Applied Probabilistic Calculus for
Financial Engineering

Applied Probabilistic Calculus for Financial Engineering

An Introduction Using R

Bertram K.C. Chan

Registered Offices
John Wiley & Sons, Inc., 111 River Street, Hoboken, NJ 07030, USA

Editorial Office
111 River Street, Hoboken, NJ 07030, USA

For details of our global editorial offices, customer services, and more information about Wiley products visit us at www.wiley.com.

Wiley also publishes its books in a variety of electronic formats and by print-on-demand. Some content that appears in standard print versions of this book may not be available in other formats.

Library of Congress Cataloguing-in-Publication Data

Name: Chan, B. K. C. (Bertram Kim-Cheong), author. Title: Applied probabilistic calculus for financial engineering : an introduction using R / by Bertram K.C. Chan. Description: Hoboken, NJ : John Wiley & Sons, Inc., 2017. | Includes bibliographical references and index. | Identifiers: LCCN 2017024496 (print) | LCCN 2017037530 (ebook) | ISBN 9781119388081 (pdf) | ISBN 9781119388043 (epub) | ISBN 9781119387619 (cloth) Subjects: LCSH: Financial engineering–Mathematical models. | Probabilities. | Calculus. | R (Computer program language) Classification: LCC HG176.7 (ebook) | LCC HG176.7.C43 2017 (print) | DDC 332.01/5192–dc23LC record available at https://lccn.loc.gov/2017024496

Dedicated to the glory of God and to my better half

Marie Nashed Yacoub Chan

Contents

Preface

The Financial Challenges and Experience of a Typical Retiring Couple – Mr. and Mrs. Smith (not their real name)

About 10 years ago, after a lifetime of steady work for some 40 years, Mr. John A. Smith and Mrs. Mary B. Smith of California were preparing for a life of active retirement, including extensive traveling worldwide. To take care of their future financial needs, they had decided to obtain the services of a local professional financial engineering and investment management company – XYZ (fictitious) – of California that conducts its transactions through a large national financial engineering corporation: LPL (Linsco – 1968 and Private Ledger – 1973).

To that end, Mr. and Mrs. Smith invested a sum of approximately $2,000,000 from their life savings, with the following twin goals:

i) The preservation of their capital of $2 M
ii) Receiving a regular net monthly cash income of at least $10,000 from XYZ

Thus, if the original capital of $2 M were to be preserved (approximately unchanged), as well as to maintain a steady withdrawal of $10,000 per month, the average annual return of the investment of the $2 M will have to be on the order of $(10,000 \times 12)/2,000,000 = 0.06$, or 6%.

The financial services management typically charges fees on the order of 1.5%. Thus, a rough estimate that the financial management should achieve would be on the order of 6% + 1.5%, or 7.5%.

How does a service such as XYZ/LPL achieve such a goal?

Approximately 10 years after their retirement, on Tuesday, November 15, 2016, the financial markets closed at

Dow (DJIA)	18,923.06/+54.03/+0.29%
Nasdaq	5,275.62/+57.22/+1.10%
S&P 500	2,180.39/+16.19/+ 0.75%
Gold	$1,229.00 per oz +0.37%

Over these 10 years, Mr. and Mrs. Smith had been receiving regularly a monthly payout from XYZ/LPL of $10,947.03! And, on the same day, the net balance of their portfolio investment account is as follows:

Portfolio ending at : $2, 111, 603.35, +$6, 152.47/ + 0.29\%$

In other words, the balance at the end of that day stood at approximately $2.1 M! And the total payout received over these past 10 years comes to $10,947.03 per month or $10,947.03 \times 12 = \$131,364.36$ per annum or $131,364.36 \times 10 = \$1,313,643.60$ over the past decade!

Exclusive of the financial management at 1.5%! How can such an investment management be achieved? Indeed, that is the central theme of this book:

The challenge in financial engineering

Whereas the nominal saving accounts of banks and credit unions in the United States have been paying at 0.1% to about 1.0%, how does a financial manager allocate the managed funds to generate, and sustain, an average return of about 7.5%? This is a typical simple example in Assets Allocation and Portfolio Optimization in Financial Engineering. It is the objective of this book to consider the underlying mathematical principles in meeting this challenge – in terms of Assets Allocation and Portfolio Optimization in Financial Engineering. This introductory text in financial engineering will include the use of the well-known and popular computer language R. Numerical worked examples are provided to illustrate the practical application of Applied Probabilistic Calculus in Financial Engineering leading to practical results in assets allocation and portfolio optimization in financial engineering using R.

About the Companion Website

This book is accompanied by a companion website:
www.wiley.com/go/chan/appliedprobabilisticcalculus

The website includes:

* Solutions to all the exercises in the body of the text, with some supportive comments

1

Introduction to Financial Engineering

1.1 What Is Financial Engineering?

In today's understanding and everyday usage, *financial engineering* is a multidisciplinary field in finance, and in theoretical and practical economics involving financial theory, the tools of applied mathematics and statistics, the methodologies of engineering, and the practice of computer programming. It also involves the application of technical methods, especially in mathematical and computational finance in the practice of financial investment and management.

However, despite its name, financial engineering does *not* belong to any of the traditional engineering fields even though many financial engineers may have engineering backgrounds. Some universities offer a postgraduate degree in the field of financial engineering requiring applicants to have a background in engineering. In the United States, ABET (the Accreditation Board for Engineering and Technology) does *not* accredit financial engineering degrees. In the United States, financial engineering programs are accredited by the International Association of Quantitative Finance.

Financial engineering uses tools from economics, mathematics, statistics, and computer science. Broadly speaking, one who uses technical tools in finance may be called a financial engineer: for example, a statistician in a bank or a computer programmer in a government economic bureau. However, most practitioners restrict this term to someone educated in the full range of tools of modern finance and whose work is informed by financial theory. It may be restricted to cover only those originating new financial products and strategies. Financial engineering plays a critical role in the customer-driven derivatives business that includes quantitative modeling and programming, trading, and risk managing derivative products in compliance with applicable legal regulations.

Applied Probabilistic Calculus for Financial Engineering: An Introduction Using R, First Edition.
Bertram K. C. Chan.
© 2017 John Wiley & Sons, Inc. Published 2017 by John Wiley & Sons, Inc.
Companion website: www.wiley.com/go/chan/appliedprobabilisticcalculus

A broader term that covers anyone using mathematics for practical financial investment purposes is "Quant", which includes financial engineers.

1.2 The Meaning of the Title of This Book

The wide use of the open-source computer software R testifies to its versatility and its concomitant increasing popularity, bearing in mind that the ubiquitous application of R is most probably due to its suitability for personal mobile-friendly desktop/laptop/panel/tablet/device computer usage. The Venn diagram that follows illustrates the interactional relationship in this context.Three subjects enunciated in this book are as follows:

1) *Applied probabilistic calculus* (APC)
2) *Assets allocation and portfolio optimization in financial engineering* (AAPOFE)
3) *The computer language* (R)

The concomitant relationship may be graphically illustrated by the mutually intersecting relationships in the following Venn diagram, **APC ∩ AAPOFE ∩** R.

Thus, this book is concerned with the distinctive subjects of importance and relevance within these areas of interest, including *Applied Probabilistic Calculus and Assets Allocation and Portfolio Optimization in Financial Engineering*, namely, **APC ∩ FE**, to be followed by critical areas of the computational and numerical aspects of *Applied Probabilistic Calculus for (Assets Allocation and Portfolio Optimization in) Financial Engineering: An Introduction Using* R, namely, **APC ∩ FE ∩** R. This is represented by the "red" area in Figure 1.1, being the common area of mutual intersection of the three areas of special interest.

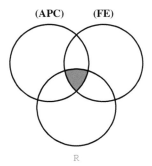

(APC)   (FE)

R

Figure 1.1 Applied Probabilistic Calculus for Financial Engineering: An Introduction Using R namely, APC ∩ FE ∩ R.

1.3   The Continuing Challenge in Financial Engineering

In the case of the investment of Mr. and Mrs. Smith (as introduced in the Preface of this book), financial engineering by the investment management company XYZ established a portfolio consisting of four accounts:

Account 1: A Family Trust

$648,857.13 (change for the day: +$150.75, +0.02%) 30.73% of Portfolio

Account 2: Individuals Trust

$504,669.30 (change for the day: +3,710.04, +0.74%)

Account 3: Traditional IRA for Person A

$476,096.53 (change for the day: $2,002.22, +0.42%)

Account 4: Traditional IRA for Person B

$457,502.17 (change for the day: $142.00, +0.03%)

The positions of these accounts are as follows:

Account 1:   A.   Cash and cash equivalents $626,331.02 (97.58%)
                 B.   Equities and options $15,839.25 (2.42%)
Account 2:   A.   Common stock – equities and options (22.71%)
                 B.   Money market – cash and cash equivalents (55.22%)
                 C.   Mutual fund, ETFs, and closed-end funds (22.07%)
Account 3:   A.   Cash and cash equivalent $247,467.33 (52.2%)
                 B.   Equities and options (3.42%)
                 C.   Money market sweeps $247,467.33 (44.38%)
                     Deposit cash account $247,447.91, account dividend 19.42
Account 4:   A.   Cash and cash equivalents (93.65%)
                 B.   Mutual funds, ETPs, and closed-end funds (44.38%)

At 11:30 p.m., Tuesday, November 15, 2016: the Smiths' account balance = $2,100,661.36 9

1.3.1   The Volatility of the Financial Market

The dynamics and volatility of the financial market is well known.

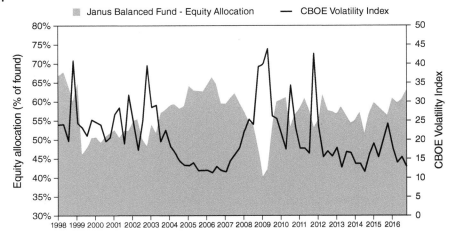

Figure 1.2 Chicago Board of Options Exchange (CBOE) Volatility Index.

For example, consider the Chicago Board of Options Exchange (CBOE) index:

CBOE Volatility Index®: Chicago Board Options Exchange index (symbol: VIX®) is the index that shows the market's expectation of 30-day volatility. It is constructed using the implied volatilities of a wide range of S&P 500 index options. This volatility is meant to be forward looking, is calculated from both calls and puts, and is a widely used measure of market risk, often referred to as the "investor fear gauge."

The VIX volatility methodology is the property of CBOE, which is not affiliated with Janus.

Clearly, Figure 1.2 reflects the dynamic nature of a typical stock market over the past 20 years. One wonders if a rational financial engineering approach may be developed to sustain the two objectives at hand simultaneously:

1) To maintain a steady level of investment
2) To produce a steady income for the investors

The remainder of this book will provide rational approaches to achieve these joint goals.

1.3.2 Ongoing Results of the XYZ–LPL Investment of the Account of Mr. and Mrs. Smith

Let us first examine the results of this investment opportunity, as seen over the past 10 years approximately.

Investment Results of the XYZ–LPL (Linsco (1968) and Private Ledger (1973)) is illustrated as follows:

LPL Financial Holdings (commonly referred to as *LPL Financial*) is the largest independent broker-dealer in the United States. The company has more than 14,000 financial advisors, over $500 billion in advisory and brokerage assets, and generated approximately $4.3 billion in annual revenue for the 2015 fiscal year. LPL Financial was formed in 1989 through the merger of two brokerage firms – Linsco (established in 1968) and Private Ledger (established in 1973) – and has since expanded its number of independent financial advisors both organically and through acquisitions. LPL Financial has main offices in Boston, Charlotte, and San Diego. Approximately 3500 employees support financial institutions, advisors, and technology, custody, and clearing service subscribers with enabling technology, comprehensive clearing and compliance services, practice management programs and training, and independent research.

LPL Financial advisors help clients with a number of financial services, including equities, bonds, mutual funds, annuities, insurance, and fee-based programs. LPL Financial does not develop its own investment products, enabling the firm's investment professionals to offer financial advice free from broker/dealer-inspired conflicts of interest.

Revenue: US$4.37 billion (2014)
Headquarters: 75 State Street, Boston, MA, USA
Traded as: NASDAQ: LPLA (https://en.wikipedia.org/wiki/LPL_Financial)

Over the past 10 years, the Smiths' received, on a monthly basis, a net income of $10,947.03. Thus, annually, the income has been

$$\$10,947.02 \times 12 = \$131,364.36.$$

And, the total income for the past 10 years has been

$$\$131,364.36 \times 10 = \$1,313,643.60.$$

Illustrated hereunder in Figure 1.3 is a snapshot of one of the four investment accounts of the Smiths'. Note the following special features of this portfolio:

1) The green area represents the investment amount: As portions of the capital were being periodically withdrawn (to satisfy U.S.A. Federal Regulations), the actual investment amount decreases in time. This loss has been more than made up by the blue area!
2) The blue area is the portfolio value of the account.

This showed that, as steady income is being generated, the portfolio value grows *more* than the amounts continually withdrawn – *periodically and regularly*.

Clearly, the goals of the investment have been achieved!

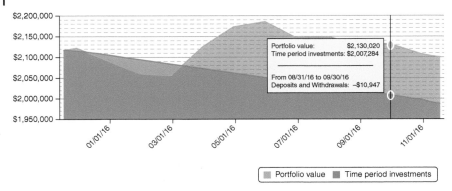

Figure 1.3 Recent time period investments and portfolio values of one of the four accounts in the portfolio of Mr. and Mrs. Smith.

While the exact algorithms used by XYZ/LPL is proprietary, the following investment strategies are clear:

1) Have each investment account placed in high-yield corporate bonds, and then SET a maximum limit of 3% draw down (loss) limit for each investment account.
2) If draw down > limit of 3%, then move investment account over to cash or money market accounts – to preserve the overall capital of the portfolio.
3) Stay in the cash or money market accounts, until the following three conditions of the market become available before returning to the high-yield corporate bonds where opportunity becomes available once again:
   a) Favorable conditions have returned – when comparing high-yield corporate bonds and the interest rates of 10-year Treasury notes, namely, when condition favoring the relative interest rates of money.
   b) Favorable conditions have returned – when comparing the volumes of money entering and leaving the market, namely, whether investors are "Pulling Back."
   c) When the direction of the market becomes clear.

It certain does not escape ones attentions that these set of conditions are rather tenuous, at best.

One should, therefore, seek more robust paths to follow.

1.4 "Financial Engineering 101": Modern Portfolio Theory

Modern portfolio theory (MPT), also known as *mean-variance analysis*, is an algorithm for building a portfolio of assets such that *the expected return is maximized for a given level of risk*, defined as variance. Its key feature is that an asset's risk and return should not be assessed by itself, *but by how it contributes to a portfolio's overall risk and return.*

(Economist Harry Markowitz introduced MPT in a 1952 paper for which he was later awarded a Nobel Memorial Prize in Economic Science!)

1.4.1 Modern Portfolio Theory (MPT)

In MPT, it is assumed that investors are *risk averse*, namely, if there exist two portfolios that offer the same expected return, rational investors will prefer the less risky one. Thus, an investor will accept increased risk only if compensated by higher expected returns. Conversely, an investor who wants higher expected returns must accept more risk. That is,

$$\text{increased risks} \Leftrightarrow \text{higher expected returns.} \qquad (1.1)$$

1.4.2 Asset Allocation and Portfolio Volatility

It is reasonable to assume that a *rational* investor will not invest in a portfolio if there exists another available portfolio with a more favorable profile. Portfolio return is the proportion-weighted combination of the constituent assets' returns.

Asset Allocation   Asset allocation is the process of organizing investments among different kinds of asset categories, such as stocks, bonds, derivatives, cash, and real estate, in order to accommodate and achieve a practical combination of risks and returns, that is consistent with an investor's specific goals. Usually, the process involves portfolio optimization, which consists of three general steps:

Step I: The investor specifies asset classes and models forward-looking assumptions for each asset classes' return and risk as well as movements among the asset classes. For a scenario-based approach, returns are simulated based on all the forward-looking assumptions.

Step II: One arrives at an optimization algorithm in which allocations to different asset classes are set. These allocations are known as the asset mix.

Step III: The asset mix return and wealth forecasts are projected over various investment probabilities, and scenarios predict and demonstrate the potential outcomes. Thus, an investor may inspect an estimate of what the portfolio value would be (say) 2 years into the future if its returns were in the bottom 10% of the projected range during this period. Mean-variance optimization (MVO) is commonly used for creating efficient asset allocation strategies. But MVO has its limitations – illustrated herein as follows.

1.4.3 Characteristic Properties of Mean-Variance Optimization (MVO)

The following are the characteristic properties of the methodology of MVO:

a) It does not take into account "fat-tailed" asset class return distributions, which matches mast real-world historical asset class returns. For example, consider the monthly total returns of the S&P 500 Index, dating back to 1926: There are 1,025 months between January 1926 and May 2011. The monthly arithmetic mean and standard deviation of the S&P 500 Index over this time period are 0.943 and 5.528%, respectively. For a normal distribution, the return that is three standard deviations away from the mean is −15.64%, calculated as (0.943%−3×5.528%).

   In a standard normal distribution, 68.27% of the data values are within one standard deviation from the mean, 95.45% within two standard deviations, and 99.73% within three standard deviations: Figure 1.4

   This implies that there is a 0.13% probability that returns would be three standard deviations below the mean, where 0.13% is calculated as

   $$(100\% - 99.73\%)/2.$$

   In other words, the normal distribution estimates that there is a 0.13% probability of returning less than −15.64%, which means that only 1.3 months out of those 1,025 months between January 1926 and May 2011 are *expected* to have returns below −15.64%, where 1.3 months is arrived at by multiplying the 0.13% probability by 1,025 months of return data.

   However, when examining historical data during this period, there are 10 months where this occurs, which is

   $$10 \text{ months}/1.3 \text{ months} = 7.69 \text{ times}$$

   or almost 8 times more than the model prediction!

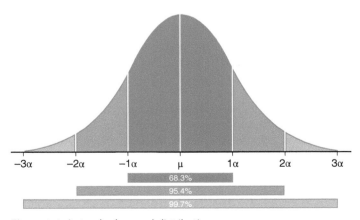

Figure 1.4 A standard normal distribution.

The following are the 10 months in question:

Month	Return (%)
Jun 1930	−16.25
Oct 2008	−16.79
Feb1933	−17.72
Oct 1929	−19.73
Apr 1932	−19.97
Oct 1987	−21.54
May 1932	−21.96
May 1940	−22.89
Mar 1938	−24.87
Sep 1931	−29.73

The normal distribution model also assumes a *symmetric bell-shaped curve*, and this seems to imply that the model is not well suited for asset classes with asymmetric return distributions. The histogram of the data, shown in Figure 1.5, plots the number of historical returns that occurred in the return range of each bar.

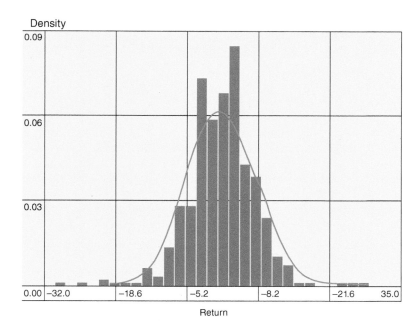

Figure 1.5 Modeling with a standard normal curve.

The curve plotted over the histogram graph shows the probability predicted by the normal distribution. In Figure 1.5, the left tail of the histogram is longer, and there are actual historical returns that the normal distribution does not predict.

Xiong and Idzorek showed that skewness (asymmetry) and excess kurtosis (larger than normal tails) in a return distribution may have a significant impact on the optimal allocations in a portfolio selection model where a downside risk measure, such as Conditional Value at Risk (CVaR), may be used as the risk parameter. Intuitively, besides lower standard deviation, investors should prefer assets with positive skewness and low kurtosis. By ignoring skewness and kurtosis, investors who rely on MVO alone may be creating portfolios that are riskier than they may realize.

b) The traditional MVO assumes that covariation of the returns on different asset classes is linear. That is, the relationship between the asset classes is consistent across the entire range of returns. However, the degree of covariation among equity markets tends to go up during global financial crises. Furthermore, a linear model may well be an inadequate representation of covariation when the relationship between two asset classes is based at least in part on optionality such as the relationship between stocks and convertible bonds. Fortunately, nonlinear covariation may be modeled using a scenario- or simulation-based approach.

c) The traditional MVO framework is limited by its ability to only optimize asset mixes for one risk metric, standard deviation. As already indicated, using standard deviation as the risk measure ignores skewness and kurtosis in return distributions. Alternative optimization models that incorporate downside risk measures may have a significant impact on optimal asset allocations.

d) The traditional MVO is a single-period optimization model that uses the arithmetic expected mean return as the measure of reward. An alternative is to use expected geometric mean return. If returns were constant, geometric mean would equal arithmetic mean. When returns vary, geometric mean is lower than arithmetic mean. Moreover, while the expected arithmetic mean is the forecasted result for the next one period, the expected geometric mean forecasts the long-term rate of return. Hence, for investors who regularly rebalance their portfolios to a given asset mix over a long period of time, the expected geometric mean is the relevant measure of reward when selecting the asset mix.

In spite of its limitations, the normal distribution has many attractive properties:

It is easy to work within a mathematical framework, as its formulas are simple.

The normal distribution is very intuitive, as 68.27% of the data values are within one standard deviation on either side of the mean, 95.45% within two standard deviations, and 99.73% within three standard deviations, and so on.

1.5 Asset Class Assumptions Modeling

Models Comparison For the first step of an asset allocation optimization process, the investor/analyst begins by specifying asset classes and then models forward-looking assumptions for each asset class's return and risk as well as relative movements among the asset classes. Generally, one may use an index, or a blended index, as a proxy to represent each asset class, although it is also possible to incorporate an investment such as a fund as the proxy or use no proxy at all. When a historical data stream, such as an index or an investment, is used as the proxy for an asset class, it may serve as a starting point for the estimation of forward-looking assumptions.

In the assumption formulation process, it is critical for the investor to ascertain what should be the return patterns of asset classes and the joint behavior among asset classes. It is common to assume these return behaviors may be modeled by a parametric return distribution function, namely, that they may be expressed by mathematical models with a small number of parameters that define the return distribution. The alternative is to directly use historical data without assuming a return distribution model – this process is known as "bootstrapping."

1.5.1 Examples of Modeling Asset Classes

1.5.1.1 Modeling Asset Classes

1.5.1.1.1 Lognormal Models
The normal distribution, also known as the Gaussian distribution, takes the form of the familiar symmetrical bell-shaped distribution curve commonly associated with MVO. It is characterized by **two** parameters: mean and standard deviation.

- Mean is the probability-weighted arithmetic average of all possible returns and is the measure of reward in MVO.
- Variance is the probability-weighted average of the square of difference between all possible returns and the mean. Standard deviation is the square root of variance and is the measure of risk in MVO.
- The prefix "log" means that the natural logarithmic form of the return relative, $\ln(1+R)$, is normally distributed. The lognormal distribution is asymmetrical, skewing to the right, because the logarithm of 0 is $-\infty$, the

lowest return possible is −100%, which reflects the fact that an unleveraged investment cannot lose more than 100%.

The lognormal distribution have the following attractive features:

1) It is very easy to work with in a mathematical framework.
2) It is scalable; therefore, mean and standard deviation can be derived from a frequency different from that of the return simulation.
3) Limitations of the model include its inability to model the skewness and kurtosis empirically observed in historical returns. That is, the lognormal distribution assumes that the skewness and excess kurtosis of $\ln(1 + R)$ are both zero.

1.5.1.1.2 Johnson Models

The Johnson model distributions are a four-parameter parametric family of return distribution functions that may be used in modeling skewness and kurtosis. Skewness and kurtosis are important distribution properties that are zero in the normal distribution and take on limited values in the lognormal model (as implied by the mean and standard deviation).

There are the following four parameters in a Johnson distribution:

- Mean
- Standard deviation
- Skewness
- Excess kurtosis

Mean and standard deviation may be described similar to their definitions. Skewness and excess kurtosis are measures of asymmetry and peakedness. Consider the following example:

Example 1

The normal distribution is a special case of the Johnson model: with skewness and excess kurtosis of *zero*. The lognormal distribution is also a special case that is generated by assigning the skewness and excess kurtosis parameters to the appropriate values.

- *Positive skewness* means that the return distribution has a longer tail on the right-hand side than the left-hand side, and *negative skewness* is the opposite.
- Excess kurtosis is *zero* for a normal distribution. A distribution with positive excess kurtosis is called *leptokurtic* and has fatter tails than a normal distribution, and a distribution with negative excess kurtosis is called *platykurtic* and has thinner tails than a normal distribution.

Besides lower standard deviation, investors should prefer assets with positive skewness and lower excess kurtosis. Skewness and excess kurtosis are often estimated from historical return data using the following formulas.

- Expected return:

$$E(R_p) = \sum_i w_i\, E(R_i) \tag{1.2}$$

where R_p is the return on the portfolio, R_i is the return on asset i, and w_i is the weighting of component asset i (that is, the proportion of asset i in the portfolio).
- Portfolio return variance σ_p is the *statistical sum* of the variances of the individual components $\{\sigma_i, \sigma_j\}$ defined as

$$\sigma_p^2 = \sum_i w_i^2 \sigma_i^2 + \sum_i \sum_{j \neq i} w_i w_j \sigma_i \sigma_j \rho_{ij} \tag{1.3}$$

where ρ_{ij} is the *correlation coefficient* between the returns on assets i and j. Also, the expression may be expressed as

$$\sigma_p^2 = \sum_i \sum_j w_i w_j \sigma_i \sigma_j \rho_{ij} \tag{1.4}$$

where $\rho_{ij} = 1$ for $i = j$.
- Portfolio return volatility (standard deviation)

$$\sigma_p = \sqrt{\sigma_p^2} \tag{1.5}$$

so that, for a *two-asset* portfolio
- Portfolio return

$$E(R_p) = w_A E(R_A) + w_B E(R_B) = w_A E(R_A) + (1 - w_A) E(R_B) \tag{1.6}$$

- Portfolio variance:

$$\sigma_p^2 = w_A^2 \sigma_A^2 + w_B^2 \sigma_B^2 + 2 w_A w_B \sigma_A \sigma_B \rho_{AB} \tag{1.7}$$

And, for a *three-asset* portfolio
- Portfolio return:

$$E(R_p) = w_A E(R_A) + w_B E(R_B) + w_C E(R_C) \tag{1.8}$$

- Portfolio variance:

$$\begin{aligned} \sigma_p^2 = {} & w_A^2 \sigma_A^2 + w_B^2 \sigma_B^2 + w_C^2 \sigma_C^2 + 2 w_A w_B \sigma_A \sigma_B \rho_{AB} \\ & + 2 w_A w_C \sigma_A \sigma_C \rho_{AC} + 2 w_B w_C \sigma_B \sigma_C \rho_{BC} \end{aligned} \tag{1.9}$$

1.5.1.1.3 Diversification of a Portfolio for Risk Reduction

To reduce portfolio risk, one may simply hold combinations of instruments that are not perfectly positively *correlated:*

$$\text{correlation } coefficient - 1 \leq \rho_{ij} < 1. \tag{1.10}$$

Thus, investors may reduce their exposure to individual asset risk by selecting a *diversified* portfolio of assets: **diversification may allow for the same portfolio expected return with reduced risk.**

(These ideas had been first proposed by Markowitz and later reinforced by other Theoreticians (applied mathematicians and economists) who had expressed ideas in the limitation of variance through portfolio theory.)

Thus, if all the asset pairs have correlations of 0 (namely, they are perfectly uncorrelated), the portfolio's return variance is the sum of all assets of the **square** of the fraction held in the asset times the asset's return variance, and the portfolio standard deviation is the square root of this sum.

Further detailed discussions of many of these ideas and approaches will be presented in Chapters 5 and 6.

1.6 Some Typical Examples of Proprietary Investment Funds

A certain investment company advertised openly investment programs such as the *Managed Payout Funds*, which claim to enable interested investors to enjoy regular monthly payments after ones retirement, without giving up control of ones funds!

Such type of funds will be briefly assessed hereunder.

One particular fund claims the following:

1) To have been designed to provide the investor with regular monthly payouts that, over time, keep pace with inflation, and to help one cover expenses in retirement.
2) To accomplish, the fund aims to strike a balance between how much income it will generate and how much principal growth it will follow.
3) To provide regular monthly payments automatically to the investor.
4) To reset the payment amounts in January every year. And payments are expected to remain the same from month to month. The fund targets future annual distribution rate of 4%.
5) Finally, the fund is NOT guaranteed: thus one may loose some of the capital invested!

Example 2: A $2,000,000 Investment with Investment Company "IC-1"

a) A quick calculation based on ones situation of investing $2M in IC-1. Using the online calculator provided by IC-1, the estimated initial monthly payout is $6,268. This compares with $10,974.03 achieved by the XYZ/LPL investment company!
b) Similarly, for a monthly payout of $10,974.00, and using the online calculator provided by IC-1, the estimated amount that one needs to invest is

$3,501,794. This compares with only $2,000,000, as required by the XYZ/LPL company!

1.7 The Dow Jones Industrial Average (DJIA) and Inflation

Many of the theoretical approaches in financial engineering depend on the time/volume inflationary characteristics of the asset values or prices of commodities. As a classical example, consider the value-time and trading-volumes characteristics of the Dow Jones Industrial Average (DJIA), illustrated in Figures 1.6 and 1.7, respectively.

Here, the following questions are obvious to be raised:

1) How high will the DJIA index go, and at what rates?
2) How high will the daily traded shares volumes go?

Figures 1.6 and 1.7 seem to indicate, in the long run, both will increase.

Recently, the following article seems to indicate that, for both parameters, "the sky is the limit"!

"Wall Street Legend Predicts "Dow 50,000"!"

- By J.L. Yastine
- November 28, 2016

A massive stock market rally is on our doorstep according to several noted economists and distinguished investors.

Ron Baron, CEO of Baron Capital, is calling for Dow 30,000.

Larry Edelson of *Money and Markets* is calling for Dow 31,000.

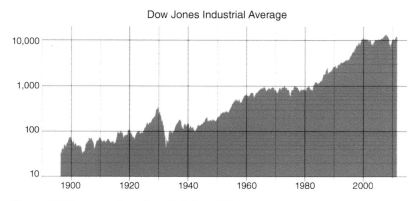

Figure 1.6 The value-time characteristics of DJIA.

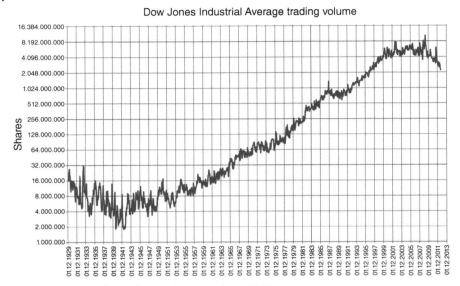

Figure 1.7 The trading volumes characteristics of DJIA.

And Jeffrey A. Hirsch, author of *Stock Trader's Almanac*, is calling for Dow 38,000.

However, Paul Mampilly's "Dow 50,000" predication is garnering national attention . . . not because it's a bold prediction, but rather because every one of Mampilly's past predictions have been spot on.

Like when he predicted the tech crash of 1999 and the time he called the financial collapse of 2008 – months before they unraveled.

Mampilly's predictions paid off big as the $6 billion hedge fund he managed was named by *Barron's* as one of the "World's Best."

And Mampilly became legendary when he won the prestigious Templeton Foundation investment competition by making a 76% return ($38 million in profit on a $50 million stake), during the 2008 and 2009 economic crisis.

In a new video presentation, Mampilly states: "Stocks are on the cusp of an historic surge. The Dow will rally to 50,000. I've never been more certain of anything in my career."

Indeed, Mampilly uses a historic chart to prove Dow 50,000 is inevitable. (In fact, one can see how the Dow could even rally to 200,000.)

The drive behind this stock market rally?

"A little-known, yet powerful economic force that has driven every bull market for the last 120 years," Mampilly explains. "I've used this same force to predict the stock market collapse of 2000 and 2008 . . . and to make personal gains of 634%, 696% and even 2,539% along the way."

And while these gains are impressive, Mampilly states they are nothing compared to what is ahead.

"The last time this scenario unfolded, it sent a handful of stocks as high as 27,000%, 28,000% and even 91,000%."

The key is to buy the right stocks before the rally.

Clearly one needs to be disciplined in investing: both in *asset allocation* as well as in *portfolio optimization*.

1.8 Some Less Commendable Stock Investment Approaches

In financial trading and investment environment today in which computerization and specialized mathematical algorithms can and do play significant roles, some investment agencies, as well as individual investors, have found it almost irresistible not to wander into and along these paths. Two such well-known paths are day trading and algorithmic trading. It should not escape once attention that serious investors should be aware of the high chances for losses in these, and similar, "get rich quick" schemes. For the sake of completeness, one should mention two such well-known schemes:

1) Day trading
2) Algorithmic trading

The following are brief description of these schemes.

1.8.1 Day Trading

Day trading is the speculation in stocks, bonds, options, currencies, and future contracts and other securities, by buying and selling financial instruments within the *same* trading day. Once, day trading was an activity exclusive to professional speculators and financial companies. The common use of "buying on margin" (using borrowed funds) greatly increases gains and losses, such that substantial gains or losses may occur in a very short period of time.

1.8.2 Algorithmic Trading

Algorithmic trading is a method of placing a large order by sending small portions of the order out to the market over time. It is an automated method of executing an order using automated preprogrammed trading instructions providing preset instructions accounting for variables such as price, volume, and time. It claims to be a way to minimize the costs, market impacts, and risks in the execution of an order – thus eliminating human intervention and decisions in the process.

1.9 Developing Tools for Financial Engineering Analysis

The remainder of this book is developed along the following steps:

- Chapter 2 will present a full discourse on the probabilistic calculus developed for modeling in financial engineering.
- Chapter 3 will discuss classical mathematical models in financial engineering, including modern portfolio theories.
- Chapter 4 will discuss the use of R programming in data analysis, providing a practical tool for the analysis.
- Chapter 5 will present the methodologies of asset allocation, using R.
- Chapter 6 will discuss financial risk modeling and portfolio optimization using R.

Review Questions

1 In finance and economics, it is tacitly assumed that inflation will continue, and will most probably continue indefinitely, with the following concomitant result:

The price of good and services such as

- food items, cars, services, and so on *will increase indefinitely,*
- the purchasing power of money, namely, the value of the $ *will decrease indefinitely,* and
- working wages and salaries *will increase indefinitely.*

  a Discuss the rationale behind each of these phenomena.

  b If these phenomena hold true indefinitely, show how one may, and should, allocate one's assets, and manage one's investment portfolios, so that one has a maximum opportunity to accumulate wealth.

2 Besides allocating one's assets in terms of buying/selling stocks, bonds, and other financial instruments, discuss the likely challenges associated with investing in the following:

  a Foreign financial entities, for example, buying stocks of overseas corporations such as financial opportunities in Europe, in China (being part of the **BRIC** group – **B**razil, **R**ussia, **I**ndia, and China). It is anticipated that, by 2050, China and India will most likely become the world's dominant suppliers of manufactured goods and services!

  b Commodities (such as grains, oils, foreign currencies, precious metals, etc.).

  c Real estates: properties and lands.

  d Services such as hospitals, retirement homes, and other commercial entities.

2

Probabilistic Calculus for Modeling Financial Engineering

2.1 Introduction to Financial Engineering

To establish useful and realistic mathematical models for financial analysis, with the objectives of assets allocation and the concomitant gains, or losses, it is useful to consider both discrete models and continuous models.

2.1.1 Some Classical Financial Data

Consider the following two typical sets of financial data:

1) **Dow Jones – 100 year historical chart** (Figure 2.1)
   Interactive chart of the Dow Jones Industrial Average (DJIA) stock market index for the last 100 years.
   This historical data is inflation-adjusted using the headline CPI and each data point represents the month-end closing value. The current month is updated on an hourly basis with today's latest value. The current value of the DJIA as of 08:48 p.m. EDT on June 3, 2016 is $17,838.56.
2) **Apple, Inc., AAPL – 5-year historical prices** (Figure 2.2)

2.2 Mathematical Modeling in Financial Engineering

2.2.1 A Discrete Model versus a Continuous Model

In general, a discrete model calls for analysis and solutions using finite difference techniques, whereas a continuous model, using partial and total derivatives, leads to a financial theory consisting of systems of partial differential equations, with total and partial derivatives. Such systems of differential equations benefit from the vast reservoir of support from the well-developed field of systems of total and partial differential equations.

Applied Probabilistic Calculus for Financial Engineering: An Introduction Using R, First Edition.
Bertram K. C. Chan.
© 2017 John Wiley & Sons, Inc. Published 2017 by John Wiley & Sons, Inc.
Companion website: www.wiley.com/go/chan/appliedprobabilisticcalculus

Figure 2.1 Dow Jones Industrial Average (DJIA) –100 year historical chart.

An outstanding example is the vast mathematical literature developed, or being developed, for the Black–Scholes model for financial derivatives.

2.2.2 A Deterministic Model versus a Probabilistic Model

2.2.2.1 Calculus of the Deterministic Model

In deterministic calculus of functions, $f(x)$, of a single variable x, there are some important results, commonly called theorems, available for relating

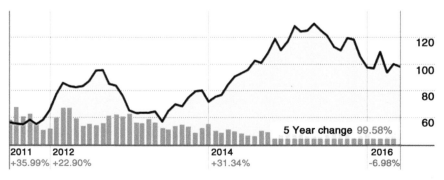

Figure 2.2 Apple, Inc., AAPL (a 5-year historical prices).

the variability characteristics of such functions. These theorems include the following:

- *The Mean Value Theorem:*

    If $f(x)$ is continuous in the closed interval $[a, b]$, and is differentiable in the open interval (a, b), then there is a value ξ of x, namely, $a < \xi < b$, such that

$$f(b) - f(a) = (b - a)f'(\xi) \tag{2.1}$$

- *The Higher Mean Value Theorems:*

    For $0 < \theta < 1$

$$f(a + h) - f(a) = hf'(a + \theta h) \tag{2.2}$$

    If $f(x)$ is continuous for $a \le x \le b$, and if $f'(x)$ exists for $a < x < b$, and if $h = b - a$, and $0 < \theta_2 < 1$, then

$$f(a + h) = f(a) + hf'(a) + \tfrac{1}{2}h^2 f''(a + \theta_2 h) \tag{2.3}$$

- *The General Mean Value Theorem (Taylor's Theorem):*

    If $f^{(n-1)}(x)$ is continuous for the closed interval $a \le x \le b$, and $f^{(n)}(x)$ exists for $a < x < b$, then

$$\begin{aligned} f(b) = f(a) + (b - a)f'(x) + \{(b - a)^2/2!\}f''(a) \\ + \cdots + \{(b - a)^{(n-1)}/(n - 1)!\}f^{(n-1)}(a) + \{(b - a)^n/n!\}f^{(n)}(\xi) \end{aligned} \tag{2.4}$$

    where $a < \xi < b$; and if $b = a + h$, then

$$\begin{aligned} f(a + h) = f(a) + hf'(a) + \tfrac{1}{2}h^2 f''(a) \\ + \cdots + \{h^{(n-1)}/(n - 1)!\}f^{(n-1)}(a) + h^n/n!\}f^{(n)}(a + \theta_n h) \end{aligned} \tag{2.5}$$

    The continuity of $f^{(n-1)}(x)$ involves that of $f(x), f'(x), f''(x), \ldots, f^{(n-2)}(x)$.

- *Taylor's Theorem for Functions of Several Variables:*

    Taylor's formula for functions of one variable, Equation (2.5), may be generalized to functions of several variables having continuous derivatives of order n in a given region of space. The following procedure is outlined for functions of two variables – the result has all the features of more general cases.

    Let $f(x, y)$ be a function of two variables (x, y) having continuous derivatives of order n in some region containing the point.

$$x = a, \quad y = b,$$

and let

$$x = a + \alpha, \quad y = b + \beta$$

be a nearby point in the region. The equations of a straight line segment joining these two points may be expressed in the form

$$x = a + \alpha t, \quad y = b + \beta t, \qquad 0 \le t \le 1 \tag{2.6}$$

and, along this segment,

$$f(x, y) = f(a + \alpha t, b + \beta t) \equiv F(t) \tag{2.7}$$

is a composite function of t having continuous derivatives of order n. Thus, $F(t)$ may be represented by the Maclaurin formula:

$$F(t) = F(0) + F'(0)t + \{F''(0)/2!\}t^2 + \cdots + \{F^{(n)}(0)/n!\}t^n, \quad 0 < \theta < t \tag{2.8}$$

The successive derivatives of $F(t)$ may be computed from Equation (2.7) and by the *chain rule* for differentiating composite functions:

If $u = f(x, y)$, then

$$du/dt = (\partial u/\partial x)(dx/dt) + (\partial u/\partial y)(dy/dt) \tag{2.9}$$

Thus,

$$\begin{aligned} F'(t) &= f_x(x, y)(dx/dt) + f_y(x, y)(dy/dt) \\ &= f_x(x, y)\alpha + f_y(x, y)\beta \end{aligned} \tag{2.10}$$

where in the last step, one recalls Equation (2.6).

Similarly,

$$\begin{aligned} F''(t) &= [f_{xx}(x, y)\alpha + f_{yx}(x, y)\beta]\{dx/dt\} + [f_{xy}(x, y)\alpha + f_{yy}(x, y)\beta](dy/dt) \\ &= f_{xx}(x, y)\alpha^2 + 2f_{xy}(x, y)\alpha\beta + f_{yy}(x, y)\beta^2 \end{aligned}$$

and

$$\begin{aligned} F^{(n)}(t) &= (\partial^n f/\partial x^n)\alpha^n + C_{n,1}(\partial^n f/\partial x^{n-1}\partial y)\alpha^{n-1}\beta + \cdots \\ &\quad + C_{n,n-1}(\partial^n f/\partial x\partial y^{n-1}) + \partial^n f/\partial y^n \beta^n \end{aligned}$$

where $C_{n,r} \equiv n!/[r! \, (n - r)!]$ are the binomial coefficients.

From Equation (2.6), it is seen that $t = 0$ corresponds to the point: $x = a, y = b$, so that

$$\begin{aligned} F(0) &= f(a, b) \\ F'(0) &= f_x(a, b)\,\alpha + f_y(a, b)\,\beta \\ F''(0) &= f_{xx}(a, b)\,\alpha^2 + 2f_{xy}(a, b)\,\alpha\beta + f_{yy}(a, b)\beta^{2t2} \end{aligned}$$

and so on.

Substituting these values in Equation (2.8) gives

$$\begin{aligned} f(x, y) &= f(a, b) + [f_x(a, b)\alpha + f_y(a, b)\beta]t \\ &\quad + (1/2!)[f_{xx}(a, b)]\alpha^2 + 2f_{xy}(a, b)\alpha\beta + f_{yy}(a, b)\beta^2] + \cdots + R_n \end{aligned} \tag{2.11}$$

where $R_n = [F^{(n)}(\theta t)/n!] \, t^n, 0 < \theta < 1$.

Now, from Equation (2.6), $\alpha t = x - a$ and $\beta t = y - b$, so that Equation (2.11) takes the form

$$f(x, y) = f(a, b) + f_x(a, b)(x - a) + f_y(a, b)(y - b) + (1/2!)[f_{xx}(a, b)(x - a)^2$$
$$+ 2f_{xy}(a, b)(x - a)(y - b) + f_{yy}(a, b)(y - b)^2] + \cdots + R_n$$

(2.12)

which is the Taylor formula (for two variables, in this case).

2.2.2.2 The Geometric Brownian Motion (GBM) Model and the Random Walk Model

Historically, the GBM had been developed in terms of the concept of the *random walk*, $W_n(t)$, which has the following characteristics:

For a positive integer n, the binomial process $W_n(t)$ may be defined by the following characteristic properties:

1) $W_n(t) = 0$, at $t = 0$
2) $W_n(t)$ has layer spacing $1/n$
3) There exist up and down jumps, equal and of magnitude $1/\sqrt{n}$
4) $W_n(t)$ has measure P, given by up and down probabilities ½ everywhere

Thus, if $X_1, X_2, X_3, \ldots, X_i$ is a sequence of independent binomial random variables taking values $+1$ or -1 with *equal* probability, then the value of W_n at the ith step is

$$W_n[i/n] = W_n[(i - 1)/n] + X_i/\sqrt{n}, \quad \text{for all } i \geq 1.$$

2.2.2.3 What Does a "Random Walk" Financial Theory Look Like?

The following chart is a simple example.

Random walk hypothesis test by increasing or decreasing the value of a fictitious stock based on the odd/even value of the decimals of π. The chart resembles a stock chart. (Figure 2.3)

Remarks:

1) A Brownian motion wanders randomly with zero mean, but a company stock usually grows at some rate (even if it just increases with normal inflation, etc.).
2) Thus one may choose to add in a drift factor μ. For example, the random walk process may be described as

$$W_t = \mu + W_{t-1} + \varepsilon_t \tag{2.13}$$

in which

W_t is the log of the asset price at time t
μ is a drift constant
ε_t is a random disturbance term for which the exceptions

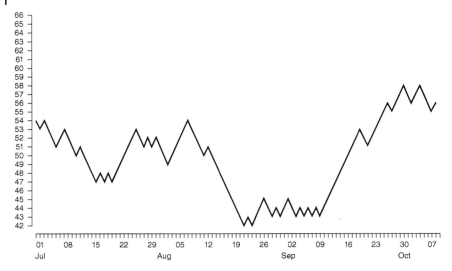

Figure 2.3 An example of a random walk based on the odd/even value of the decimals of $\pi = 3.14159$. (Ordinates are arbitrary) (Available at https://en.wikipedia.org/wiki/Random_walk_hypothesis)

$$E[\varepsilon_t] = 0 \quad \text{and} \quad E[\varepsilon_t \varepsilon_\tau] = 0, \quad \text{for} \quad \tau \neq t$$

It had been demonstrated that the random walk hypothesis is inaccurate, there are trends in the stock market, and the stock market is somewhat predictable.

For progressing from a pure GBM to a modifiable deterministic–probabilistic model for the stock market, the following path may be followed:

- A deterministic model versus a probabilistic model
- A combined deterministic and probabilistic model

2.3 Building an Effective Financial Model from GBM via Probabilistic Calculus

The attractive characteristics of the GBM are quite powerful. They may be retained if some of its attractive features are retained in a more realistic environment, even if any such hybrid result is considerably complex. This may well be a worthwhile endeavor!

For any smooth, namely, differentiable, path, consider a small time segment, it is acceptable that this time segment consists of a continuous and differentiable function df_t within this time slot dt such that

$$df_t = \mu_t \, dt \tag{2.14}$$

where μ_t is the slope or drift scaling function.

From this simple beginning, one may consider all other possible Newtonian functions. A likely possibility is one in which the drift μ_t may depend on the existing value of the function f_t, *namely*,

$$df_t = \mu_t(df_t, t) \, dt \tag{2.15}$$

2.3.1 A Probabilistic Model for the Stock Market

In a stock market that reflects a realistic probabilistic characteristic, the stock price X_t, at any time t, may be considered as one having both

  i) a Browning motion term $\sigma_t \, dW_t$ and
 ii) a Newtonian characteristic term based upon dt: $\mu_t(X_t, t) \, dt$,

so that the infinitesimal change of X may be represented by

$$dX_t = \sigma_t \, dW_t + \mu_t(X_t, t) \, dt \tag{2.16}$$

in which

a) the drift μ_t can depend on the time t,
b) the drift μ_t can depend on the value of X and W, up to time t, and
c) the noisiness or volatility, σ_t, of the process X can depend on the history of the process.

This approach, however, does provide a practical (though not universal) definition of a probabilistic process (sometimes known as a stochastic process).

2.3.2 Probabilistic Processes for the Stock Market Entities

A probabilistic process X is a continuous process such that

$$X_t = X_0 + \int_0^t \sigma_s \, dW_s + \int_0^t \mu_s \, ds \tag{2.17}$$

in which σ and μ are random processes such that $\int_0^t \sigma_s^2 + |\mu_s| ds$ is finite for all times t, with probability 1. The differential form of Equation (2.17) is

$$dX_t = \sigma_t dW_t + \mu_t \, dt \tag{2.18}$$

2.3.3 Mathematical Modeling of Stock Prices

In modeling stock prices, it certainly should not escape ones attention that, in practice, *a price can and may change at any instant,* and not at some fixed times when a portfolio may be considered for rebalancing. Hence, these observations call for the requirements that

$$\sigma_t = \sigma(X_t, t) \qquad \text{and} \qquad \mu_t = \mu(X_t, t)$$

where both σ_t and μ_t depend on W, and Equation (2.18) may be written as

$$dX_t = \sigma(X_t, t)\, dW_t + \mu(X_t, t)\, dt \tag{2.19}$$

which is the *probabilistic (or stochastic) differential equation (PDE)* for X_t.

As with any deterministic differential equation, a PDE

  i) need not have an explicit or implicit solution and
 ii) if a solution should exist, it need not be unique.

Moreover, probabilistic differentials of systems whose volatility and drift depend *not only* on t and X_t, *but also* on other factors in the history of the system behavior.

2.3.4 A Simple Case

When σ and μ are *both constants:*

If X has constant volatility σ and drift μ, then the system PDE for X is

$$dX_t = \sigma\, dW_t + \mu\, dt \tag{2.20a}$$

then (if $X_0 = 0$) the corresponding solution is clearly

$$X_t = \sigma\, W_t + \mu\, t \tag{2.20b}$$

And, if the differential form of $\sigma\, W_t$ may be assumed to be $\sigma\, dW_t$, then it is seen that Equation (2.20b) is the unique solution of Equation (2.20a).

Nontrivial Cases: For a somewhat more complex case,

$$dX_t = X_t(\sigma\, dW_t + \mu\, dt) \tag{2.21}$$

It seems that the solution to the equation (2.21) is nontrivial.

2.4 A Continuous Financial Model Using Probabilistic Calculus: Stochastic Calculus, Ito Calculus

Deterministic Calculus

Probabilistic Calculus

For a continuous model representing the real-life market of financial entities such as stocks, bonds, derivatives, and so on, the following basic model characteristics should be included:

1) The value of the entity can change at any time, and indeed, from moment to moment.
2) The actual values may be expressed in arbitrarily small fractions in terms of a real number.
3) The process changes continuously, and the value cannot make – abrupt – jumps. Thus, if the security values changes from 10 to 11, it must have passed through all the values in between, albeit quickly.

Historically, the motion of the stock price, variations with time, had been likened to one well-known physical process – that followed by a random movement of gas particles – namely, the geometric Brownian motion (GBM).

2.4.1 A Brief Observation of the Geometric Brownian Motion

Figure 2.4a shows two sample paths of GBM, with different parameters. The blue line has larger *drift*, the green line has larger *variance*.

A GBM may be considered as a continuous-time probabilistic motion in which the logarithm of the randomly varying quantity follows a Brownian motion (also known as a Wiener process) with drift. Thus, it is an important example of probabilistic processes satisfying a probabilistic differential equation. It is used in mathematical finance to model stock prices in the Black–Scholes model.

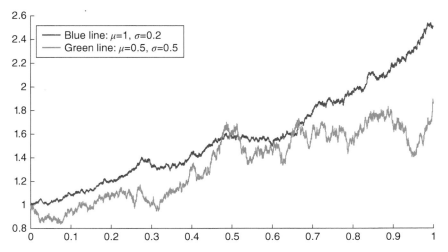

Figure 2.4 (a) Geometric Brownian motion (See: Wikipedia).

One may approach the solution of equations such as (2.20), in steps, now known as the *Ito calculus*, as follows:

2.4.2 Ito Calculus

Here one needs to establish the tools for solving a probabilistic (or stochastic) differential equation such as Equation (2.22)

$$dX_t = X_t(\sigma\, dW_t + \mu\, dt) \tag{2.22}$$

to provide for steps similar tools developed for deterministic or Newtonian calculus: along the lines of integration by parts, the chain rule, the product rule, and so on.

2.4.2.1 The Ito Lemma

The Ito Lemma is named after its discoverer, the brilliant Japanese mathematician Prof. Dr. Kiyoshi Ito, who passed away on November 10, 2008, at the age of 93. His work created a field of mathematics that is a calculus of probabilistic, or stochastic, variables (http://www.sjsu.edu/faculty/watkins/ito.htm).

Changes in a variable, such as the price of a stock, involve two components:

1) A deterministic component that is a function of time.
2) A probabilistic, or stochastic, component that depends upon a random variable.

Let S be the stock price at time t, and let dS be the infinitesimal change in S over the infinitesimal interval of time dt. The change in the random variable z over this interval of time is dz. Hence, the change in stock price is given by

$$dS = a\,dt + b\,dz \tag{2.23}$$

where a and b may be functions of S and t as well as other variables; that is,

$$dS = a(S, t, x)dt + b(S, t, x)dz \tag{2.24}$$

The *expected* value of dz is zero, so the expected value of dS is equal to the deterministic component $a\,dt$. Concomitantly, the random variable dz represents an accumulation of random contributions over the interval dt. Applying the central limit theorem, which implies that dz has a *normal* distribution and hence is completely characterized by its mean and standard deviation:

- The mean, or expected value, of dz is zero.
- The variance of a random variable that is the accumulation of independent effects over a given interval of time is proportional to the length of the interval: in this case dt. Hence, the standard deviation of dz is proportional to the square root of dt, namely, $(dt)^{1/2}$.

All of this means that the random variable dz is equivalent to a random variable $w(dt)^{1/2}$, where w is a standard normal variable with mean zero and standard deviation equal to unity.

Now consider another variable C, such as the price of a call option: it is a function of S and t, say $C = f(S, t)$: because C is a function of the probabilistic variable S, it will have a probabilistic component S as well as a deterministic component. Thus, C will have a representation of the form

$$dC = pdt + qdz \tag{2.25}$$

where the coefficients p and q may be functions of S, t, and possibly other variables; that is,

$$p = p(S, t, x), \quad \text{and} \quad q = q(S, t, x) \tag{2.26}$$

It remains to be determined how the functions p and q are related to the functions a and b in the equation

$$dS = adt + bdz \tag{2.27}$$

The answer may be found using the *Ito Lemma*:

The *deterministic* component p and *probabilistic* (viz., *stochastic*) component q of dC are given by

$$p = \partial f/\partial t + (\partial f/\partial S)a + \tfrac{1}{2}(\partial^2 f/\partial S^2)\, b^2 \tag{2.28a}$$

$$q = (\partial f/\partial S)b \tag{2.28b}$$

respectively.

Remark: The Ito Lemma is crucial in deriving differential equations for the value of derivative securities such as stock options.

A Proof of the Ito Lemma:
A derivation of the Ito Lemma is provided herein.

The Taylor series for $f(S, t)$ gives the increment in C as

$$dC = (\partial f/\partial t)dt + (\partial f/\partial S)dS + \tfrac{1}{2}(\partial^2 f/\partial S^2)(dS)^2 + (\partial^2 f/\partial S\partial t)(dS)(dt)$$
$$+ \tfrac{1}{2}(\partial^2 f/\partial t^2)(dt)^2$$
$$+ \text{Higher Order Terms} \tag{2.29}$$

For example: The increment in stock price dS is given by

$$dS = adt + bdz, \quad \text{but} \quad dz = vw[dt]^{1/2} \tag{2.30}$$

where w is a standard normal random variable and v is the scale of the variability of the random element; that is, its standard deviation. Substitution of $adt +$

$bvw(dt)^{\frac{1}{2}}$ for dS in Equation (2.27) yields

$$dC = (\partial f/\partial t)dt + (\partial f/\partial S)adt + \partial f/\partial S)bvw(dt)^{\frac{1}{2}}$$
$$+ \tfrac{1}{2}(\partial^2 f/\partial S^2)(adt + bvw(dt)^{\frac{1}{2}})^2$$
$$+ (\partial^2 f/\partial S\partial t)(adt + bvw(dt)^{\frac{1}{2}})(dt) + \tfrac{1}{2}(\partial^2 f/\partial t^2)(dt)^2$$
$$+ \text{higher order terms} \qquad (2.31)$$

With the expansion of the squared term and the product term the result is

$$dC = (\partial f/\partial t)dt + (\partial f/\partial S)adt + \partial f/\partial S)bvw(dt)^{\frac{1}{2}}$$
$$+ \tfrac{1}{2}(\partial^2 f/\partial S^2)(a^2\,dt^2 + 2abvw(dt)^{3/2} + b^2v^2w^2dt)$$
$$+ (\partial^2 f/\partial S\partial t)(a(dt)^2 + bvw(dt)^{3/2}) + \tfrac{1}{2}(\partial^2 f/\partial t^2)(dt)^2$$
$$+ \text{higher order terms} \qquad (2.32)$$

Taking into account the infinitesimal nature of dt so that dt to *any power higher than unity* vanishes, namely, *all* the terms in red in Equation (2.32) vanish, so that Equation (2.32) may be reduced to

$$dC = (\partial f/\partial t)dt + (\partial f/\partial S)adt + (\partial f/\partial S)bvw(dt)^{\frac{1}{2}}$$
$$+ \tfrac{1}{2}(\partial^2 f/\partial S^2)(b^2v^2w^2dt) \qquad (2.33)$$

Now, the expected value of w^2 is unity, hence the expected value of dC is

$$[\partial f/\partial t + (\partial f/\partial S)a + \tfrac{1}{2}(\partial^2 f/\partial S^2)b^2]dt \qquad (2.34)$$

This, (2.32), is the *deterministic* component of dC.

The probabilistic, or stochastic, component is the term that depends upon dz, which in (2.31) is represented as $vw(dt)^{\frac{1}{2}}$. Therefore, the probabilistic, or stochastic, component is

$$[(\partial f/\partial S)b]dz \qquad (2.35)$$

In the foregoing derivation, it would seem that there is an additional stochastic term that arises from the random deviations of w^2 from its expected value of 1; that is, the additional term

$$\tfrac{1}{2}(\partial^2 f/\partial S^2)(b^2v^2w^2dt) \qquad (2.36)$$

However, the variance of this additional term is proportional to $(dt)^2$, whereas the variance of the stochastic term given in Equation (2.33) is proportional to (dt). Thus the stochastic term given in Equation (2.34) vanishes in comparison with the stochastic term given in Equation (2.33).

Worked Example 2.1

If $X_t = \exp(W_t)$, what is dX_t?

Solution:

Since $X_t = \exp(W_t)$,

$$
\begin{aligned}
dX_t &= d[\exp(W_t)] && \text{by direct substitution} \\
&= \exp(W_t)dW_t + \tfrac{1}{2}\exp(W_t)\,dt && \text{by the Ito Lemma} \\
&= X_t dW_t + \tfrac{1}{2}X_t dt && \text{by direct substitution}
\end{aligned}
$$

Worked Example 2.2

On the differential of the *product* of a nonvolatile process *and* a stochastic process: If B_t is a nonvolatile process and X_t is a stochastic process, then the differential of the product $B_t X_t$ is

$$
d(B_t X_t) = B_t dX_t + X_t dB_t
$$

Solution:

Two approaches may be used:

First Method:

Applying the product rule in the differential of a functions two variables, the result is

$$
\begin{aligned}
d(B_t X_t)/dt &= B_t(dX_t/dt) + X_t(dB_t/dt) \\
&= (B_t dX_t + X_t dB_t)/dt \\
\therefore d(B_t X_t) &= B_t dX_t + X_t dB_t
\end{aligned}
$$

as required.

Second Method:

Applying the Ito Lemma first, and then simplifying the results:

Since X_t is a probabilistic (stochastic) process: $dX_t = \sigma_t\, dW_t + \mu_t\, dt$, and since B_t is a nonvolatile process: $dB_t = B_t\, dt$

Moreover, since

$$
(B_t + X_t)^2 \equiv B_t^2 + 2B_t X_t + X_t^2
$$

one obtains

$$
B_t X_t \equiv \tfrac{1}{2}[(B_t + X_t)^2 - B_t^2 - X_t^2]
$$

hence,

$$
\begin{aligned}
d(B_t X_t) &\equiv d\{\tfrac{1}{2}[(B_t + X_t)^2 - B_t^2 - X_t^2]\} \\
&\equiv \tfrac{1}{2}\,d[(B_t + X_t)^2 - B_t^2 - X_t^2] \\
&= \tfrac{1}{2}\,d\{(B_t + X_t)^2 - \tfrac{1}{2}\,dB_t^2 - \tfrac{1}{2}\,dX_t^2 \\
&= (B_t + X_t)\,d(B_t + X_t) - B_t\,dB_t + X_t dX_t
\end{aligned}
$$

and upon applying the Ito Lemma

$$
\begin{aligned}
&= (B_t + X_t)\, d(B_t + X_t) + \tfrac{1}{2}\sigma_t^2 dt - B_t dB_t - X_t dX_t - \tfrac{1}{2}\sigma_t^2 dt \\
&= (B_t + X_t)\, d(B_t + X_t) - B_t dB_t - X_t dX_t \\
&= (B_t\, dB_t + B_t dX_t + X_t dB_t + X_t dX_t) - B_t\, dB_t - X_t dX_t \\
&= B_t\, dX_t + X_t dB_t
\end{aligned}
$$

as required.

Remarks:

1) *The Ito Lemma Is Essential in the Derivation of the Black–Scholes Equation*
   An immediate question is whether Black–Scholes equation is an extension of Ito's Lemma for stable distributions of z other than the normal distribution. This question may be investigated in the study of stable distributions.

2) *Probabilistic (or Stochastic) Calculus and the Ito Lemma*
   Now, a probabilistic process S_t is a GBM if it satisfies the following probabilistic differential equation:

$$
dS_t = \mu S_t\, dt + \sigma S_t\, dW_t \tag{2.37}
$$

   where W_t is the Wiener or Brownian motion, and μ the percentage drift and σ the percentage volatility are constants.
   Here, μ is used to model deterministic behavior and σ is used to model a set of unpredictable events occurring during the motion.

3) *Solving the Probabilistic Differential Equation*
   For an arbitrary initial value S_0 the above probabilistic differential Equation (2.37) has the analytic solution

$$
S_t = S_0 \exp\{(\mu - \sigma^2/2)t + \sigma W_t\} \tag{2.38}
$$

This result may be obtained as follows:

- First divide the probabilistic differential Equation (2.37) by S_t in order to have the choice random variable on only one side.
- Next, write the equation in integral form (known as the Ito form):

$$
\int_0^t dS/S_t = \mu t + \sigma W_t, \text{ assuming } W_0 = 0 \tag{2.39}
$$

- At this point, it might appear that dS/S_t may be related to $\ln S_t$. *However, S_t is an Ito process, requiring the use of Ito Calculus.* Applying the Ito Lemma leads to

$$
d(\ln S_t) = dS_t/S_t - \tfrac{1}{2}(1/S_t^2)dS_t\, dS_t \tag{2.40}
$$

where $dS_t \, dS_t$ is the quadratic variation of the SDE, which may be written as $d[S]t$ or $<S.>_t$. In this case, one has

$$dS_t dS_t = \sigma 2 \, S_t^2 dt \tag{2.41}$$

- Substituting the value of dS_t in Equation (2.41), and simplifying, one obtains

$$\ln (S_t/S_0) = \{\mu - (\sigma^2/2)\}t + \sigma W_t \tag{2.42}$$

- Taking the exponential, and multiplying both sides by S_0 gives the solution (2.40).

2.5   A Numerical Study of the Geometric Brownian Motion (GBM) Model and the Random Walk Model Using R

2.5.1   Modeling Real Financial Data

2.5.1.1   The Geometric Brownian Motion (GBM) Model and the Random Walk Model

Historically, the GBM had been developed in terms of the concept of the *random walk*, $W_n(t)$, which has the following characteristics:

For a positive integer n, the binomial process $W_n(t)$ may be defined by the following characteristic properties:

1) $W_n(t) = 0$, at $t = 0$
2) $W_n(t)$ has layer spacing $1/n$
3) There exist up and down jumps, equal and of magnitude $1/\sqrt{n}$
4) $W_n(t)$ has measure P, given by up and down probabilities ½ everywhere.

Thus, if $X_1, X_2, X_3, \ldots, X_i$ is a sequence of independent binomial random variables taking values $+1$ or -1 with *equal* probability, then the value of W_n at the ith step is

$$W_n[i/n] = W_n[(i-1)/n] + X_i/\sqrt{n}, \quad \text{for all } i \geq 1 \tag{2.43}$$

What does a "Random Walk" look like?

The chart that follows is a simple example.

Random walk hypothesis test by increasing or decreasing the value of a fictitious stock based on the odd/even value of the decimals of π. The chart resembles a stock chart (Figure 2.5a).

Notes:

1) A Brownian motion wanders randomly with zero mean, but a company stock usually grows at some rate (even if it just increases with normal inflation).
2) Thus one may choose to add in a drift factor. For example, the process with a constant noise factor σ may be represented by

$$S_t = \sigma W_t + \mu t \tag{2.44}$$

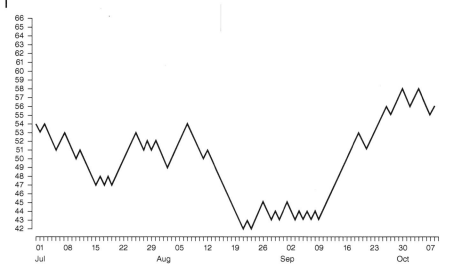

Figure 2.5 (a) An example of a random walk based on the odd/even values of the decimals of π.

- **A Deterministic Model versus a Probabilistic Model**
- **A *Combined* Deterministic and Probabilistic Model**

2.5.1.2 Other Models for Simulating Random Walk Systems Using R

The following are the two, among many, efficient models for simulating random walk systems that have been found to be useful in financial engineering and econometric data modeling:

1) The Cox–Ingersoll–Ross (CIR) model
2) The Chan–Karolyl– Longstaff–Sanders (CKLS) model

The *CIR model* is generally used to describe the evolution of interest rates. It is a type of "one-factor model" (aka a short rate model) as it models interest rate movements as driven by only *one* source of market risk. The model may be used in the valuation of interest rate derivatives. It was introduced in 1985 by J.C. Cox, J.E. Ingfersoll, and S.A. Ross. The CIR model assumes that the instantaneous interest rate follows the stochastic (probabilistic) differential equation. The CIR is an ergodic process, and possesses a stationary distribution.

The *CKLS models* belong to a class of parametric stochastic differential equations widely used in many finance applications, in particular to model interest rates or asset prices. This model nests a class of asset pricing models such as the CIR. This flexibility of the CKLS model spawns many empirical applications.

2.5.2   Some Typical Numerical Examples of Financial Data Using R

To demonstrate the GBM model as well as the random walk model, the R Program `Sim.DiffProc`, available from CRAN, is chosen. A typical worked example is selected, and the associated R program is run, outputting the associated numerical and graphical results:

Package:	`Sim.DiffProc`
Type:	Package
Title:	Simulation of Diffusion Processes
Version:	3.2
Date:	February 9, 2016
Author	A. C. Guidoum and K. Boukhetala
Maintainer	A. C. Guidoum <acguidoum@usthb.dz>
Encoding	UTF-8
Depends	R (>= 2.15.1)
Imports	scatterplot3d, rgl
Description	This R program provides the functions for simulation and modeling of stochastic differential equations (SDEs): the Ito type. This package contains many objects, the numerical methods to find the solutions to SDEs (1, 2, and 3 dimemsion/s), for simulating the corresponding flow trajectories, with satisfactory accuracy. Many theoretical problems on the SDEs have become the object of research, as statistical analysis and simulation of solution of SDE's, enabling workers in different domains to use these equations to modeling and to analyze practical problems, in financial and actuarial modeling and other areas of application.
License	GPL (>= 3) | file LICENCE
Classification	/MSC 37H10, 37M10, 60H05, 60H10, 60H35, 60J60, 68N15
Needs Compilation	yes
Repository	CRAN
Date/ Publication	2016-02-09 09:46:41

R **topics documented**:

```
Sim.DiffProc-package
bconfint
BM
bridgesde1d
bridgesde2d
bridgesde3d
fitsde
fptsde1d
fptsde2d
fptsde3d
HWV
Irates
plot2d
rsde1d
rsde2d
rsde3d
snssde1d
snssde2d
snssde3d
st.int
```

The following numerical examples are selected, all being typical representations of the geometric Brownian motion (GBM) model and the random walk model:

Worked Example 2-I in R:

BM: Brownian motion, Brownian bridge, geometric Brownian motion, and arithmetic Brownian motion simulators

Description
The (S3) generic function for simulation of Brownian motion, Brownian bridge, geometric Brownian motion, and arithmetic Brownian motion.

Usage
```
BM(N, . . . )
BB(N, . . . )
GBM(N, . . . )
ABM(N, . . . )
```

Default S3 method:

```
BM(N =100,M=1,x0=0,t0=0,T=1,Dt, . . . )
```

Default S3 method:

```
BB(N =100,M=1,x0=0,y=1,t0=0,T=1,Dt, . . . )
```

Default S3 method:

```
GBM(N =100,M=1,x0=1,t0=0,T=1,Dt,theta=1,sigma=1, . . . )
```

Default S3 method:

```
ABM(N =100,M=1,x0=0,t0=0,T=1,Dt,theta=1,sigma=1, . . . )
```

Arguments

N	Number of simulation steps.
M	Number of trajectories.
x0	Initial value of the process at time *t0*.
y	Terminal value of the process at time *T* of the BB.
t0	Initial time.
T	Final time.
Dt	Time step of the simulation (discretization). If it is missing a default $\Delta t = (T - t_0)/N$
theta	The interest rate of the ABM and GBM.
sigma	The volatility of the ABM and GBM.
. . .	Further arguments for (nondefault) methods.

Details

The function BM returns a trajectory of the standard Brownian motion (Wiener process) in the time interval $[t_0, T]$. Indeed, for $W(dt)$ it holds true that

$$W(dt) \rightarrow W(dt) - W(0) \rightarrow N(0, dt)$$

where $N(0, 1)$ is normal distribution Normal.

The function BB returns a trajectory of the Brownian bridge starting at x_0 at time t_0 and ending at y at time T; that is, the diffusion process solution of stochastic differential equation:

$$dX_t = \{(y - X_t)/(T - t)\} \, dt + dW_t$$

The function GBM returns a trajectory of the geometric Brownian motion starting at x_0 at time t_0; that is, the diffusion process solution of stochastic differential equation:

$$dX_t = \theta X_t \, dt + \sigma X_t \, dW_t$$

The function GBM returns a trajectory of the arithmetic Brownian motion starting at x_0 at time t_0; that is, the diffusion process solution of stochastic differential equation:

$$dX_t = \theta dt + \sigma dW_t$$

Value

X   a visible t_s object

Author(s)

A.C. Guidoum and K. Boukhetala.

Examples

```
op <- par(mfrow = c(2, 2))
##
## Brownian motion
set.seed(1234)
X <- BM(N = 1000, M = 50)
plot(X,plot.type="single")
lines(as.vector(time(X)),rowMeans(X),col="red")
##
## Brownian bridge
set.seed(1234)
X <- BB(N = 1000, M =50)
plot(X,plot.type="single")
lines(as.vector(time(X)),rowMeans(X),col="red")
##
## Geometric Brownian motion
set.seed(1234)
X <- GBM(N = 1000, M = 50)
plot(X,plot.type="single")
lines(as.vector(time(X)),rowMeans(X),col="red")
##
## Arithmetic Brownian motion
```

```
set.seed(1234)
X <- ABM(N = 1000, M = 50)
plot(X,plot.type="single")
lines(as.vector(time(X)),rowMeans(X),col="red")
##
par(op)
```

In the R domain:

```
>
> install.packages("Sim.DiffProc ")
Installing package into 'C:/Users/Bert/Documents/R/win-
library/3.2'
(as 'lib' is unspecified)
--- Please select a CRAN mirror for use in this session ---
Warning message:
```

A CRAN mirror is selected.

```
package 'Sim.DiffProc ' is not available (for R version 3.2.2)
> library(Sim.DiffProc)
Package 'Sim.DiffProc' version 3.2 loaded.
help(Sim.DiffProc) for summary information.
Warning message:
package 'Sim.DiffProc' was built under R version 3.2.5
> ls("package:Sim.DiffProc")
```

```
[1] "ABM"                   "ABM.default"        "AIC.fitsde"
[4] "BB"                    "BB.default"         "bconfint"
[7] "bconfint.bridgesde1d"  "bconfint.           "bconfint.
                             bridgesde2d"         bridgesde3d"
[10] "bconfint.default"     "bconfint.fptsde1d"  "bconfint.fptsde2d"
[13] "bconfint.fptsde3d"    "bconfint.rsde1d"    "bconfint.rsde2d"
[16] "bconfint.rsde3d"      "bconfint.snssde1d"  "bconfint.snssde2d"
[19] "bconfint.snssde3d"    "bconfint.st.int"    "BIC.fitsde"
[22] "BM"                   "BM.default"         "bridgesde1d"
[25] "bridgesde1d.          "bridgesde2d"        "bridgesde2d.
default"                                          default"
[28] "bridgesde3d"          "bridgesde3d.        "coef.fitsde"
                             default"
[31] "confint.fitsde"       "fitsde"             "fitsde.default"
[34] "fptsde1d"             "fptsde1d.default"   "fptsde2d"
[37] "fptsde2d.default"     "fptsde3d"           "fptsde3d.default"
```

```
 [40] "GBM"                  "GBM.default"          "HWV"
 [43] "HWV.default"          "kurtosis"             "kurtosis.bridgesde1d"
 [46] "kurtosis.            "kurtosis.             "kurtosis.default"
      bridgesde2d"           bridgesde3d"
 [49] "kurtosis.fptsde1d"   "kurtosis.fptsde2d"    "kurtosis.fptsde3d"
 [52] "kurtosis.rsde1d"     "kurtosis.rsde2d"      "kurtosis.rsde3d"
 [55] "kurtosis.snssde1d"   "kurtosis.snssde2d"    "kurtosis.snssde3d"
 [58] "kurtosis.st.int"     "lines.bridgesde1d"    "lines.bridgesde2d"
 [61] "lines.bridgesde3d"   "lines.snssde1d"       "lines.snssde2d"
 [64] "lines.snssde3d"      "lines.st.int"         "lines2d"
 [67] "lines2d.             "lines2d.default"      "lines2d.snssde2d"
      bridgesde2d"
 [70] "logLik.fitsde"       "mean.bridgesde1d"     "mean.bridgesde2d"
 [73] "mean.bridgesde3d"    "mean.fptsde1d"        "mean.fptsde2d"
 [76] "mean.fptsde3d"       "mean.rsde1d"          "mean.rsde2d"
 [79] "mean.rsde3d"         "mean.snssde1d"        "mean.snssde2d"
 [82] "mean.snssde3d"       "mean.st.int"          "median.
                                                    bridgesde1d"
 [85] "median.             "median.               "median.fptsde1d"
      bridgesde2d"           bridgesde3d"
 [88] "median.fptsde2d"     "median.fptsde3d"      "median.rsde1d"
 [91] "median.rsde2d"       "median.rsde3d"        "median.snssde1d"
 [94] "median.snssde2d"     "median.snssde3d"      "median.st.int"
 [97] "moment"              "moment.               "moment.
                             bridgesde1d"           bridgesde2d"
[100] "moment.             "moment.default"       "moment.fptsde1d"
      bridgesde3d"
[103] "moment.fptsde2d"     "moment.fptsde3d"      "moment.rsde1d"
[106] "moment.rsde2d"       "moment.rsde3d"        "moment.snssde1d"
[109] "moment.snssde2d"     "moment.snssde3d"      "moment.st.int"
[112] "OU"                  "OU.default"           "plot.bridgesde1d"
[115] "plot.bridgesde2d"    "plot.bridgesde3d"     "plot.fptsde1d"
[118] "plot.fptsde2d"       "plot.fptsde3d"        "plot.rsde1d"
[121] "plot.rsde2d"         "plot.rsde3d"          "plot.snssde1d"
[124] "plot.snssde2d"       "plot.snssde3d"        "plot.st.int"
[127] "plot2d"              "plot2d.               "plot2d.default"
                             bridgesde2d"
[130] "plot2d.snssde2d"     "plot3D"               "plot3D.
                                                    bridgesde3d"
[133] "plot3D.default"      "plot3D.snssde3d"      "points.
                                                    bridgesde1d"
```

```
[136] "points.          "points.             "points.snssde1d"
      bridgesde2d"       bridgesde3d"

[139] "points.snssde2d"  "points.snssde3d"   "points.st.int"

[142] "points2d"         "points2d.           "points2d.default"
                         bridgesde2d"

[145] "points2d.         "print.bridgesde1d" "print.bridgesde2d"
      snssde2d"

[148] "print.            "print.fitsde"      "print.snssde1d"
      bridgesde3d"

[151] "print.snssde2d"   "print.snssde3d"    "print.st.int"

[154] "quantile.         "quantile.           "quantile.
      bridgesde1d"       bridgesde2d"         bridgesde3d"

[157] "quantile.         "quantile.fptsde2d" "quantile.fptsde3d"
      fptsde1d"

[160] "quantile.rsde1d"  "quantile.rsde2d"   "quantile.rsde3d"

[163] "quantile.         "quantile.snssde2d" "quantile.snssde3d"
      snssde1d"

[166] "quantile.st.int"  "rsde1d"            "rsde1d.default"

[169] "rsde2d"           "rsde2d.default"    "rsde3d"

[172] "rsde3d.default"   "skewness"          "skewness.
                                             bridgesde1d"

[175] "skewness.         "skewness.          "skewness.default"
      bridgesde2d"       bridgesde3d"

[178] "skewness.         "skewness.fptsde2d" "skewness.fptsde3d"
      fptsde1d"

[181] "skewness.rsde1d"  "skewness.rsde2d"   "skewness.rsde3d"

[184] "skewness.         "skewness.snssde2d" "skewness.snssde3d"
      snssde1d"

[187] "skewness.st.int"  "snssde1d"          "snssde1d.default"

[190] "snssde2d"         "snssde2d.default"  "snssde3d"

[193] "snssde3d.default" "st.int"            "st.int.default"

[196] "summary.fitsde"   "summary.fptsde1d"  "summary.fptsde2d"

[199] "summary.fptsde3d" "summary.rsde1d"    "summary.rsde2d"

[202] "summary.rsde3d"   "summary.snssde1d"  "summary.snssde2d"

[205] "summary.snssde3d" "summary.st.int"    "time.bridgesde1d"

[208] "time.bridgesde2d" "time.bridgesde3d"  "time.snssde1d"

[211] "time.snssde2d"    "time.snssde3d"     "time.st.int"

[214] "vcov.fitsde"

>
> op <- par(mfrow = c(2, 2))
> utils:::menuShowCRAN()
```

> *# Output:* Figure 2.4b **Sim.DiffProc-1**

Figure 2.4 (b) `Sim.DiffProc-1`.

```
>
> ##
> ## Brownian motion
> set.seed(1234)
> X <- BM(N = 1000, M = 50)
> plot(X,plot.type="single")
```

> *# Output:* Figure 2.5b **Sim.DiffProc-2**

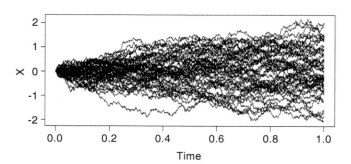

Figure 2.5 (b) `Sim.DiffProc-2` (Ordinates are arbitrary) (Available at https://en.wikipedia.org/wiki/Random_walk_hypothesis).

```
>
> lines(as.vector(time(X)),rowMeans(X),col="red")
```

> # *Output:* Figure 2.6 **Sim.DiffProc-3**

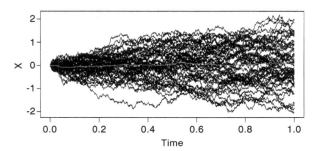

Figure 2.6 Sim.DiffProc-3 (Ordinates are arbitrary)

```
>
> ##
> ## Brownian bridge
> set.seed(1234)
> X <- BB(N = 1000, M = 50)
> plot(X,plot.type="single")
```

> # *Output:* Figure 2.7 **Sim.DiffProc-4**

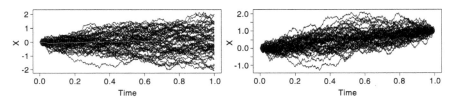

Figure 2.7 Sim.DiffProc-4 (Ordinates are arbitrary)

```
>
> lines(as.vector(time(X)),rowMeans(X),col="red")
```

> # *Output:* Figure 2.8 **Sim.DiffProc-5**

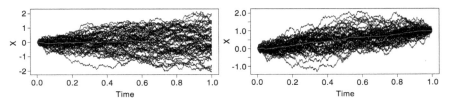

Figure 2.8 Sim.DiffProc-5 (Ordinates are arbitrary)

```
>
> ##
> ## Geometric Brownian motion
> set.seed(1234)
> X <- GBM(N = 1000, M = 50)
> plot(X,plot.type="single")
```

> # *Output:* Figure 2.9 **Sim.DiffProc-6**

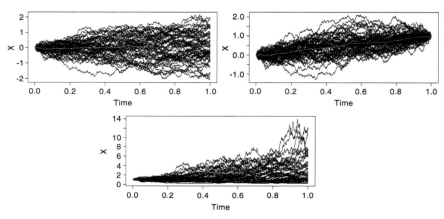

Figure 2.9 Sim.DiffProc-6 (Ordinates are arbitrary)

```
>
> lines(as.vector(time(X)),rowMeans(X),col="red")
```

> # *Output:* Figure 2.10 **Sim.DiffProc-7**

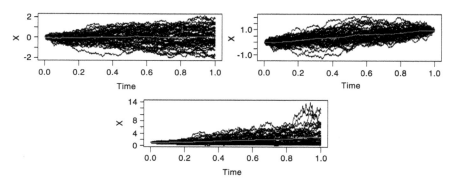

Figure 2.10 Sim.DiffProc-7 (Ordinates are arbitrary)

```
>
> ##
> ## Arithmetic Brownian motion
> set.seed(1234)
> X <- ABM(N = 1000, M = 50)
> plot(X,plot.type="single")
```

> # *Output:* Figure 2.11 **Sim.DiffProc-8**

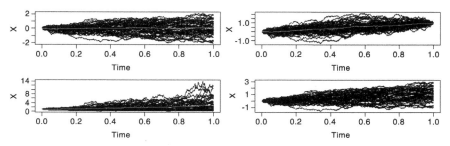

Figure 2.11 Sim.DiffProc-8 (Ordinates are arbitrary)

```
>
> lines(as.vector(time(X)),rowMeans(X),col="red")
> ##
```

> # *Output:* Figure 2.12 **Sim.DiffProc-9**

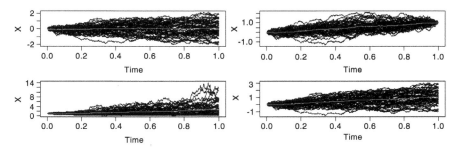

Figure 2.12 Sim.DiffProc-9 (Ordinates are arbitrary)

```
> ##
> par(op)
>
```

Worked Example 2 in R:

Irates: Monthly interest rates

Description
The CKLS Model of Short-term Interest Rates
   Chan, K. C., Karolyi, G. A., Longstaff, F. A., Sanders, A. B. (1992) An empirical comparison of alternative models of the short-term interest rate. Journal of Finance, **47** (3), 1209–1227.
   Listing monthly observations from December 1946 to February 1991
   Number of observations: 531
   Observation: country

Values
Usage
   data(Irates)

Format
   A time series containing the following:

r1 interest rate for a maturity of 1 months (% per year).
r2 interest rate for a maturity of 2 months (% per year).
r3 interest rate for a maturity of 3 months (% per year).
r5 interest rate for a maturity of 5 months (% per year).
r6 interest rate for a maturity of 6 months (% per year).
r11 interest rate for a maturity of 11 months (% per year).
r12 interest rate for a maturity of 12 months (% per year).
r36 interest rate for a maturity of 36 months (% per year).
r60 interest rate for a maturity of 60 months (% per year).
r120 interest rate for a maturity of 120 months (% per year).

Source
   McCulloch, J.H. and Kwon, H.C. (1993) U.S. term structure data, 1947–1991, Ohio State Working paper 93–6, Ohio State University, Columbus, Ohio
   The datasets Irates are in package Ecdat.

References
   Croissant, Y. (2014) Ecdat: Data sets for econometrics.
   R package version 0.2–5.

Examples

```
data(Irates)
rates <- Irates[,"r1"]
rates <- window(rates, start=1964.471, end=1989.333)
·## CKLS modele vs CIR modele
```

```
## CKLS : dX(t) = (theta1+theta2* X(t))* dt + theta3 *
X(t)^theta4 * dW(t)
fx <- expression(theta[1]+theta[2]*x)
gx <- expression(theta[3]*x^theta[4])
fitmod <- fitsde(rates,drift=fx,diffusion=gx,pmle="euler",
start = list(theta1=1,theta2=1,
theta3=1,theta4=1),optim.method = "L-BFGS-B")
theta <- coef(fitmod)
N <- length(rates)
res <- snssde1d(drift=fx,diffusion=gx,M=200,t0=time(rates)[1],
T=time(rates)[N],
Dt=deltat(rates),x0=rates[1],N)
plot(res,plot.type="single",ylim=c(0,50))
lines(rates,col=2,lwd=2)
legend("topleft",c("real data","CKLS modele"),
inset = .01,col=c(2,1),lwd=2,cex=0.8)
dev.new()
plot(res,plot.type="single",type="n",ylim=c(0,35))
lines(rates,col=2,lwd=2)
lines(time(res),mean(res),col=3,lwd=2)
lines(time(res),bconfint(res,level=0.95)[,1],col=4,lwd=2)
lines(time(res),bconfint(res,level=0.95)[,2],col=4,lwd=2)
legend("topleft",c("real data","mean path",
paste("bound of", 95,"percent confidence")),
inset = .01,col=2:4,lwd=2,cex=0.8)
```

In the R domain:

```
>
> data(Irates)
> rates <- Irates[,"r1"]
> rates <- window(rates, start=1964.471, end=1989.333)
> ## CKLS modele vs CIR modele
> ## CKLS : dX(t) = (theta1+theta2* X(t))* dt + theta3 *
>##              X(t)^theta4 * dW(t)
> fx <- expression(theta[1]+theta[2]*x)
> gx <- expression(theta[3]*x^theta[4])
> fitmod <- fitsde(rates,drift=fx,diffusion=gx,pmle="euler",
+ start = list(theta1=1,theta2=1,
+ theta3=1,theta4=1),optim.method = "L-BFGS-B")
> theta <- coef(fitmod)
> N <- length(rates)
> res <- snssde1d(drift=fx,diffusion=gx,
+ M=200,t0=time(rates)[1],
```

```
+ T=time(rates)[N],
+ Dt=deltat(rates),x0=rates[1],N)
> plot(res,plot.type="single",ylim=c(0,50))

> # Output: Figure 2.13 Irates-1
```

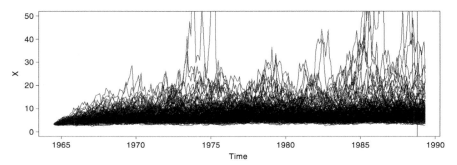

Figure 2.13 Irates-1 (Ordinates are arbitrary) The CKLS model of short-term interest rates: interest rate for a maturity of 1 month.

```
>
> lines(rates,col=2,lwd=2)

> # Output: Figure 2.14 Irates-2
```

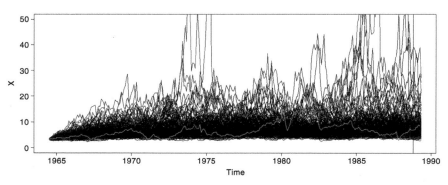

Figure 2.14 Irates-2 the CKLS model of short-term interest rates: interest rate for a maturity of 1 month Irates-1 Comparing with the mean interest rate.

```
>
> legend("topleft",c("real data","CKLS modele"),
+ inset = .01,col=c(2,1),lwd=2,cex=0.8)

> # Output: Figure 2.15 Irates-3
>
> dev.new()
```

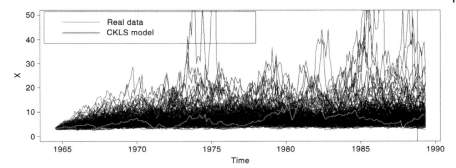

Figure 2.15 `Irates`-3 The CKLS model of short-term interest rates: interest rate for a maturity of 1 month `Irates`-1 Comparing with the mean interest rate – with legend.

```
> # Output: Figure 2.16 Irates-4
```

Figure 2.16 `Irates`-4: Preparation for new data to follow (pattern only).

```
>
> plot(res,plot.type="single",type="n",ylim=c(0,35))

> # Output: Figure 2.17 Irates-5
```

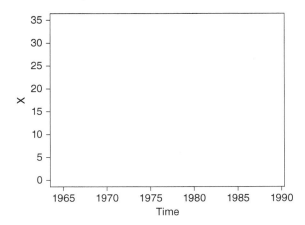

Figure 2.17 `Irates`-5-5: With axes coordinates added.

```
>
> lines(rates,col=2,lwd=2)

> # Output: Figure 2.18 Irates-6
```

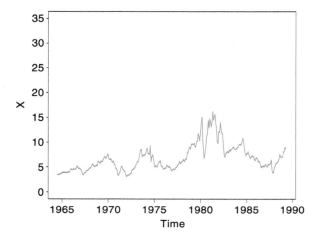

Figure 2.18 `Irates-6`: `lines(rates,col=2,lwd=2)` Mean interest rates added (red).

```
>
> lines(time(res),mean(res),col=3,lwd=2)

> # Output: Figure 2.19 Irates-7
```

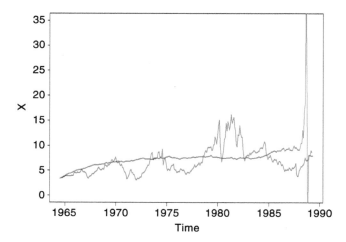

Figure 2.19 `Irates-7`: Line for mean path added (green).

>
> lines(time(res),bconfint(res,level=0.95)[,1],col=4,lwd=2)

> # *Output:* Figure 2.20 Irates-8

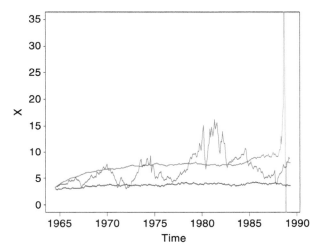

Figure 2.20 Irates-8: Lower bound of 95% confidence level added (blue).

>
> lines(time(res),bconfint(res,level=0.95)[,2],col=4,lwd=2)

> # *Output:* Figure 2.21 Irates-9

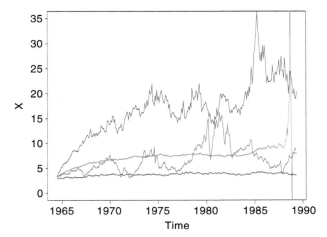

Figure 2.21 Irates-9: Upper bound of 95% confidence level added (blue) (Ordinates are arbitrary).

```
>
> legend("topleft",c("real data","mean path",paste("bound of",
+ 95,"percent confidence")),
+ inset = .01,col=2:4,lwd=2,cex=0.8)

> # Output: Figure 2.22 Irates-10
>
```

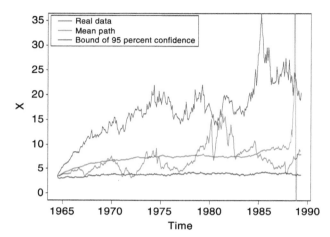

Figure 2.22 `Irates`-10: Legend added for real data, mean path, and upper and lower bounds of 95% confidence levels (Ordinates are arbitrary).

Review Questions and Exercises

1 Define differential calculus.

   a Define Ito calculus.

   b What are the essential differences between these two concepts? Explain

2 **a** Why is the Ito calculus suitable for analyzing GBM (*Geometric Brown motion*)?

   b i) What are the advantages of using Ito calculus in this application?

   ii) What are the disadvantages? Any limitations?

3 GBM: Let Y be a normal $N(0, 1)$ function, then the process $X_t = Y\sqrt{t}$ is continuous, and is marginally distributed as a normal $N(0, t)$.

   Is X_t a Brownian motion? Explain.

4   GBM: Let Y_t and $\underline{Y}_t$ be two independent Brownian motions, and λ is a constant such that $-1 < \lambda < 1$. Then the process

$$X_t = \lambda\, Y_t + \{\sqrt{(1 - \lambda^2)}\}\ \underline{Y}_t$$

is continuous and has marginal distributions $N(0, t)$.
Is the process X_t a GBM? Explain.

5   The Black–Scholes formula for pricing European call options is

$$V(s, T) = s\,\Phi[\{\log(s/k) + (r + \tfrac{1}{2}\,\sigma^2)T\}/\sigma\sqrt{T}] - ke^{-\tau T}\Phi[\{\log(s/k)$$
$$+(r - \tfrac{1}{2}\sigma^2)T\}/\sigma\sqrt{T}]$$

where $\Phi(x) = (2\pi)^{-\frac{1}{2}}\int_{-\infty}^{x}\exp(y^2/2)dy$, the probability that a normal distribution
$N(0, 1)$ has values less than x, then one may calculate that $V_0 = V(s, T)$.
Consider the change of variable

$$v = -(x + \tfrac{1}{2}\sigma^2 T)/\sigma\sqrt{T},$$

and use it to establish the common form of the Black–Scholes formula, namely,

$$V_0 = s\Phi(\alpha + \sigma\sqrt{T}) - ke^{-\tau T}\Phi(\alpha)$$

3

Classical Mathematical Models in Financial Engineering and Modern Portfolio Theory

3.1 An Introduction to the Cost of Money in the Financial Market

The Prevailing Financial Scenery   Much of financial engineering depends on an understanding of the perceived behavior of the financial market, particularly the stock markets that include daily trading of corporate and governmental stocks, bonds, mutual funds, derivatives, and many other commodities.

Some Current Observations in the United States: As of August 11, 2016

A) *Financial Interest Rates: The Cost of Money*
A typical local (California) federally insured credit union is advertising the following *typical* savings and lending rates for the month of July, 2016:

Money market savings account, dividend rate: 0.85% per annum

Free checking account, dividend rate: 0.25% per annum

Certificate account rate, dividend rate: 1.00% per annum

Secured real estate loan rate: 4.00% per annum

Credit card loan rate: 8.75% per annum

Remark: It certainly should not escape one's attention that such low levels of dividend rates in ordinary savings accounts may well be a prime incentive for one's interests in seeking higher returns in other avenues within the financial market or elsewhere!

B) *The Financial News Media*
CNN reported in http://money.cnn.com that there are "5 Reasons Why Stocks May Keep Going Higher":

"Most major stock market indexes in the U.S. are trading near their all-time highs. So why do so many investors . . . and even Wall Street experts . . . feel so lousy?"

Applied Probabilistic Calculus for Financial Engineering: An Introduction Using R, First Edition.
Bertram K. C. Chan.
© 2017 John Wiley & Sons, Inc. Published 2017 by John Wiley & Sons, Inc.
Companion website: www.wiley.com/go/chan/appliedprobabilisticcalculus

"There is a growing skepticism that stocks can keep hitting new records. Yet, the market keeps grinding higher. Investors may be talking about how nervous they are. But the numbers tell a different story. The VIX, a measure of volatility often dubbed Wall Street's 'Fear Gauge,' is near its lowest level in a year. And *CNN Money's Fear & Greed Index*, which looks at the VIX and six other indicators of investor sentiment, has been showing signs of Extreme Greed in the market for the past month. Concerns about the U.K. Brexit (viz., United Kingdom of Britain exiting the European Union) vote initially rocked the market . . . (for less than a week) . . . but such worries, that Brexit could wind up being the 2016 equivalent of Lehman Brothers, almost turned out to be short-lived. Investors do not seem terribly concerned about the impact that the U.S. presidential election (2016 being such an election year in the United States) will have on stocks either – even though the gains may be modest –"

C) *For the Foreseeable Future.*
Here are five outstanding reasons:

1) Corporate America is getting healthier.
2) Stocks could stay pricy for a while.
3) The Fed (viz., U.S. Federal Reserve Banks—the central banking system) is still your friend!
4) The (corporate) merger boom is not over yet!
5) "Slow and steady wins the race!"

Conclusion: "Patient investors should be rewarded if they don't do anything crazy. We are invested conservatively and believe it is better to make money slowly than just take speculative bets. There are still ways to make money and do it smartly."

Creative Financing   Indeed, there are virtually unlimited ways to create opportunities of financing. Here are two examples:

I) *Fixed Rate Mortgage versus Adjustable Rate Mortgage (ARM) versus LIBOR ARM*

- A fixed rate mortgage has the same payment for the entire term of the loan.
- An *A*djustable *R*ate *M*ortgage (ARM) has a rate that can change, causing the monthly payment to increase or decrease or remain unchanged.
- LIBOR (*L*ondon *I*nter*B*ank *O*ffered *R*ate) is an index set by a group of London-based banks, and sometimes used as a base for U.S. adjustable rate mortgages.

II) *Seller-Financing*
As an example, the owner of a real asset may offer his own line of credit to any acceptable buyers of the said asset.

3.2 Modern Theories of Portfolio Optimization

In making investment decisions in asset allocation, the modern portfolio theory focuses on potential return in relation to the *concomitant* potential risk. The strategy is to select and then evaluate individual investments (securities, bonds, funds, derivatives, commodities, etc.) *as part of an overall portfolio rather than solely for their own strengths or weaknesses as an investment.*

Asset allocation is therefore a primary tactic—because it allows investors to create portfolios—to obtain the strongest possible return without assuming a greater level of risk than they would like to bear!

Another critical feature of an acceptable portfolio theory is that investors must be rewarded, in terms of realizing a greater return, for assuming greater risk. Otherwise, there would be little motivation and incentive to make investments that might result in a loss of principal!

With such preconditions, two outstanding theories of portfolio allocation are presented herein:

1) The Markowitz model
2) The Black–Litterman model

There are numerous modifications/improvements for these "standard bearers" that maintain a fruitful area in research and development in mathematical finance and financial engineering. In this chapter, these two approaches will be presented, followed by the more favored modifications of these theories!

3.2.1 The Markowitz Model of Modern Portfolio Theory (MPT)

Modern Portfolio Theory   Modern portfolio theory, or mean variance analysis, is a mathematical model for building a portfolio of financial assets such that the *expected return* is maximized for *a given level of risk*, defined as the *variance*. Its key feature is as follows:

> "The risks of the asset and the return of profits should not be assessed by themselves, but by how it contributes to an overall risk and return of the portfolio."

3.2.1.1   Risk and Expected Return

The MPT assumes that investors are risk-averse, meaning that given the two portfolios that offer the same expected return, investors will prefer the *less* risky one: Thus, *an investor will take on increased risk* only *if compensated by higher expected returns. Conversely, an investor who wants higher expected returns must accept more risk.*

This trade-off will be the same for all investors, but different investors will evaluate the trade-off differently based on individual risk-aversion characteristics. The implication is that a *rational* investor will not invest in a portfolio if a second portfolio exists with a more favorable *risk-expected return profile*—that is, if for that level of risk, an alternative portfolio exists that has better expected returns.

Under the model:

- Portfolio return is the *proportion-weighted combination* of the constituent assets' returns.
- Portfolio volatility is a function of the *correlations* ρ_{ij} of the component assets, for all asset pairs (i, j).

In general:

- Expected return:

$$E(R_p) = \sum_i w_i E(R_i) \tag{3.1}$$

where

R_p is the return on the portfolio,
R_i is the return on asset i, and w_i is the weighting of component asset i (i.e., the proportion of asset i in the portfolio).

- Portfolio return variance σ_p is the *statistical sum* of the variances of the individual components $\{\sigma_i, \sigma_j\}$ defined as follows:

$$\sigma_p^2 = \sum_i w_i^2 \sigma_i^2 + \sum_i \sum_{j \neq i} w_i w_j \sigma_i \sigma_j \rho_{ij} \tag{3.2}$$

where ρ_{ij} is the *correlation coefficient* between the returns on assets i and j. Alternatively, the expression can be written as follows:

$$\sigma_p^2 = \sum_i \sum_j w_i w_j \sigma_i \sigma_j \rho_{ij} \tag{3.3}$$

where $\rho_{ij} = 1$ for $i = j$.

- Portfolio return volatility (standard deviation):

$$\sigma_p = \sqrt{\sigma_p^2} \tag{3.4}$$

Thus, for a *two-asset* portfolio:

- Portfolio return:

$$E(R_p) = w_A E(R_A) + w_B E(R_B) = w_A E(R_A) + (1 - w_A) E(R_B). \tag{3.5}$$

- Portfolio variance:

$$\sigma_p^2 = w_A^2 \sigma_A^2 + w_B^2 \sigma_B^2 + 2 w_A w_B \sigma_A \sigma_B \rho_{AB} \tag{3.6}$$

And, for a *three-asset* portfolio:

- Portfolio return:

$$E(R_p) = w_A E(R_A) + w_B E(R_B) + w_C E(R_C) \tag{3.7}$$

- Portfolio variance:

$$\sigma_p^2 = w_A^2 \sigma_A^2 + w_B^2 \sigma_B^2 + w_C^2 \sigma_C^2 + 2w_A w_B \sigma_A \sigma_B \rho_{AB} + 2w_A w_C \sigma_A \sigma_C \rho_{AC}$$
$$+ 2w_B w_C \sigma_B \sigma_C \rho_{BC} \tag{3.8}$$

3.2.1.2 Diversification

An investor may reduce portfolio risk simply by holding combinations of instruments that are not perfectly positively *correlated:*

$$\text{correlation coefficient} - 1 \leq \rho_{ij} < 1 \tag{3.9}$$

In other words, investors can reduce their exposure to individual asset risk by selecting a *diversified* portfolio of assets: *Diversification may allow the same portfolio expected return with reduced risk.* These ideas had been first proposed by Markowitz and later reinforced by other economists and mathematicians who have expressed ideas in the limitation of variance through portfolio theory.

Thus, if all the asset pairs have correlations of 0 (viz., they are perfectly uncorrelated), the portfolio's return variance is the sum of all assets of the square of the fraction held in the asset times the asset's return variance (and the portfolio standard deviation is the square root of this sum).

3.2.1.3 Efficient Frontier with No Risk-Free Assets

As shown in the graph in Figure 3.1, every possible combination of the risky assets, without including any holdings of the risk-free asset, may be plotted in

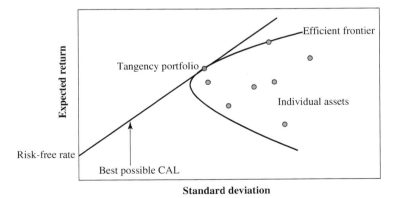

Standard deviation

Figure 3.1 *Efficient Frontier:* The hyperbola, popularly known as the "Markowitz Bullet," is the efficient frontier if no risk-free asset is available. (For a risk-free asset, the straight line is the efficient frontier.)

risk versus expected-return space, and the collection of all such possible portfolios defines a characteristic region in this space.

The left boundary of this region is a hyperbola, and the upper edge of this region is the *efficient frontier* in the absence of a risk-free asset (called "the Markowitz Bullet"). Combinations along this upper edge represent portfolios (including no holdings of the risk-free asset) for which there is *lowest* risk for a given level of expected return. Equivalently, a portfolio lying on the efficient frontier represents the combination offering the best possible expected return for the given risk level. The tangent to the hyperbola at the tangency point indicates the best possible Capital Allocation Line (CAL).

In the description and mathematical development of the MPT, matrices are generally preferred for calculations of the efficient frontier.

Remarks:

In matrix form, for a given "risk tolerance" $q \in (0, \infty)$, the efficient frontier may be obtained by minimizing the following expression:

$$w^T \sum w - q^* R^T w \tag{3.10}$$

Here

1) w is a vector of portfolio weights and $\sum_i w_i = 1$. (The weights may be negative, which means investors can *short* a security.)
2) $\sum$ is the *covariance matrix* for the returns on the assets in the portfolio.
3) $q \geq 0$ is a *risk tolerance* factor, where 0 results in the portfolio with minimal risk and ∞ results in the portfolio infinitely far out on the frontier with both expected return and risk unbounded.
4) R is a vector of expected return.
5) $w^T \sum w$ is the variance of portfolio return.
6) $R^T w$ is the expected return on the portfolio.

The above optimization finds the point on the frontier at which the *inverse* of the slope of the frontier would be q if portfolio return variance instead of standard deviation were plotted horizontally. The frontier is parametric on q.

3.2.1.4 The Two Mutual Fund Theorem

An important result of the above analysis is the *Two Mutual Fund Theorem* that states:

> "Any portfolio on the efficient frontier may be generated by using a combination of any two given portfolios on the frontier; the latter two given portfolios are the 'mutual funds' in the theorem's name."

Thus, in the absence of a risk-free asset, an investor may achieve any desired efficient portfolio even if all that is available is a pair of efficient mutual funds:

1) If the location of the desired portfolio on the frontier is between the locations of the two mutual funds, then both mutual funds may be held in positive quantities.
2) If the desired portfolio is outside the range spanned by the two mutual funds, then one of the mutual funds must be sold short (held in negative quantity) while the size of the investment in the other mutual fund must be greater than the amount available for investment (the excess being funded by the borrowing from the other fund).

3.2.1.5   Risk-Free Asset and the Capital Allocation Line

A risk-free asset is the asset that pays a *risk-free rate*. In practice, short-term government securities (such as U.S. *Treasury Bills)* are *considered* a risk-free asset, because they pay a fixed rate of interest and have *exceptionally low default risks* (none, in the United States, so far!). The risk-free asset has zero variance in returns (being risk-free). It is also uncorrelated with any other asset (by definition, since its variance is zero).

As a result, when it is combined with any other asset or portfolio of assets, the change in return is linearly related to the change in risk as the proportions in the combination vary.

3.2.1.6   The Sharpe Ratio

In mathematical finance, the *Sharpe Ratio* (or the *Sharpe Index*, or the *Sharpe Measure*, or the *Reward-to-Variability Ratio*) is an index for examining the performance (or risk premium) per unit of deviation in an investment asset or a trading strategy, typically referred to as risk (and is a deviation risk measure). It is named after W.F. Sharpe (recipient of the 1990 Nobel Memorial Prize in Economic Sciences).

In application, the Sharpe Ratio is similar to the Information Ratio; whereas the Sharpe Ratio is the *excess* return of an asset over the return of a risk-free asset divided by the variability or standard deviation of returns, the Information Ratio is the *active* return to the most relevant benchmark index divided by the standard deviation of the *active* return.

3.2.1.7   The Capital Allocation Line (CAL)

When a risk-free asset is included, the half-line shown in Figure 3.2 becomes the new efficient frontier: It is tangent to the hyperbola at the pure risky portfolio with the highest Sharpe Ratio. Its vertical intercept represents a portfolio with 100% of holdings in the risk-free asset; the tangency with the hyperbola

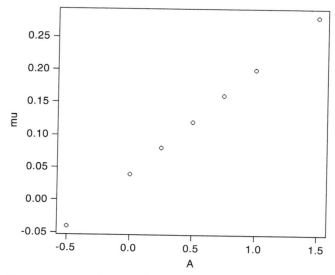

Figure 3.2 mu <- A muA + B muB.

represents a portfolio with no risk-free holdings and 100% of assets held in the portfolio occurring at the point of contact, namely, the tangency point:

1) Points between these two positions are portfolios containing positive amounts of both the risky tangency portfolio and the risk-free asset.
2) Points on the half-line beyond the tangency point are leveraged portfolios involving negative holdings of the risk-free asset (the latter has been sold short—in other words, the investor has borrowed at the risk-free rate) and an amount invested in the tangency portfolio equal to more than 100% of the investor's initial capital. This efficient half-line is called the *Capital Allocation Line* (CAL), and its equation may be shown to be

$$E(R_C) = R_F + (\sigma_C[\{E(R_P) - R_F\}/\sigma_P]) \qquad (3.11)$$

In Equation (3.11),

P is the subportfolio of risky assets at the tangency with the Markowitz bullet.
F is the risk-free asset.
C is the combination of portfolios P and F.

Using the diagram, the introduction of the risk-free asset as a likely component of the portfolio has improved the range of risk-expected return combinations available because everywhere (except at the tangency portfolio) the half-line has a higher expected return than the hyperbola does at every possible risk level. That all points on the linear efficient locus can be achieved by a combination of holdings of the risk-free asset and the tangency portfolio is known as the "One Mutual Fund Theorem," in which the mutual fund referred to is the tangency portfolio.

3.2.1.8 Asset Pricing

Up to this point, the analysis describes the optimal behavior of an individual investor. Asset pricing theory depends on this analysis in the following way:

Since each investor holds the risky assets in identical proportions to each other, namely, in the proportions given by the tangency portfolio, in market equilibrium the risky assets' prices, and therefore their expected returns, will adjust so that the ratios in the tangency portfolio are the same as the ratios in which the risky assets are supplied to the market. Thus, relative supplies will equal relative demands:

Modern Portfolio Theory *derives the required expected return for a correctly priced asset in this context.*

3.2.1.9 Specific and Systematic Risks

- *Specific risks* are the risks associated with individual assets. Within a portfolio, these risks may be reduced through diversification, namely, canceling out each other. Specific risk is also called diversifiable, unique, unsystematic, or idiosyncratic risk.
- *Systematic risks*, namely, portfolio risks or market risks are risks common to all securities—except for *selling short*. Systematic risk cannot be diversified away within one market. Within the market portfolio, asset-specific risk may be diversified away to the extent possible. Systematic risks are, therefore, equated with the risks of the market portfolio.

Since a security will be purchased only if it improves the risk-expected return characteristics of the market portfolio, *the relevant measure of the risk of a security is the risk it adds to the market portfolio*, and *not* its risk in isolation. In this context, the volatility of the asset and its correlation with the market portfolio are historically observed and are, therefore, available for consideration. Systematic risks within one market can be managed through a strategy of using both long and short positions within one portfolio, creating a "market-neutral" portfolio. Market-neutral portfolios will have a correlation of zero.

3.2.2 Capital Asset Pricing Model (CAPM)

For a given asset allocation, the return depends on the price and total amount paid for the asset. The goal of the investment is that the price paid should ensure that the market portfolio's risk-return characteristics *improve* when the asset is added to it. The *CAPM* is an approach that derives the theoretical required expected return (i.e., discount rate) for an asset in a market, given the risk-free rate available to investors and the risk of the market as a whole.

The CAPM is usually expressed as follows:

$$E(R_i) = R_f + \beta_i[E(R_m) - R_f] \tag{3.12}$$

where β is the *asset sensitivity* to a change in the overall market, and is usually found via correlations on historical data, noting that

a) $\beta > 1$: signifying more than average "riskiness" for the asset's contribution to overall portfolio risk;
b) $\beta < 1$: signifying a lower than average risk contribution to the portfolio risk.
c) $[E(R_m) - R_f]$ is the market premium, the expected excess return of the market portfolio's expected return *over the risk-free rate.*

The above conclusions may be established as follows:

i) The initial risks of the portfolio $\mathbf{m} = w_m^2 \sigma_m^2$
When an additional risky asset **a** is added to the market portfolio **m**, the incremental impact on risk and expected return follows from the formulas for a two-asset portfolio. The results may then be used to derive the asset-appropriate discount rate as follows:

ii) The market portfolio's risk $= \left(w_m^2 \sigma_m^2 + \left[w_a^2 \sigma_a^2 + 2w_m w_a \rho_{am} \sigma_a \sigma_m\right]\right)$
Hence, the risk added to the portfolio $= \left[w_a^2 \sigma_a^2 + 2w_m w_a \rho_{am} \sigma_a \sigma_m\right]$ but the weight of the asset will be relatively low, namely, $w_a^2 \approx 0$, hence $w_a^2 \sigma_a^2 \approx 0$, so that the additional risk $\approx 2w_m w_a \rho_{am} \sigma_a \sigma_m$.

iii) Since the market portfolio's expected return $= (w_m E(R_m) + [w_a E(R_a)])$, the additional expected return $= [w_a E(R_a)]$.

iv) On the other hand, if an asset a is accurately priced, the improvement in *its risk-to-expected* return ratio obtained by adding it to the market portfolio m will at least match the gains of spending that money on an increased stake in the market portfolio. The assumption here is that the investor will buy the asset with funds borrowed at the risk-free rate R_f. This is reasonable if

$$E(R_a) > R_f \tag{3.13}$$

Hence,

$$[w_a(E(R_a) - R_f)]/\{2w_m w_a \rho_a \sigma_m\} = [w_a(E(R_m) - R_f)]/[2w_m w_a \sigma_m \sigma_m] \tag{3.14}$$

that is,

$$[E(R_a)] = R_f + [E(R_m) - R_f]^*[\rho_{am} \sigma_a \sigma_m]/[\sigma_m \sigma_m] \tag{3.15}$$

or

$$[E(R_a)] = R_f + [E(R_m) - R_f]^*[\sigma_{am}]/[\sigma_{mm}] \tag{3.16}$$

where $[\sigma_{am}]/[\sigma_{mm}]$ is the "beta", β_{return}: the *covariance* between the asset's return and the market's return divided by the variance of the market return, which is the sensitivity of the asset price to movement in the market value of the portfolio.

3.2.2.1 The Security Characteristic Line (SCL)

Equation (3.11) may be computed statistically using the following regression equation, known as the SCL regression equation:

$$R_{i,t} - R_f = \alpha_i + \beta_i(R_{M,t} - R_f) + \in_{i,t} \tag{3.17}$$

in which

α_i is the asset's *alpha* coefficient,
β_i is the asset's *beta* coefficient, and
SCL is the Security Characteristic Line.

If the Expected Return $E(R_i)$ is calculated using *CAPM*, then the future *cash flow* of the asset may be discounted to their present value using this rate to establish the correct price for the asset.

Remarks:

1) A riskier stock will have a higher beta and will be discounted at a higher rate.
2) Less-sensitive stocks may have lower betas and may be discounted at a lower rate.
3) Theoretically, an asset is correctly priced when its observed price is the *same* as its value calculated using the CAPM-derived discount rate.
4) If the observed price is higher than the valuation, then the asset is over-valued; and it is undervalued for a too low price.
5) Despite its theoretical importance, critics of the Modern Portfolio Theory (MPT) question whether it is an ideal investment tool, because its model of financial markets does *not* match the real world in many ways.
6) The risk, return, and correlation measures used by MPT are based on *expected values*, namely, they are mathematical statements about the future; the expected value of returns is
    - *explicit* in the foregoing equations, and
    - *implicit* in the definitions of variance and covariance.

    In practice, financial analysts must rely on predictions based on historical records of asset return and *volatility* for these values in the model equations.

    Often such expected values are at variance with the situations of the prevailing circumstances that may not exist when the historical data were generated.
7) *Probabilistic Characteristics:* It should not escape one's attention that when using the Modern Portfolio Theory (MPT), financial analysts may need to

estimate some key parameters from *past* market data because MPT attempts to model risk in terms of the likelihood of losses, without indicating why those losses might occur. Thus, the risk measurements used are *probabilistic* in nature, not structural. This is a major difference as compared to many alternative approaches to risk management.

8) *Estimation of Errors:* This is critical in the Modern Portfolio Theory (MPT). In an MPT or mean-variance optimization analysis, accurate estimation of the *Variance–Covariance* matrix is critical. In this context, numerical forecasting with Monte Carlo simulation with the Gaussian copula and well-specified marginal distributions may be effective. The modeling process may be expected to adjust for empirical characteristics in stock returns such as autoregression, asymmetric volatility, skewness, and kurtosis. Neglecting to account for these factors may result in severe estimation errors occurring in the correlation of the *Variance Covariance*, *resulting in high* negative biases.

9) *MPT:* This has one important conceptual difference from the *Probabilistic Risk Assessment* (PRA) used in the risk assessment of (say) nuclear power plants. In classical econometrics, a PRA is considered as a *structural model* in which the components of a system and their relationships are modeled using Monte Carlo simulations: if valve X fails, it causes a loss of back pressure on pump Y, which, in turn, causing a drop in flow to vessel Z, etc. However, in the Black–Scholes model and *MPT*, there is no attempt to explain an underlying structure to price changes. Various outcomes are simply given as probabilities. And, unlike the PRA, if there is no history of a particular system-level event like a *liquidity crisis*. Thus, *there is no way to compute its odds!*

   (If nuclear safety engineers compute risk management this way, they would *not* be able to compute the odds of a nuclear meltdown at a particular plant until several similar events occurred in the same reactor design—to produce realistic nuclear safety data!)

10) Mathematical risk measurements are useful only to the extent that they reflect investors' true concerns—There is no point minimizing a variable that nobody cares about in practice. Modern Portfolio Theory (*MPT*) uses the mathematical concept of *variance* to quantify risk, and this might be acceptable if the assumptions of *MPT* such as the assumption of returns may be taken as normally distributed returns. For general returns, distributions of other risk measures might better reflect investors' true preference.

3.2.3 Some Typical Simple Illustrative Numerical Examples of the Markowitz MPT Using R

To demonstrate the Markowitz MPT model, two numerical examples are selected:

1) An illustrative example that may be treated with simple arithmetical calculations.

2) An example selected from the R Package MarkowitzR, available from CRAN, is chosen. For this example, the associated R program is run, outputting the concomitant associated numerical and graphical results.

3.2.3.1 Markowitz MPT Using R: A Simple Example of a Portfolio Consisting of Two Risky Assets

This example introduces modern portfolio theory in a simple setting of only a single risk-free asset and two risky assets.

A Portfolio of Two Risky Assets   Consider an investment problem in which there are two nondividend-paying stocks: Stock A and Stock B, and over the next month let

R_A be the monthly simple return on the Stock A, and
R_B be the monthly simple return on the Stock B.

Since these returns will not be realized until the end of each month, their returns may be treated as random variables.

Assume that the returns R_A and R_B are *jointly normally distributed*, and that the following information are available regarding the means, variances, and covariances of the probability distribution of the two returns:

The means μ, standard deviations σ, covariance σ_{AB}, and correlation coefficient ρ_{AB} are defined as follows:

$$\mu_A = E[R_A] \tag{3.18a}$$

$$\sigma_A^2 = \text{var}(R_A) \tag{3.18b}$$

$$\mu_B = E[R_B] \tag{3.19a}$$

$$\sigma_B^2 = \text{var}(R_B) \tag{3.19b}$$

$$\sigma_{AB} = \text{cov}(R_A, R_B) \tag{3.20}$$

$$\rho_{AB} = \text{cor}(R_A, R_B) = \sigma_{AB}/(\sigma_A \sigma_B) \tag{3.21}$$

where E is the expectation and *var* is the variance.

Generally, these values are *estimated* from historical return data for the two stocks. On the other hand, they may also be subjective estimates by an analyst!

For this exercise, one may assume that these values are taken as given:

Remarks:

1) For the monthly returns on each of the two stocks, the expected returns μ_A and μ_B are considered to be best estimated expectations. However, since the investment returns are random variables, one must recognize that the actual realized returns may be different from these expectations. The variances σ_A^2 and σ_B^2 provide some estimated measures of the uncertainty associated with these monthly returns.

2) One may also consider the variances as measuring the risk associated with the investments:
   - Assets with high return variability (or volatility) are often considered— understandably—to be risky.
   - Assets with low return volatility are often thought to be safe.
3) The covariance σ_{AB} may provide some probabilistic information about the direction of any linear dependence between returns:
   - If $\sigma_{AB} > 0$, the two returns tend to move in the same direction.
   - If $\sigma_{AB} < 0$, the two returns tend to move in opposite directions.
   - If $\sigma_{AB} = 0$, the two returns tend to move independently.
4) The strength of the dependence between the returns is provided by the correlation coefficient ρ_{AB}:

   - If $|\rho_{AB}| \rightarrow 1$, the returns approach each other very closely.
   - If $|\rho_{AB}| \rightarrow 0$, the returns show very little relationship.

Example 3.1 A Numerical Example on Assets Allocation, Given the Portfolio Information of Two Risky Assets—Stock A and Stock B:

Table 3.1 shows the annual return distribution parameters for two hypothetical assets A and B:

- Whereas asset A is a comparatively high-risk asset with an annual return of $\mu_A = 0.2 = 20\%$, asset B is a relatively low-risk asset with an annual return of $\mu_B = 0.04 = 4\%$, and an annual standard deviation of $\sigma_A = \sqrt{(\sigma_A^2)} = \sqrt{0.05} = 0.2236 = 22.36\%$.
- Asset B is a comparatively lower risk asset with annual return of $\mu_B = 4\%$, and annual standard deviation of $\sigma_B = \sqrt{(\sigma_B^2)} = \sqrt{0.01} = 0.1000 = 10.00\%$.
- These assets are correlated with correlation coefficient $\rho_{AB} = -0.17$.
- The covariance σ_{AB}, given the standard deviations σ_A and σ_B, and the correlation coefficient ρ_{AB} may be calculated from (3.21):

$$\sigma_{AB} = \rho_{AB}(\sigma_A \sigma_B)$$

$$= (-0.17)(0.2236)(0.1000)$$

$$= -0.0038$$

Table 3.1 Example data for a two-asset portfolio.

μ_A	σ_A^2	μ_B	σ_B^2	σ_{AB}	ρ_{AB}
0.2	0.05	0.04	0.01	−0.005	−0.17

Table 3.2 Example data for a two-asset portfolio.

μ_A	σ_A^2	μ_B	σ_B^2	σ_A	σ_B	σ_{AB}	ρ_{AB}
0.2	0.05	0.04	0.01	0.2236	0.1000	−0.0038	−0.17

And the characteristics of the two-asset portfolio data may be expanded, to be given by Table 3.2, as follows:

Given the available investment data as expressed in Table 3.2, the asset allocation problem may be stated as follows:

Assets Allocation For a given amount of initial liquid asset, totaling T_0, the task at hand is that one will fully invest *all* this amount in the two stocks A and B. The investment problem is the asset allocation decision: how much to allocate in asset A and how much in asset B?

Let x_A be the fraction of T_0 to be invested in stock A and x_B be the fraction of T_0 to be invested in stock B.

The values of x_A and x_B may be positive, zero, or negative:

- Positive values imply *long* positions, in the *purchase*, of the asset.
- Negative values imply *short* positions, in the *sales*, of the asset.
- Zero values imply *no* positions in the asset.

As all the wealth is invested in these two assets, it follows that

$$x_A + x_B = 1 \tag{3.22}$$

Thus, if asset A is "shorted," then the proceeds of the short sale will be used to buy more of asset B and vice versa. (To "short" an asset, one borrows the asset, usually from a broker, and then sells it: and the proceeds from the short sale are usually kept on account with a broker—often there may be some restrictions that prevent the use of these funds for the purchase of other assets. The short position is closed out when the asset is repurchased, and then returned to original owner. Should the asset drops in value, then a gain is made on the short sale; and if the asset increases in value, then a loss occurred!) Hence, the investment problem is

to ascertain the values of x_A and x_B in such a way that the overall profit of the investment will always be maximized!

The present investment in the two stocks forms a *portfolio*, and the shares of A and B are the *portfolio shares* or *weights*.

In this example, the return on the portfolio, R_p, over the following month is a random variable, given by

$$R_p = x_A R_A + x_B R_B \tag{3.23}$$

that is, a weighted average or linear combination of the random variables R_A and R_B. Moreover, as both R_A and R_B are assumed to be normally distributed, it follows from (3.23) that R_p is also normally distributed. Moreover, using the properties of linear combinations of random variables, (3.23) may be used to determine the mean and variance of the distribution of R_p, that is, the return of the entire portfolio.

Probable Expected Return and Variance of the Portfolio   Again, according to (3.23), the mean, variance, and standard deviation of the distribution of the return on the portfolio may be readily derived as follows:

- The mean of the portfolio μ_p is given by

$$\begin{aligned} \mu_p &= E[R_p] \\ &= E[x_A R_A + x_B R_B], \quad \text{upon direct subsitution from (3.23)} \\ &= x_A E[R_A] + x_B E[R_B], \quad \text{by the linearity property of the expectation operator } E \\ &= x_A \mu_A + x_B \mu_B \end{aligned}$$

namely,

$$\mu_p = E[R_p] = x_A \mu_A + x_B \mu_B \tag{3.24}$$

- The variance of the portfolio σ_p^2 is given by

$$\begin{aligned} \sigma_p^2 &= \text{var}(R_p) \\ &= E[(R_p - \mu_p)^2], \quad \text{by definition of the variance property} \\ &= E[\{(x_A R_A + x_B R_B) - (x_A \mu_A + x_B \mu_B)\}^2] \end{aligned}$$

upon direct substitutions, respectively, from (3.23) and (3.24):

$$= E[(x_A R_A + x_B R_B)^2 + (x_A \mu_A + x_B \mu_B)^2 - 2(x_A R_A + x_B R_B)(x_A \mu_A + x_B \mu_B)]$$

upon squaring the terms

$$\begin{aligned} = E[&(x_A^2 R_A^2 + x_B^2 R_B^2 + 2x_A R_A x_B R_B) \\ &+ (x_A^2 \mu_A^2 + x_B^2 \mu_B^2 + 2x_A \mu_A x_B \mu_B) \\ &- (2x_A^2 R_A \mu_A + 2x_A x_B R_A \mu_B + 2x_A x_B \mu_A R_B + 2x_B^2 R_B \mu_B)] \end{aligned}$$

upon expanding the multiplication of terms

$$\begin{aligned} = E[&x_A^2 R_A^2 + x_B^2 R_B^2 + 2x_A R_A x_B R_B \\ &+ x_A^2 \mu_A^2 + x_B^2 \mu_B^2 + 2x_A \mu_A x_B \mu_B \\ &- 2x_A^2 R_A \mu_A - 2x_A x_B R_A \mu_B - 2x_A x_B \mu_A R_B - 2x_B^2 R_B \mu_B] \end{aligned}$$

upon removing all the parentheses

$$\begin{aligned} = E[&(x_A^2 R_A^2 - 2x_A^2 R_A \mu_A + x_A^2 \mu_A^2) + (x_B^2 R_B^2 - 2x_B^2 R_B \mu_B + x_B^2 \mu_B^2) \\ &+ (2x_A x_B R_A R_B + 2x_A \mu_A x_B \mu_B - 2x_A x_B R_A \mu_B - 2x_A x_B \mu_A R_B)] \end{aligned}$$

upon rearranging and recollecting terms

$$= E[x_A^2(R_A^2 - 2R_A\mu_A + \mu_A^2) + x_B^2(R_B^2 - 2R_B\mu_B + \mu_B^2)$$
$$+ 2x_Ax_B(R_AR_B + \mu_A\mu_B - R_A\mu_B - \mu_AR_B)]$$

upon further rearranging and recollecting terms

$$= E[x_A^2(R_A^2 - 2R_A\mu_A + \mu_A^2) + x_B^2(R_B^2 - 2R_B\mu_B + \mu_B^2)$$
$$+ 2x_Ax_B(R_AR_B - R_A\mu_B - \mu_AR_B + \mu_A\mu_B)]$$

upon further rearranging terms

$$= E[x_A^2(R_A^2 - 2R_A\mu_A + \mu_A^2) + x_B^2(R_B^2 - 2R_B\mu_B + \mu_B^2)$$
$$+ 2x_Ax_B\{R_A(R_B - \mu_B) - \mu_A(R_B - \mu_B)\}]$$

upon further factoring terms

$$= E[x_A^2(R_A - \mu_A)^2 + x_B^2(R_B - \mu_B)^2 + 2x_Ax_B(R_A - \mu_A)(R_B - \mu_B)]$$

upon further factoring terms

$$= E[x_A^2(R_A - \mu_A)^2] + E[x_B^2(R_B - \mu_B)^2] + E[2x_Ax_B(R_A - \mu_A)(R_B - \mu_B)]$$
$$= x_A^2 E[(R_A - \mu_A)^2] + x_B^2 E[(R_B - \mu_B)^2] + 2x_Ax_B E[(R_A - \mu_A)(R_B - \mu_B)]$$

upon further factoring the expression

$$= x_A^2\sigma_A^2 + x_B^2\sigma_B^2 + 2x_Ax_B\sigma_{AB}$$

namely,

$$\sigma_p^2 = \text{var}(R_p) = x_A^2\sigma_A^2 + x_B^2\sigma_B^2 + 2x_Ax_B\sigma_{AB} \tag{3.25}$$

Finally,

$$\sigma_p = \text{SD}(R_p) = \sqrt{\text{var}(R_p)} = \sqrt{(x_A^2 s_A^2 + x_B^2\sigma_B^2 + 2x_Ax_B\sigma_{AB})} \tag{3.26}$$

• These relationships imply that

$$R_p \sim N(\mu_p, \sigma_p^2) \tag{3.27}$$

Remarks:

1) Equation (3.25) shows that the variance of the portfolio is given as a weighted average of the variances of the individual component assets plus twice the product of the portfolio weights times the covariance between the component assets.
2) Thus, if the portfolio weights are both positive, then a positive covariance will likely increase the portfolio variance since both returns tend to vary in the same direction. Likewise, a negative covariance may reduce the portfolio variance.

3) Hence, assets with negatively correlated returns may be beneficial when building up a portfolio since the concomitant risk, as indicated by portfolio standard deviation, is reduced.
4) Forming portfolios with positively correlated assets may reduce risk as long as the correlation is small.

Example 3.2 Some Preliminary Model Portfolios

Using the asset information in Table 3.2, one may create some preliminary portfolios:

Portfolio 1: An equally weighted portfolio with $x_A = 0.500 = x_B$.
Given:
Here one may use Equations (3.23–3.26):

$$R_p = x_A R_A + x_B R_B \tag{3.23}$$

$$\mu_p = E[R_p] = x_A \mu_A + x_B \mu_B \tag{3.24}$$

$$\sigma_p^2 = \text{var}(R_p) = x_A^2 \sigma_A^2 + x_B^2 \sigma_B^2 + 2x_A x_B \sigma_{AB} \tag{3.25}$$

$$\sigma_p = \text{SD}(R_p) = \sqrt{\text{var}(R_p)} = \sqrt{(x_A^2 \sigma_A^2 + x_B^2 \sigma_B^2 + 2x_A x_B \sigma_{AB})} \tag{3.26}$$

to obtain

$$
\begin{aligned}
\mu_p &= E[R_p] = x_A \mu_A + x_B \mu_B, \quad \text{using (3.24)} \\
&= (0.500)(0.2) + (0.500)(0.04), \quad \text{using Table 3.2} \\
&= 0.100 + 0.020 \\
&= 0.120
\end{aligned}
$$

$$
\begin{aligned}
\sigma_p^2 &= \text{var}(R_p) = x_A^2 \sigma_A^2 + x_B^2 \sigma_B^2 + 2x_A x_B \sigma_{AB}, \quad \text{using (3.25) and Table 3.2} \\
&= (0.500)^2 (-0.2236)^2 + (0.500)^2 (-0.1000)^2 \\
&\quad + 2(0.500)(0.500)(-0.0038)(-0.1000) \\
&= 0.0125 + 0.0025 + 0.00019 \\
&= 0.01519
\end{aligned}
$$

$$
\begin{aligned}
\sigma_p &= \sqrt{\sigma_p^2} = \text{SD}(R_p) = \sqrt{\text{var}(R_p)}, \quad \text{using (3.26)} \\
&= \sqrt{(0.01519)} \\
&= 0.12325
\end{aligned}
$$

$$\mu_p = E[R_p] = x_A \mu_A + x_B \mu_B$$

Remark:
The following R-code* segment may be used for this computations:

```
mu <- A*muA + B*muB
```
$$\sigma_p^2 < -\text{var}(R_p) < -x_A^2 \sigma_A^2 + x_B^2 \sigma_B^2 + 2x_A x_B \sigma_{AB}$$
```
var <- A*A*sigA*sigA + B*B*sigB*sigB + 2*A*B*sigAB)
```

*A comprehensive presentation of the R computer code is provided in Chapter 4. Here, one may have recourse to using the R code to facilitate repetitive computations, in order to investigate results for other choices of x_A, and hence x_B.

Thus, for

$$x_A = 1, 0.75, 0.50, 0.25, 0, 1.5, \text{ and } -0.5,$$

the following R code segment would undertake the computation. The results follow, together with some graphic presentations of the results.

In the R domain:

```
>
> muA <- 0.2
> sigmasqA <- 0.05
> muB <- 0.04
> sigmasqB <- 0.01
> sigA <- -0.2236
> sigB <- -0.1000
> sigAB <- -0.0038
> rhoAB <- -0.17
>
> A <- c(1, 0.75, 0.50, 0.25, 0, 1.5, -0.5)
> B <- c(0, 0.25, 0.50, 0.75, 1, -0.5, 1.5)
> mu <- A*muA + B*muB
> var <- A*A*sigA*sigA + B*B*sigB*sigB + 2*A*B*sigAB
>
> mu # Outputting:
[1]  0.20  0.16  0.12  0.08  0.04  0.28 -0.04
>
> var # Outputting:
[1] 0.04999696   0.02732329   0.01309924   0.00732481
[2] 0.01000000   0.12069316   0.04069924
>
> plot(A, mu) # Outputting: Figure 3.2
> plot(A, var) # Outputting: Figure 3.3
>
```

Representing:

$$\mu_p = E[R_p] = x_A\mu_A + x_B\mu_B \tag{3.24}$$

Representing:

$$\sigma_p^2 = \text{var}(R_p) = x_A^2\sigma_A^2 + x_B^2\sigma_B^2 + 2x_Ax_B\sigma_{AB} \tag{3.25}$$

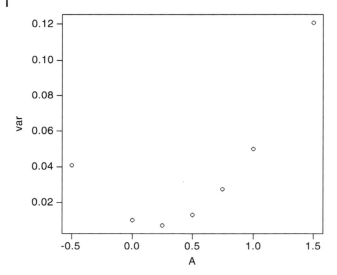

Figure 3.3 `var<- A*A*sigA*sigA + B*B*sigB*sigB + 2*A*B*sigAB`.

Remarks:

1) The variance of the portfolio is a weighted average of the variance of the individual assets *plus* twice the product of the portfolio weights times the covariance between the assets.

2) If the portfolio weights are both positive, then a positive covariance will tend to *increase* the portfolio variance, because both returns tend to move in the *same* direction; likewise, a negative covariance will tend to reduce the portfolio variance.

3) Hence, choosing assets with negatively correlated returns may be beneficial when forming portfolios because risk, as measured by portfolio standard deviation, is reduced.

4) Note also that what forming portfolios with positively correlated assets can also reduce risk as long as the correlation is not too large.

Example 3.3 Further Remarks on the Two-Asset Portfolio

- When forming a portfolio using the asset information in Table 3.2, if the first portfolio is an equally weighted portfolio with

$$x_A = x_B = 0.5$$

then using (3.24–3.26) and Table 3.2, one obtains

$$
\begin{aligned}
\mu_p &= E[R_p] \\
&= x_A\mu_A + x_B\mu_B \\
&= (0.5)(0.2) + (0.5)(0.04) \\
&= 0.1 + 0.02 \\
&= 0.12
\end{aligned}
$$

$$\sigma_p^2 = \text{var}(R_p)$$
$$= x_A^2\sigma_A^2 + x_B^2\sigma_B^2 + 2x_Ax_B\sigma_{AB}$$
$$= (0.5)^2(0.05) + (0.5)^2(0.01) + 2(0.5)(0.5)(-0.0038)$$
$$= 0.0125 + 0.0025 + (-0.0019)$$
$$= 0.0131$$
$$\sigma_p = \sqrt{0.0131}$$
$$= 0.11446$$

- This portfolio has an expected return of $\mu_p = 0.12$, or 12%, which is halfway between the expected returns on asset A ($\mu_A = 0.20$ or 20%) and on asset B ($\mu_B = 0.04$ or 4%), but the portfolio standard deviation ($\sigma_p = 0.11446$ or 11.446%) is *less than halfway* between the two asset standard deviations, namely,

$$|(-0.2236 + -0.1000|/2 = 0.1618, \text{ or } 16.18\%.$$

This reflects *risk reduction via diversification*.
- In the "long–short combination" portfolio with $x_A = 1.5$ and $x_B = -0.5$, in which asset B is sold short and the proceeds of this short sale are used to leverage the investment in asset A. The portfolio characteristics are then given by

$$\mu_p = E[R_p]$$
$$= x_A\mu_A + x_B\mu_B$$
$$= (1.5)(0.2) + (-0.5)(0.04)$$
$$= 0.3 - 0.02$$
$$= 0.28$$
$$\sigma_p^2 = \text{var}(R_p)$$
$$= x_A^2\sigma_A^2 + x_B^2\sigma_B^2 + 2x_Ax_B\sigma_{AB}$$
$$= (1.5)^2(0.05) + (-0.5)^2(0.01) + 2(1.5)(-0.5)(-0.0038)$$
$$= 0.1125 + 0.0025 + 0.0057$$
$$= 0.1207$$
$$\sigma_p = \sqrt{0.1207}$$
$$= 0.34742$$

Thus, comparing with the "equally weighted" portfolio in which $x_A = x_B = 0.5$, the "long–short combination" portfolio with $x_A = 1.5$ and $x_B = -0.5$ shows both a *higher* expected return ($\mu_p = 0.28$ compared with 0.2) and a *higher* standard deviation asset A ($\sigma_p = 0.34742$ compared with $|-0.2236| = 0.2236$)! Moreover, the relative expected return (RER) is given by

$$\text{RER} = (\mu_p) - \text{"equally weighted"}/(\mu_p) - \text{"unequally weighted"}$$
$$= 0.28/0.20$$
$$= 1.4$$

Thus, for $x_A = 1$, 0.75, 0.50, 0.25, 0, 1.5, and −0.5,
the following R code segment would undertake the computation. The results
follow, together with some graphic presentations of the results:

In the R domain:

```
>
> muA <- 0.2
> sigmasqA <- 0.05
> muB <- 0.04
> sigmasqB <- 0.01
> sigA <- -0.2236
> sigB <- -0.1000
> sigAB <- -0.0038
> rhoAB <- -0.17
>
> A <- c(1, 0.75, 0.50, 0.25, 0, 1.5, -0.5)
> B <- c(0, 0.25, 0.50, 0.75, 1, -0.5, 1.5)
> mu <- A*muA + B*muB
> var <- A*A*sigA*sigA + B*B*sigB*sigB + 2*A*B*sigAB
> sigma <- sqrt(var)
>
> mu
[1]  0.20 0.16 0.12 0.08 0.04 0.28 -0.04
>
> sigma
[1] 0.2236000    0.1652976    0.1144519   0.0855851
[5] 0.1000000    0.3474092    0.2017405
>
> plot(A, sigma) # Outputting: Figure 3.4
> RER <- mu/sigma
> RER
[1]  0.8944544    0.9679513    1.0484753    0.9347421

[5]  0.4000000    0.8059660   -0.1982745
>
> plot(A, RER) # Outputting: Figure 3.5
```

3.2.3.2 Evaluating a Portfolio

Consider an initial investment I_0 in a portfolio of assets A and B, for which

a) the return is to be given by (3.23),

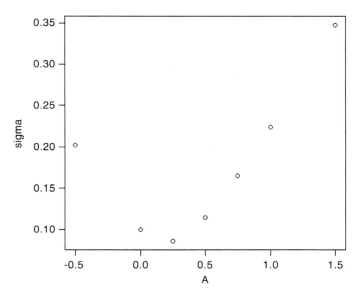

Figure 3.4 `plot(A, sigma)`.

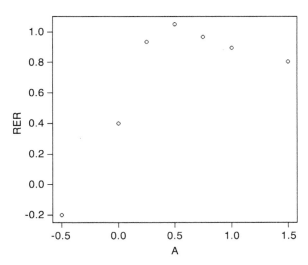

Figure 3.5 `plot(A, RER)`.

b) the expected return is to be given by (3.24), and

c) the variance is to be given by (3.25).

$$R_p = x_A R_A + x_B R_B \tag{3.23}$$

$$\mu_p = E[R_p] = x_A \mu_A + x_B \mu_B \tag{3.24}$$

$$\sigma_p^2 = \mathrm{var}(R_p) = x_A^2 \sigma_A^2 + x_B^2 \sigma_B^2 + 2 x_A x_B \sigma_{AB} \tag{3.25}$$

$$\sigma_p = \mathrm{SD}(R_p) = \sqrt{\mathrm{var}(R_p)} = \sqrt{(x_A^2 \sigma_A^2 + x_B^2 \sigma_B^2 + 2 x_A x_B \sigma_{AB})} \tag{3.26}$$

These relationships show that

$$R_p \sim N(\mu_p, \sigma^2) \tag{3.27}$$

Now, for $\alpha \in (0, 1)$, the $100\alpha\%$ portfolio value-at-risk is

$$\mathrm{VaR}_{p,\alpha} = q_{p,\alpha}^R I_0 \tag{3.28}$$

where q_α^R is the α quantile of the distribution of R_p and is given by

$$q_{p,\alpha}^R = \mu_p + \sigma_p q_\alpha^z \tag{3.29}$$

where q_α^z is the α quantile of the standard normal distribution. If R_p is a continuously compounded return, then the implied simple return quantile is $q_{p,\alpha}^R = \exp(\mu_p + \sigma_p q_\alpha^z) - 1$.

Relationship between the Portfolio *VaR* and the Individual Asset *VaR* In general, the Portfolio VaR is *not* the weighted average of the Individual Asset VaRs. They may be seen in the following counterexample: Consider the portfolio weighted average of the individual asset return *quantiles* for a two-asset system (A, B) for which the weighted average of the asset return quantiles may be expressed as

$$
\begin{aligned}
x_A q_{A,\alpha}^R + x_B q_{B,\alpha}^R &= x_A(\mu_A + \sigma_A q_\alpha^z) + x_B(\mu_B + \sigma_B q_\alpha^z) \\
&= x_A \mu_A + x_A \sigma_A q_\alpha^z + x_B \mu_B + x_B \sigma_B q_\alpha^z \\
&= x_A \mu_A + x_B \mu_B + x_A \sigma_A q_\alpha^z + x_B \sigma_B q_\alpha^z \\
&= (x_A \mu_A + x_B \mu_B) + (x_A \sigma_A + x_B \sigma_B) q_\alpha^z \\
&= \mu_p + (x_A \sigma_A + x_B \sigma_B) q_\alpha^z
\end{aligned}
\tag{3.30}
$$

Remark:

The weighted asset quantile (3.30) is *not* equal to the portfolio quantile (3.29) unless $\sigma_p = \rho_{AB} = 1$. Hence, weighted asset VaR, *in general*, is not equal to portfolio VaR because the quantile (3.30) ignores the correlation between R_A and R_B.

μ_A	σ_A^2	μ_B	σ_B^2	σ_A	σ_B	σ_{AB}	ρ_{AB}
0.2	0.05	0.04	0.01	0.2236	0.1000	-0.0038	-0.17

Example 3.4 A Numerical Example of a Portfolio Consisting of Two Risky Assets

Consider an initial investment of $I_0 = \$1,000,000$, and assume that returns are simple, the 5% VaRs on assets A and B are

$$
\begin{aligned}
\text{VaR}_{A,0.05} &= q_{0.05}^{R_A} I_0 \\
&= (\mu_A + \sigma_A q_{0.05}^z) I_0 \\
&= \{0.2 + (0.2236)(-1.645)\}(1,000,000) \\
&= (0.2 - 0.36782)(1,000,000) \\
&= (-0.167822)(1,000,000) \\
&= -167,822
\end{aligned}
\tag{3.31}
$$

$$
\begin{aligned}
\text{VaR}_{B,0.05} &= q_{0.05}^{R_B} I_0 \\
&= (\mu_B + \sigma_B q_{0.05}^z) I_0 \\
&= \{0.04 + (0.1000)(-1.645)\}(1,000,000) \\
&= (0.04 - 0.16540)(1,000,000) \\
&= (-0.12540)(1,000,000) \\
&= -125,400
\end{aligned}
\tag{3.32}
$$

Now, the 5% VaR on the equally weighted portfolio with $x_A = 0.5 = x_B$ is

$$
\begin{aligned}
\text{VaR}_{p,0.05} &= q_{0.05}^{R_p} I_0 \\
&= (\mu_P + \sigma_P q_{0.05}^z) I_0 \\
&= \{0.28 + (0.34742)(-1.645)\}(1,000,000) \\
&= (0.28 - 0.57151)(1,000,000) \\
&= (-0.29151)(1,000,000) \\
&= -291,510
\end{aligned}
\tag{3.33}
$$

The weighted average of the individual asset VaRs is

$$
\begin{aligned}
x_A \text{VaR}_{A,0.05} + x_B \text{VaR}_{B,0.05} &= (0.5)(-167,822) + (0.5)(-125,400) \\
&= (-83,911) + (-62,700) \\
&= -146,611
\end{aligned}
\tag{3.34}
$$

The 5% VaR on the "long–short" portfolio in which $x_A = 1.5$ and $x_B = -0.5$ is

$$
\text{VaR}_{p,0.05} = q_{0.05}^{R_p} I_0 = -167,822, \quad \text{from (3.31)}
$$

And the weighted-average of the individual asset VaRs is

$$
\begin{aligned}
x_A \text{VaR}_{A,0.05} + x_B \text{VaR}_{B,0.05} &= (1.5)(-167,822) + (0.5)(-125,400) \\
&= (-251,733) - (-62,700) \\
&= -189,033
\end{aligned}
\tag{3.35}
$$

Remark: Note that

$$\text{VaR}_{p,0.05} = -291,510, \text{ from } (3.33)$$

and

$$x_A \text{VaR}_{A,0.05} + x_B \text{VaR}_{B,0.05} = -189,033, \text{ from } (3.35)$$

namely,

$$\text{VaR}_{p,0.05} \neq x_A \text{VaR}_{A,0.05} + x_B \text{VaR}_{B,0.05}$$

because

$$\rho_{AB} \neq 1.$$

3.2.4 Management of Portfolios Consisting of Two Risky Assets

In the management of portfolios consisting of two risky assets, the approach may begin by constructing portfolios that are mean-variance efficient, using the following distribution of asset returns and the investors' behavior:

1. All returns are	a) covariance stationary,
	b) ergodic, and
	c) jointly normally distributed over the investment horizon.

This implies that means, variances, and covariances of returns are constant over the investment horizon and completely characterize the joint distribution of returns.

1) Investors know the values of asset return means, variances, and covariances.
2) Investors are only concerned about the portfolio expected return and portfolio variance. Investors prefer portfolios with high expected return but not portfolios with high return variance.

With the above assumptions, it is possible to characterize the set of efficient portfolios, that is, those portfolios that have the highest expected return for a given level of risk as measured by portfolio variance. These portfolios are the investors' most likely choice.

For a numerical example of the management of portfolios consisting of two risky assets, use the data set in Table 3.2.

Example 3.5 Portfolios Consisting of Two Risky Assets

The following R code segment would undertake the computation as outlined in Section 3.1.3.3. The results follow, together with a graphical presentations of the results:

```
mu.A <- 0.2
sig2.A <- 0.05
mu.B <- 0.04
sig2.B <- 0.01
sig.A <- 0.2236
sig.B <- 0.1000
sig.AB <- -0.0038
rhoAB <- -0.17
x.A = seq(from=-0.4, to=1.4, by=0.1)
x.B = 1 - x.A
mu.p = x.A*mu.A + x.B*mu.B
sig2.p = x.A^2 * sig2.A + x.B^2 * sig2.B + 2*x.A*x.B*sig.AB
sig.p = sqrt(sig2.p)
plot(sig.p, mu.p, type="b", pch=16,ylim=c(0, max(mu.p)),
xlim=c(0, max(sig.p)),xlab=expression(sigma[p]),
ylab=expression(mu[p]),
col=c(rep("red", 6), rep("green", 13)))
plot(sig.p, mu.p, type="b", pch=16,
ylim=c(0, max(mu.p)), xlim=c(0, max(sig.p)),
xlab=expression(sigma[p]), ylab=expression(mu[p]),
col=c(rep("red", 6), rep("green", 13)))
text(x=sig.A, y=mu.A, labe
ls="Asset A", pos=4)
text(x=sig.B, y=mu.B, labels="Asset B", pos=4)
```

In the R domain:

```
> mu.A <- 0.2
> sig2.A <- 0.05
> mu.B <- 0.04
> sig2.B <- 0.01
> sig.A <- 0.2236
> sig.B <- 0.1000
> sig.AB <- -0.0038
> rhoAB <- -0.17
>
> x.A = seq(from=-0.4, to=1.4, by=0.1)
> x.B = 1 - x.A
> mu.p = x.A*mu.A + x.B*mu.B
> sig2.p = x.A^2 * sig2.A + x.B^2 * sig2.B + 2*x.A*x.B*sig.AB
> sig.p = sqrt(sig2.p)
> plot(sig.p, mu.p, type="b", pch=16,
+ ylim=c(0, max(mu.p)), xlim=c(0, max(sig.p)),
+ xlab=expression(sigma[p]), ylab=expression(mu[p]),
```

```
+ col=c(rep("red", 6), rep("green", 13)))
> plot(sig.p, mu.p, type="b", pch=16,
+ ylim=c(0, max(mu.p)), xlim=c(0, max(sig.p)),
+ xlab=expression(sigma[p]), ylab=expression(mu[p]),
+ col=c(rep("red", 6), rep("green", 13)))
> text(x=sig.A, y=mu.A, labels="Asset A", pos=4)
> text(x=sig.B, y=mu.B, labels="Asset B", pos=4)
>
> # Outputting: Figure 3.6 Frontier of the Two-Asset
Portfolio (A, B)
>
```

Figure 3.6 shows the frontier of the two-asset portfolio (A, B) of the example data in Table 1.1. The set of all feasible portfolios, or the investment possibilities set in the case of two assets, is simply all possible portfolios that can be formed by varying the portfolio weights x_A and x_B such that the weights sum to 1: $x_A + x_B = 1$. One may summarize the expected return-risk (viz., mean variance) properties of the feasible portfolios in a plot with the portfolio expected return μ_p on the ordinate and portfolio standard deviation σ_p on the abscissa. The portfolio standard deviation is used instead of variance because standard deviation is measured in the same units as the expected value.

Example 3.6 Options in Asset Allocation for a Given Set of Investment Data

For the given set of investment data such as the set given in Table 3.2, Example 3.5, and Figure 3.6, the portfolio weight on asset A, x_A, is varied from -0.4 to 1.4

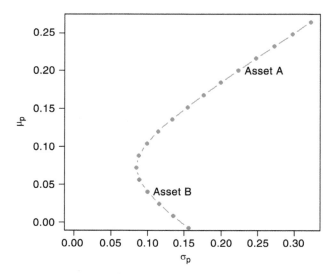

Figure 3.6 Frontier of the two-asset portfolio (A, B).

in increments of 0.1 and, since $x_A + x_B = 1$, the weight on asset B then varies from 1.4 to -0.4, providing a total of 18 portfolios with weights:

$$(x_A, x_B) = (-0.4, 1.4), (-0.3, 1.3), (-0.2, 1.2), \ldots, (1.2, -0.2), (1.3, -0.3), (1.4, -0.4)$$

For each of these portfolios, one uses Equations (3.24) and (3.26) to compute μ_p and σ_p and plotting the results in Figure 3.6 shows that the plot of μ_p versus σ_p resembles a parabola lying on its side (actually, it is one side of a hyperbola). Assuming that investors prefer portfolios with the highest expected return μ_p for a given level of risk σ_p:

- Combinations that are in the *upper left corner* are the best portfolios.
- Combinations that are in the *lower right corner* are the worst portfolios.

Notice that the portfolio at the bottom of the parabola has the smallest variance among all feasible portfolios. Thus, this portfolio is called the *Global Minimum-Variance* portfolio.

For a given level of risk, portfolios with the highest expected return are colored green in Figure 3.6. These are known as efficient portfolios.

Inefficient portfolios are such that there is another feasible portfolio that has the same risk (σ_p) but a higher expected return (μ_p)—they are colored red in Figure 3.6. Thus, Figure 3.6 shows that the inefficient portfolios are the feasible portfolios that lie below the global minimum-variance portfolio, and the efficient portfolios are those that lie above the global minimum-variance portfolio.

3.2.4.1 The Global Minimum-Variance Portfolio

Using elementary calculus, finding the global minimum-variance portfolio is a simple exercise in calculus. This constrained optimization may be defined as

$$\underset{x_A, x_B}{\text{minimize}} \ \{\sigma_p^2 = x_A^2 \sigma_A^2 + x_B^2 \sigma_B^2 + 2 x_A x_B \sigma_{AB}\}, \quad \text{such that } x_A + x_B = 1$$

This is a constrained optimization problem and may be solved in the following two ways:

First Method: The *method of substitution*—using the constraint relationship to substitute one of the two variables (x_A, x_B) to transform the constrained optimization problem (in two variables) into an unconstrained optimization problem in only *one* variable. Thus, substituting, from $x_A + x_B = 1$, by inserting $x_B = 1 - x_A$ into the formula for σ_p^2 reduces the optimization problem to

$$\underset{x_A}{\text{minimize}} \ \{\sigma_p^2 = x_A^2 \sigma_A^2 + (1 - x_A)^2 \sigma_B^2 + 2 x_A (1 - x_A) \sigma_{AB}\} \tag{3.36}$$

The condition for a local stationary point in the expression in (3.36) is that

$$d(\sigma_p^2)/dx_A = 0 \tag{3.37}$$

The differentiation may be achieved using the chain rule:

$$d(\sigma_p^2)/dx_A = (d/dx_A)[x_A^2\sigma_A^2 + (1-x_A)^2\sigma_B^2 + 2x_A(1-x_A)\sigma_{AB}]$$
$$= [2x_A\sigma_A^2 - 2(1-x_A)\sigma_B^2 + 2(1-2x_A)\sigma_{AB}]$$

Using (3.37), one obtains $x_A^{\min} = x_A$ given by

$$2x_A\sigma_A^2 - 2(1-x_A)\sigma_B^2 + 2(1-2x_A)\sigma_{AB} = 0$$

namely,

$$x_A\sigma_A^2 - (1-x_A)\sigma_B^2 + (1-2x_A)\sigma_{AB} = 0$$

or

$$x_A\sigma_A^2 - \sigma_B^2 + x_A\sigma_B^2 + \sigma_{AB} - 2x_A\sigma_{AB} = 0$$

or

$$x_A\sigma_A^2 + x_A\sigma_B^2 - 2x_A\sigma_{AB} = \sigma_B^2 - \sigma_{AB}$$

or

$$x_A(\sigma_A^2 + \sigma_B^2 - 2x_A\sigma_{AB}) = (\sigma_B^2 - \sigma_{AB})$$

namely,

$$x_A^{\min} = (\sigma_B^2 - \sigma_{AB})/(\sigma_A^2 + \sigma_B^2 - 2x_A\sigma_{AB}) \tag{3.38a}$$

and

$$x_B^{\min} = 1 - x_A^{\min} \tag{3.38b}$$

Second Method: The *Method of Auxiliary Lagrange Multipliers λ—*

First, one puts the constraint: $x_A + x_B = 1$ into a homogeneous form,
and writing: $F_1 = x_A + x_B - 1 = 0$
as well as: $F_2 = \sigma_p^2 = x_A^2\sigma_A^2 + x_B^2\sigma_B^2 + 2x_Ax_B\sigma_{AB}$

The Lagrangian Multiplier Function L is formed by adding to F_2 the homogeneous constraint F_1, multiplied by an auxiliary variable λ, the Lagrangian Multiplier, to give

$$L(x_A, x_B, \lambda) = F_2 + \lambda F_1$$
$$= (x_A^2\sigma_A^2 + x_B^2\sigma_B^2 + 2x_Ax_B\sigma_{AB}) + \lambda(x_A + x_B - 1) \tag{3.39}$$

Next, (3.39) is minimized, with respect to x_A, x_B, and λ, leading to three auxiliary conditions:

$$\partial L/\partial x_A = 0, \quad \partial L/\partial x_B = 0, \quad \partial L/\partial \lambda = 0 \tag{3.40}$$

With the Lagrangian Multiplier Function L given by (3.39), one has

$$
\begin{aligned}
0 = \partial L/\partial x_A &= \partial L(x_A, x_B, \lambda)/\partial x_A \\
&= \partial\{(x_A^2 \sigma_A^2 + x_B^2 \sigma_B^2 + 2x_A x_B \sigma_{AB}) + \lambda(x_A + x_{B-1})\}/\partial x_A \\
&= (2x_A \sigma_A^2 + 0 + 2x_B \sigma_{AB}) + \lambda(1 + 0 - 0) \\
&= 2x_A \sigma_A^2 + 2x_B \sigma_{AB} + \lambda
\end{aligned}
\tag{3.41a}
$$

$$
\begin{aligned}
0 = \partial L/\partial x_B &= \partial L(x_A, x_B, \lambda)/\partial x_B \\
&= \partial\{(x_A^2 \sigma_A^2 + x_B^2 \sigma_B^2 + 2x_A x_B \sigma_{AB}) + \lambda(x_A + x_{B-1})\}/\partial x_B \\
&= (0 + 2x_B \sigma_B^2 + 2x_A \sigma_{AB}) + \lambda(1 + 0 - 0) \\
&= 2x_B \sigma_B^2 + 2x_A \sigma_{AB} + \lambda
\end{aligned}
\tag{3.41b}
$$

$$
\begin{aligned}
0 = \partial L/\partial \lambda &= \partial L(x_A, x_B, \lambda)/\partial \lambda \\
&= \partial\{(x_A^2 \sigma_A^2 + x_B^2 \sigma_B^2 + 2x_A x_B \sigma_{AB}) + \lambda(x_A + x_{B-1})\}/\partial \lambda \\
&= (0 + 0 + 0) + (x_A + x_B - 1) \\
&= x_A + x_B - 1
\end{aligned}
\tag{3.41c}
$$

Combining (3.41a) and (3.41b), one obtains

$$2x_A \sigma_A^2 + 2x_B \sigma_{AB} + \lambda = 0 = 2x_B \sigma_B^2 + 2x_A \sigma_{AB} + \lambda$$

namely,

$$2x_A \sigma_A^2 + 2x_B \sigma_{AB} + \lambda = 2x_B \sigma_B^2 + 2x_A \sigma_{AB} + \lambda$$

namely,

$$2x_A \sigma_A^2 + 2x_B \sigma_{AB} = 2x_B \sigma_B^2 + 2x_A \sigma_{AB}$$

namely,

$$x_A \sigma_A^2 + x_B \sigma_{AB} = x_B \sigma_B^2 + x_A \sigma_{AB}$$

namely,

$$x_A \sigma_A^2 - x_A \sigma_{AB} = x_B \sigma_B^2 - x_B \sigma_{AB}$$

namely,

$$x_A(\sigma_A^2 - \sigma_{AB}) = x_B(\sigma_B^2 - \sigma_{AB})$$

or,

$$x_B = x_A\{(\sigma_A^2 - \sigma_{AB})/(\sigma_B^2 - \sigma_{AB})\} \tag{3.42}$$

Finally, to obtain the required values $(x_A^{\min}, x_B^{\min})$, one may incorporate the third Lagrangian Multiplier condition (3.41c), namely, combining (3.42) and (3.41c):

From (3,42), $\quad x_B^{\min} = x_A^{\min}\{(\sigma_A^2 - \sigma_{AB})/(\sigma_B^2 - \sigma_{AB})\} \tag{3.43a}$

From (3,41C), $\quad x_B^{\min} = 1 - x_A^{\min} \tag{3.43b}$

Combining (3.43a) and (3.43b):

$$x_A^{\min}\{(\sigma_A^2 - \sigma_{AB})/(\sigma_B^2 - \sigma_{AB})\} = 1 - x_A^{\min}$$

$$=> x_A^{\min}\{(\sigma_A^2 - \sigma_{AB}) = (\sigma_B^2 - \sigma_{AB})(1 - x_A^{\min})$$

$$= (\sigma_B^2 - \sigma_{AB}) - (\sigma_B^2 - \sigma_{AB})x_A^{\min}$$

$$=> x_A^{\min}\{(\sigma_A^2 - \sigma_{AB}) + (\sigma_B^2 - \sigma_{AB})x_A^{\min} = (\sigma_B^2 - \sigma_{AB})$$

$$=> x_A^{\min}\{(\sigma_A^2 - \sigma_{AB}) + (\sigma_B^2 - \sigma_{AB}) = (\sigma_B^2 - \sigma_{AB})$$

$$=> x_A^{\min}(\sigma_A^2 - \sigma_{AB} + \sigma_B^2 - \sigma_{AB}) = (\sigma_B^2 - \sigma_{AB})$$

$$=> x_A^{\min}\{(\sigma_A^2 - \sigma_{AB}) + (\sigma_B^2 - \sigma_{AB}) = (\sigma_B^2 - \sigma_{AB})$$

$$=> x_A^{\min}(\sigma_A^2 - \sigma_{AB} + \sigma_B^2 - \sigma_{AB}) = (\sigma_B^2 - \sigma_{AB})$$

$$=> x_A^{\min}(\sigma_A^2 + \sigma_B^2 - 2\sigma_{AB}) = (\sigma_B^2 - \sigma_{AB})$$

$$=> x_A^{\min} = (\sigma_B^2 - \sigma_{AB})/(\sigma_A^2 + \sigma_B^2 - 2\sigma_{AB}) \tag{3.44a}$$

which is, of course, the same as (3.38a):
Finally,

$$x_B^{\min} = 1 - x_A^{\min} \tag{3.44b}$$

which is, of course (again) (3.38b).

Remarks:

1) In the *Method of Auxiliary Lagrange Multipliers*, the Lagrangian Multiplier Function, *λ*, *is introduced at the beginning.* And, after *λ* facilitated in solving the problem, in a rather elegant but indirect way, it is no longer required in further analysis!
2) In the *mathematics of optimization*, the Method of Lagrange.

Langrangian Multipliers (named after the French mathematician Joseph Lagrange, 1811) are used as a methodology for locating the minima and the maxima of a function subject to a set of equality constraints. For instance, in Figure 3.7, consider the following optimization problem:

"To maximize $f(x, y)$, subject to the constraint $g(x, y) = 0$."

It is understood that both f and g have continuous first partial derivatives. The Method of Lagrangian Multipliers
Introduce a new variable $λ$, called a *Lagrange Multiplier* and study the *Lagrange Function* (or the *Lagrangian*) defined by

$$L(x, y, \lambda) = f(x, y) - \lambda g(x, y)$$

where the $λ$ term may be either added or subtracted. If $f(x_0, y_0)$ is a maximum of $f(x, y)$ for the original constrained problem, then there exists λ_0 such that

$f(x,y)$

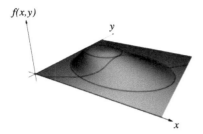

Figure 3.7 Method of Lagrange Multipliers: Finding x and y to maximize (or minimize), subject to a constraint (in red): g(x, y) = constant.

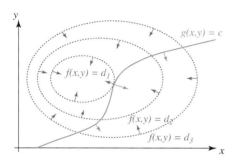

Figure 3.8 Contour Map of Figure 3.7. The red line represents the constraint g(x, y) = c. The blue lines are contours of $f(x, y)$. The solution is the point where the red line *tangentially touches* a blue contour. (Since $d_1 > d_2$, the solution is a *maximization* of the function $f(x, y)$. (Both ordinates and abscissae are arbitrary.)

(x_0, y_0, λ_0) is a *stationary point* for the Lagrange function (stationary points are those points where the partial derivatives of L are zero). However, *not* all stationary points yield a solution of the original problem. Thus, the method of Lagrange multipliers yields a *necessary condition* for optimality in constrained problems. Sufficient conditions for a minimum or maximum also exist (Figure 3.8).

Example 3.7 Estimation of Global Minimum-Variance: a numerical example of the management of portfolios consisting of two risky assets, using the data set in Table 3.2.

Using the data in Table 3.2 (restated above) and Equations (3.44a) and (3.44b):

$$
\begin{aligned}
x_A^{\min} &= (\sigma_B^2 - \sigma_{AB})/(\sigma_A^2 + \sigma_B^2 - 2\sigma_{AB}) \\
&= \{0.01 - (-0.0038)\}/\{0.05 + 0.01 - 2(-0.0038)\} \\
&= 0.0138/0.0676 \\
&= 0.2041
\end{aligned}
\tag{3.44a}
$$

$$
\begin{aligned}
x_B^{\min} &= 1 - x_A^{\min} \\
&= 1 - 0.2041 \\
&= 0.7959
\end{aligned}
\tag{3.44b}
$$

For these values of the investment portfolio position, (x_A^{min}, x_B^{min}), the corresponding values of the expected return μ_p, variance $var(R_p)$, and standard deviation σ_p may be calculated from the following relationships:

$$\mu_p = E[R_p] = x_A\mu_A + x_B\mu_B \tag{3.24}$$

$$\sigma_p^2 = var(R_p) = x_A^2\sigma_A^2 + x_B^2\sigma_B^2 + 2x_Ax_B\sigma_{AB} \tag{3.25}$$

$$\sigma_p = SD(R_p) = \sqrt{var(R_p)} = \sqrt{(x_A^2\sigma_A^2 + x_B^2\sigma_B^2 + 2x_Ax_B\sigma_{AB})} \tag{3.26}$$

Hence,

$$
\begin{aligned}
\mu_p = E[R_p] &= x_A^{min}\mu_A + x_B^{min}\mu_B \\
&= (0.2041)(0.2) + (0.7959)(0.04) \\
&= 0.04082 + 0.03184 \\
&= 0.07266
\end{aligned}
$$

$$
\begin{aligned}
\sigma_p^2 = var(R_p) &= (x_A^{min})^2\sigma_A^2 + (x_B^{min})^2\sigma_B^2 + 2(x_A^{min})(x_B^{min})\sigma_{AB} \\
&= (0.2041)^2(0.05) + (0.7959)^2(0.01) \\
&\quad + 2(0.2041)(0.7959)(-0.0038) \\
&= 0.00208 + 0.00633 - 0.00123 \\
&= 0.00718
\end{aligned}
$$

$$
\begin{aligned}
\sigma_p = SD(R_p) &= \sqrt{var(R_p)} \\
&= \sqrt{(0.00718)} \\
&= 0.08473
\end{aligned}
$$

Suggested Exercise for Computation Using R**:**   As an exercise, write a program in R to undertake all the computations.

3.2.4.2   Effects of Portfolio Variance on Investment Possibilities

For a portfolio of any two assets, A and B, the *correlation* between A and B may strongly affect the investment possibilities of the portfolio. Thus,

a) If ρ_{AB} is close to 1, then the investment set approaches a linear relationship such that the return is close to a straight line connecting the portfolio with all the funds invested in asset B, namely, $(x_A, x_B) = (0, 1)$, to the portfolio with all the funds placed in asset A only, that is, $(x_A, x_B) = (1, 0)$.

b) In a plot of μ_p (as the ordinate) versus σ_p (as the abscissa) (see Figure 1.2), as ρ_{AB} approaches 0, the set bows toward the μ_p-axis, and the power of diversification starts to make its presence felt! If $\rho_{AB} = -1$, the set will actually tangentially touch the μ_p-axis. This implies that the assets A and B will be perfectly negatively correlated, and there exists a portfolio of A and

B that has positively expected return but zero variance! And to determine the portfolio with

$\sigma_p^2 = 0$ when $\rho_{AB} = -1$, one may use (3.44a) and (3.44b), and the condition that

$$\sigma_{AB} = \rho_{AB}\sigma_A\sigma_B \tag{3.27}$$

to give

$$x_A^{\min} = (\sigma_B^2 - \sigma_{AB})/(\sigma_A^2 + \sigma_B^2 - 2\sigma_{AB}) \tag{3.44a}$$

Since $\rho_{AB} = -1$

$$\sigma_{AB} = \rho_{AB}\sigma_A\sigma_B = (-1)\sigma_A\sigma_B = -\sigma_A\sigma_B \tag{3.45}$$

Now, from (3.44),

$$
\begin{aligned}
x_A^{\min} &= (\sigma_B^2 - \sigma_{AB})/(\sigma_A^2 + \sigma_B^2 - 2\sigma_{AB}) \\
&= \{\sigma_B^2 - (-\sigma_A\sigma_B)/\{\sigma_A^2 + \sigma_B^2 - 2(\sigma_A\sigma_B)\}, \quad \text{from (3.45)} \\
&= (\sigma_B^2 + \sigma_A\sigma_B)/(\sigma_A^2 + \sigma_B^2 + 2\sigma_A\sigma_B) \\
&= \sigma_B(\sigma_A + \sigma_B)/(\sigma_A^2 + 2\sigma_A\sigma_B + \sigma_B^2) \\
&= \sigma_B(\sigma_B + \sigma_A)/(\sigma_B + \sigma_A)^2 \\
&= \sigma_B/(\sigma_B + \sigma_A)
\end{aligned} \tag{3.44a}
$$

And using (3.44b),

$$x_B^{\min} = 1 - x_A^{\min} \tag{3.44b}$$

3.2.4.3 Introduction to Portfolio Optimization

For the efficient set of portfolios as described in Figure 3.6, the critical question will be

"Which portfolio should an investor choose? And why?"

Of the efficient portfolios, investors will select the one that closely supports their risk preferences. In general, risk-averse investors will prefer a portfolio that has low risk, namely, low volatility, and will prefer a portfolio close to the global minimum-variance portfolio.

On the other hand, risk-tolerant investors will ignore volatility and seek portfolios with high expected returns. These investors will choose portfolios with large amounts of asset A that may involve short-selling asset B.

3.2.5 Attractive Portfolios with Risk-Free Assets

In Section 3.1.4, an efficient set of portfolios in the absence of a risk-free asset was constructed. Now consider what happens when a risk-free asset is being introduced.

A risk-free asset is equivalent to default-free pure discount bond that matures at the end of the assumed investment period. The risk-free rate r_f, may be represented by the nominal return on a low-risk bond. For example, if the investment period is 1 month, then the risk-free asset is a 30-day U.S. Treasury bill (T-bill), and the risk-free rate is the nominal rate of return on the T-bill. (The "default-free" assumption of U.S. national debt has been questioned owing to the possible, and probable, inability of the U.S. Congress to address the long-term debt problems of the U.S. government.)

If the portfolio holding of the risk-free assets is *positive*, then it is equivalent to "lending money" at the risk-free rate. If the portfolio holding is negative, at "risk-free" rates, then one is "borrowing" money that is risk-free!

3.2.5.1 An Attractive Portfolio with a Risk-Free Asset

Consider an arbitrary portfolio with asset B, one may examine the consequences if one should introduce a risk-free asset (say, a T-Bill) into this portfolio.

Now the risk-free rate is constant (fixed) over the investment portfolio that has the following properties:

$$\mu_f = E[r_f] = r_f \tag{3.45a}$$

$$\text{var}(r_f) = 0 \tag{3.45b}$$

$$\text{cov}(R_B, r_f) = 0 \tag{3.45c}$$

Then, within this portfolio,

- let the proportion of the investment in asset $B = x_B$,
- then the proportion of the investment in T-Bills $= (1 - x_B)$
  and hence the return of this portfolio is

$R_p = $ Return owing to the asset $B + $ return owing to the risk-free T-Bills $= R_B x_B$. Hence, the portfolio return is given by

$$
\begin{aligned}
R_p &= \text{Return from the asset B} + \text{return from the risk} - \text{free asset} \\
&= x_B R_B + (1 - x_B) r_f \\
&= r_f + (R_B - r_f) x_B
\end{aligned}
$$

$$\tag{3.46}$$

The term $(R_B - r_f)$ is the *Excess Return* over the T-Bills return on asset B.

Hence, the *Expected Return* of the portfolio is

$$
\begin{aligned}
\mu_p &= r_f + (E[R_B] - r_f) x_B \\
&= r_f + (\mu_B - r_f) x_B
\end{aligned}
\tag{3.47}
$$

In (3.47), the factor $(\mu_B - r_f)$ is the *Expected Excess Return* or *Risk Premium* on asset B.

Remarks:

1) The risk premium is generally *positive* for risky assets, showing that investors will expect a *higher* return on the risky assets than the safe assets. This "risk premium" on the portfolio may be expressed in terms of the risk premium on asset B as follows, using (3.47):

$$\mu_p = r_f + (E[R_B] - r_f)x_B$$
$$= r_f + (\mu_B - r_f)x_B \qquad (3.47)$$

$$\text{Risk Premium} = (\mu_p - r_f)$$
$$= (\mu_B - r_f)x_B \qquad (3.48)$$

2) Thus, the more one invests in asset B, the higher the risk premium on the portfolio.

3) Since the risk-free rate is constant, the portfolio variance depends only on the variability of asset B, and is given by

$$\sigma_p^2 = x_B^2 \sigma_B^2 \qquad (3.49)$$

4) Hence, the portfolio standard deviation is proportional to the standard deviation on asset B:

$$\sigma_p = x_B \sigma_B \qquad (3.50a)$$

which may be solved for x_B:

$$x_B = \sigma_p / \sigma_B \qquad (3.50b)$$

5) The *Capital Allocation Line* (CAL)—Combining (3.47) and (3.50b), one obtains the set of Efficient Portfolios:

$$\mu_p = r_f + (E[R_B] - r_f)x_B$$
$$= r_f + (\mu_B - r_f)x_B \qquad (3.47)$$

$$\text{Risk Premium} = (\mu_p - r_f)$$
$$= (\mu_B - r_f)x_B \qquad (3.48)$$

or

$$(\mu_B - r_f) = (\mu_p - r_f)/x_B$$
$$= (\mu_B - r_f)(\sigma_p/\sigma_B), \quad \text{from (3.50b)}$$

Hence,

$$\mu_B = \{(\mu_B - r_f)/\sigma_B\}\sigma_p + r_f \qquad (3.51)$$

which is a straight line in the (σ_p, μ_p)-space, with slope $\{(\mu_B - r_f)/\sigma_B\}$, and ordinate intercept r_f.

This is the Capital Allocation Line (CAL)—see Section 3.1.1.7.

The slope of this line is the Sharpe Ratio (SR)—see Section 3.1.1.6. It is a measure of the risk premium on the asset per unit risk—as measured by the standard deviation of the asset.

6) Characteristics of a two-asset portfolio with one asset taking short positions to appreciate the economic characteristics of a two-asset portfolio; one may consider the effects of combining two assets to form a portfolio. One can extend the analysis to cases in which short positions may be taken.

First, assume that one may either go long on the risk-free asset lend) or take a short position in it borrow) at the same interest rate, e_1.

Let x_A and x_B be the proportions invested in assets A and B, respectively, and let e_A and e_B be their respective returns.

Then, the Expected Return of the *whole* portfolio, e_p, will be given by

$$e_p = x_A e_A + x_B e_B \tag{3.52}$$

and since

$$x_A + x_B = 1,$$

namely,

$$x_A = 1 - x_B$$

(3.52) may be written as

$$e_p = (1 - x_B)e_A + x_B e_B$$

namely,

$$e_p = e_A + x_B(e_B - e_A) \tag{3.53}$$

The variance of the portfolio, var_p, will be a function of

a) the proportions, x_A and x_B, invested in these assets,
b) their return variances: var_A and var_B, and
c) the covariance between their returns: covar_{A-B}:

$$\text{covar}_{A-B} = x_A^2 \text{var}_A + x_B^2 \text{var}_B + 2x_A x_B \text{covar}_{A-B}$$

Upon substituting $(1 - x_B)$ for x_A to derive an expression relating the variance of the portfolio to the amount invested in asset B:

$$
\begin{aligned}
\text{var}_{A-B} &= (1 - x_B)^2 \, \text{var}_A + x_B^2 \, \text{var}_B + 2(1 - x_B)x_B \, \text{covar}_{A-B} \\
&= (1 - 2x_B + x_B^2) \, \text{var}_A + x_B^2 \, \text{var}_B + 2x_B \, \text{covar}_{A-B} - 2x_B^2 \, \text{covar}_{A-B} \\
&= \text{var}_A - 2x_B \text{var}_A + x_B^2 \, \text{var}_A + x_B^2 \, \text{var}_B + 2x_B \, \text{covar}_{A-B} - 2x_B^2 \, \text{covar}_{A-B}
\end{aligned}
$$

namely,

$$\text{var}_{A-B} = \text{var}_A + 2x_B(\text{covar}_{A-B} - \text{var}_A) + x_B^2(\text{var}_A - 2\text{covar}_{A-B} + \text{var}_B) \quad (3.54)$$

and the standard deviation of return, SD_R, is the square root of this variance:

$$SD_R = \sqrt{(\text{var}_{A-B})}$$

For any two given assets in a portfolio, and without loss of generality, one may assume that Asset 1 has less risk and, concomitantly, smaller expected return.

Now, consider the risk-return trade-offs associated with different combinations of the two assets. Also, consider the shape of the curves for mean-variance and mean-standard deviation plots that result as more investment is added to the risky asset, namely, as x_2 (for the risky asset) is increased and x_1 (for the risk-free asset) is decreased.

3.2.5.1.1 Investments with One Risk-Free Asset and One Risky Asset

When the investment consists only asset A and a risk-free asset (T-Bills), the Sharpe Ratio, SR_A (viz., the *excess* return of an asset over the return of a risk-free asset divided by the variability or standard deviation of returns, see Section 3.1.1.6) is given by

$$
\begin{aligned}
SR_A &= (\mu_A - r_f)/\sigma_A \\
&= (0.2 - 0.03)/0.2236 \quad (3.55a) \\
&= 0.7603
\end{aligned}
$$

On the other hand, when the investment consists only asset B and T-Bills, the Sharpe Ratio, SR_B, is

$$
\begin{aligned}
SR_B &= (\mu_B - r_f)/\sigma_B \\
&= (0.04 - 0.03)/0.1000 \quad (3.55b) \\
&= 0.1000
\end{aligned}
$$

Here the data are taken from Table 3.2, and the risk-free rate, $r_f = 0.03$, is the nominal return on the bond. Thus, for example, if the investment horizon is 1 month, then the risk-free asset is a "30-day U.S. Treasury Bill (T-Bill), and the risk-free rate r_f is the nominal rate of return on the T-Bill.

Remarks:

1) The expected return-risk trade-off for these portfolios are linear.

Figure 3.9 plots the locus of mean-standard deviation combinations for values.

If x_A and x_B are the proportions invested in assets A and B, respectively, and μ_A and μ_B are their respective expected returns, then the expected return of the portfolio of these two assets, μ_p, will be given by

$$\mu_p = x_A\mu_A + x_B\mu_B$$

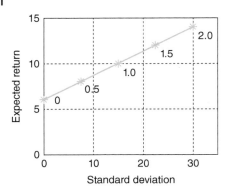

Figure 3.9 Locus of mean standard deviations.

Since

$$x_A + x_B = 1, \quad x_A = 1 - x_B$$

hence,

$$\mu_p = (1 - x_B)\mu_A + x_B\mu_B$$
$$= \mu_A - x_B\mu_A + x_B\mu_B$$
$$= \mu_A + x_B(\mu_B - \mu_A)$$

for x_B between 0 and 1 when $x_A = 6$, $\mu_B = 10$, $\sigma_A = 0$, $\sigma_B = 15$, and covar$_{A-B} = 0$.

In this case, as in every case involving a riskless and a risky asset, the relationship is linear. This is easily seen. Recall that e_p is always linear in x_B as shown earlier. If asset A is risk-free, s_p ($= x_A\sigma_A + x_B\sigma_B$) will also be linear in x_B, since both σ_A^2 and covar$_{A-B}$ will equal zero. In such a case,

$$\sigma_{AB}^2 = (x_B^2)\sigma_B^2$$
$$s_p(= x_A\sigma_A + x_B\sigma_B) = \sqrt{\{(x_B^2)\sigma_B^2\}}$$
$$= |(x_B)|\sqrt{(\sigma_B^2)}$$
$$= |(x_B)|\sigma_B$$

where $|(x_B)|$ denotes the absolute value of x_B.

This result may be applied to cases where it is possible to take short positions. First, assume one can either go long on the risk-free asset (lend) or take a short position in it (borrow) at the same interest rate (e_A). The above equations may then be applied. Figure 3.9 shows the results obtained by using *leverage* in this way. Thus, the point marked 1.5 is associated with $x_B = 1.5$ and $x_A = -0.5$. It shows that by "leveraging up" an investment in asset B by 50%, one may obtain a probability distribution of return on initial capital with an expected value of 12% and a standard deviation of 22.5%. The other points in the figure correspond to the indicated values of x_B. Those above the original point involve borrowing ($x_A < 0$) while those below it involve lending ($x_A > 0$).

2) The portfolios that are combinations of asset A and T-Bills have *Expected Returns* uniformly higher than the portfolios that are combinations of asset B and T-Bills. This result due to the Sharpe Ratio of asset A, 0.7603, is higher than the Sharpe Ratio of Asset B, 0.1000.

3) The portfolios of asset A and T-Bills are more efficient relative to the portfolios of asset B and T- Bills.

4) The Sharpe Ratio may be used as an index for ranking the return efficiencies of individual assets: Assets with higher Sharpe Ratios have superior risk-returns than assets with lower Sharpe Ratios. Thus, in financial engineering, analysts do rank assets on the basis of their Sharpe Ratios.

5) For an investment portfolio consisting of a risky asset A and a risk-free asset (such as T-Bills), one may assume that the risk-free rate is constant (viz., fixed) over the investment period during which it has the following special properties: for the *risk-free asset*, it is assumed that

$$\mu_f = E[r_f] = r_f \tag{3.56a}$$

$$\text{var}(r_f) = 0 \tag{3.56b}$$

$$\text{cov}(R_A r_f) = 0 \tag{3.56c}$$

If x_A denote the share of investment in asset A and x_f denote the share of investment in the T-Bill portion of the portfolio, then

$$x_A + x_f = 1 \tag{3.57}$$

For this investment, the portfolio return, R_p, is given by

$$
\begin{aligned}
R_p &= \text{Return from the asset } A + \text{return from the risk} - \text{free investment} \\
&= x_A R_A + x_f r_f \\
&= x_A \mu_A + (1 - x_A) r_f
\end{aligned}
$$

namely,

$$R_p = x_A(\mu_A - r_f) + r_f \tag{3.58}$$

Remarks:

1) The quantity ($\mu_A - r_f$), in (3.55), is the *Expected Excess Return*, or *Risk Premium on asset A* (over and above a risk-free asset). For risky assets, the Expected Excess Return is typically positive. That is, investors will expect a higher return on more risky investments.

2) The Expected Excess on the portfolio may be expressed in terms of Expected Excess on the risky asset A as follows:

$$(\mu_p - r_f) = x_A(\mu_A - r_f) \tag{3.59}$$

Equation (3.59) shows that for this category of investment portfolio, the greater the *proportion* of risky asset A invested, the higher the Expected Excess.

3) Since the risk-free return rate is constant, the variance of the whole portfolio will depend only on the variability of the risky asset A, and is given by

$$\sigma_p^2 = x_A^2 \sigma_A^2 \tag{3.60a}$$

Hence, the portfolio standard deviation σ_p is proportional to the standard deviation of asset A:

$$\sigma_p = x_A \sigma_A \tag{3.60b}$$

from which one may solve for x_A:

$$x_A = \sigma_p / \sigma_A \tag{3.60c}$$

4) The feasibility and efficiency of the portfolio is given by the follow relationship: From (3.59):

$$\mu_p = r_f + x_A(\mu_A - r_f)$$

and from (3.60c):

$$\mu_p = \sigma_p\{(\mu_A - r_f)/\sigma_A\} + r_f \tag{3.61}$$

which is a straight line in the (μ_p, σ_p) space with slope $\{(\mu_A - r_f)/\sigma_A\}$ and intercept r_f.

The straight line (3.61) is the *Capital Allocation Line* (CAL) and the slope of CAL is called the *Sharpe Ratio* (SR) (formerly the *Reward-to-Variability Ratio*). It measures the ratio of the Expected Value of a zero-investment strategy to the Standard Deviation of that strategy. A special case of this approach involves a *zero-investment strategy* where funds are borrowed at a fixed rate of interest and invested in a risky asset!

Figure 3.10 plots the locus of mean-standard deviation combinations for values of x_B between 0 and 1 when $\mu_A = 6$, $\mu_B = 10$, $\sigma_A = 0$, $\sigma_B = 15$ and $\text{covar}_{A-B} = 0$.

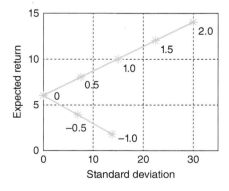

Figure 3.10 Effects of shorting the risky asset B.

As shown earlier,
a) $e_p \ (= x_A\mu_A + x_B\mu_B)$ is always linear in x_B, and
b) If asset A is risk-free, $s_p \ (= x_A\,\sigma_A + x_B\,\sigma_B)$ will also be linear in x_B, since both σ_A^2 and covar$_{A-B}$ will equal zero. In such a case,

$$s_p (= x_A\sigma_A + x_B\sigma_B) = \sqrt{\{(x_B^2)\sigma_B^2\}} = |(x_B)|\sigma_B$$

5) If an investor short the risky asset ($x_B < 0$) and invest the proceeds obtained from the short sale in the risk-free asset A, the standard equations apply. However, note that the variance will be positive, as will the standard deviation, since a negative number (x_B) squared is always positive. Figure 3.11 shows the effects of negative x_B values.

3.2.5.1.2 Investments with One Risk-Free Asset and Two Risky Assets
Next consider a portfolio of two risky assets, A and B, and some T-Bills. In this case, the efficient set will still be a straight line in the (μ_p, σ_p) space with intercept r_j. The slope of the efficient set, namely, the maximum Sharpe Ratio, is such that *it is tangential to the efficient set constructed just using the two risky assets A and B.*

Remarks:

1) If one invests only in asset A and T-Bills, then this gives a Sharpe Ratio of

$$SR_A = (\mu_A - r_f)/\sigma_A = (0.2 - 0.03)/0.2236 = 0.7603$$

and the Capital Allocation Line (CAL) will *intersect* the efficiency parabola (say, at point A). Thus, this is certainly *not* the efficient set of portfolios.
2) On the other hand, if one invests only in asset B and T-Bills, then this gives a Sharpe Ratio of

$$SR_B = (\mu_B - r_f)/\sigma_B = (0.04 - 0.03)/0.1000 = 0.1000$$

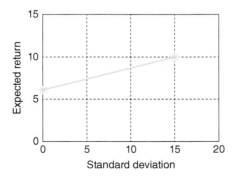

Figure 3.11 A two-asset portfolio: 1 risk-free and 1 risky.

and the Capital Allocation Line (CAL) will *intersect* the efficiency parabola (say, at point B). Again, this is certainly *not* the efficient set of portfolios.

3) Indeed, one could do better if one invests some *combination* of assets A and B, together with some T-Bills. And clearly, the most efficient portfolio would be one such that the CAL is *tangential* to the parabola. This Tangency Portfolio will consist of assets A and B such that the CAL is just tangential to the parabola.

4) And so the set of efficient portfolios will consist of such sets of assets A and B, together with some risk-free assets, such as T-Bills. These portfolios are Tangency Portfolio of assets A and B.

Example 3.8 A Two-Asset Portfolio: 1 Risk-Free and 1 Risky

Figure 3.11 is a plot of Expected Returns, e_i, (on the ordinate: vertical axis), versus Standard Deviations, SD_i (on the abscissa: horizontal axis), showing the locus of the following two-asset portfolio ($A = $ risk-free, and $B = $ risky), given their Expected Returns μ_A and μ_B, their variances σ_A^2 and σ_B^2, as well as their correlation coefficient $covar_{A-B}$. In this example,

- their Expected Returns: $\mu_A = 6$ and $\mu_B = 10$,
- their Variances: $var_A = 0$ and $var_B = 15$, and
- their correlation coefficient: $covar_{i-B} = 0$.

In cases of portfolios consisting of a risk-free asset and a risky asset, in proportions x_A and x_B, respectively, this relationship is linear, because the Expected Return e_p is always linear in x_B (see (3.53)):

$$e_p = e_A + x_B(e_B - e_A) \tag{3.61}$$

Now, if asset A is risk-free (by hypothesis), the portfolio standard deviation of return (σ_p) will also be linear in x_B, since

$$\sigma_p = SD(R_p) = \sqrt{var(R_p)} = \sqrt{(x_A^2\sigma_A^2 + x_B^2\sigma_B^2 + 2x_Ax_B\sigma_{AB})}$$
$$= \sqrt{(0 + x_B^2\sigma_B^2 + 0)}, \quad \text{since} \quad \sigma_A^2 = 0 = \sigma_{AB}$$
$$= \sqrt{(x_B^2\sigma_B^2)}$$
$$= x_B\sigma_B \tag{3.26}$$

Example 3.9 Use of Leveraging in a Two-Asset Portfolio: 1 Risk-Free and 1 Risky

The result in Example 3.8 may be extended to portfolios in which leveraging positions may be taken where one may

i) either go "long" in the risk-free asset A (viz., to "lend")
ii) or take a "short" in the risk-free asset A (viz., to "borrow")

at the same rate of interest (e_A).

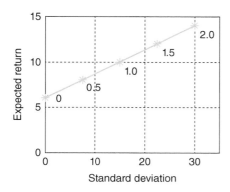

Figure 3.12 Combining a risk-free asset and a risky asset: use of leveraging.

The foregoing equations do apply.
Figure 3.12 shows the results for this use of leverage:

- The point marked "1.5" represents the case of

$$x_B = 1.5 \quad \text{and} \quad x_A = 1 - x_B = 1 - 1.5 = -0.5$$

- It shows that by "leveraging up" a portfolio in asset B by 50%, an investor can obtain a probability distribution of return-on-initial capital with an expected value of 12% and a standard deviation of 22.5%.
- The other points in Figure 3.12 correspond to the indicated values of the risky asset B, with proportions $x_2 = 0.0, 0.5, 1.0, 1.5$, and 2.0. Those above the point marked "1.0" involve "borrowing," namely, $x_A < 0$, and those below the point marked "1.0" involve "lending", namely, $x_A > 0$.

Figure 3.13 plots the locus of mean-standard deviation combinations for values of x_B between 0 and 1, when

$$\mu_A = 6, \quad \mu_B = 10, \quad \sigma_A = 0, \quad \sigma_B = 15, \quad \text{covar}_{A-B} = 0$$

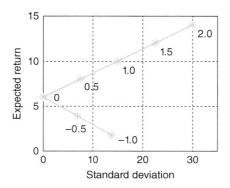

Figure 3.13 Effects of shorting the risky asset *B*.

Clearly, the endpoints A and B of the straight line are

$$(\sigma_A = 0, \quad \mu_A = 6) \quad \text{and} \quad (\sigma_B = 15, \quad \mu_B = 10)$$

Since

$$\mu_p = E[R_p] = x_A\mu_A + x_B\mu_B \tag{3.24}$$

μ_p is always linear in x_B.

If asset A is risk-free, then both var $A = \sigma_A^2 = 0$ and $\sigma_{AB} = \text{covar}_{A-B} = 0$, hence

$$\sigma_A = 0$$

and since

$$s_p = x_A\sigma_A + x_B\sigma_B = 0 + x_B\sigma_B = x_B\sigma_B$$

Hence, s_p also *always will be linear in x_B*.

And starting with the relationship

$$\begin{aligned}
\sigma_p^2 &= \text{var}\,(R_p) = x_A^2\sigma_A^2 + x_B^2\sigma_B^2 + 2x_Ax_B\sigma_{AB} \\
&= x_A^2(0) + x_B^2\sigma_B^2 + 2x_Ax_B(0)
\end{aligned} \tag{3.25}$$

Hence,

$$\sigma_p^2 = x_B^2\sigma_B^2$$

and

$$\sigma_p = \sqrt{(\sigma_p^2)} = \sqrt{(x_B^2\sigma_B^2)} = |x_B|\sqrt{(\sigma_B^2)} = |x_B|\sigma_B$$

that is,

$$\sigma_p = |x_B|\sigma_B \tag{3.26}$$

where $|x_B|$ is the absolute value of x_B.

Remarks:

1) If, during the process of allocation of assets, one takes short positions, and then the foregoing result can be extended to such cases as well.
2) For example, if in one investment portfolio, one goes long on the risk-free asset (viz., "lend"), or chooses a short position instead (viz., "borrow") at the same rate of interest, e_1, then the above analysis may be applied.
3) The results obtained from applying this leverage method is illustrated in Figure 3.10. Thus, the point labeled 1.5 is associated with

$$x_A = 0.5 \quad \text{and} \quad x_B = 1.5$$

This shows that by the use of the leverage method applying to asset B by 50% (from $x_B = 1.0–1.5$) in an investment, the investor may obtain a probability distribution of return on initial investment capital with

a) an *expected return* value of 12% (as indicated in the ordinate of the "Expected return" axis), and

b) a predicted standard deviation of 22.5% (as read on the abscissa of the "Standard Deviation" axis.

4) In this figure, the other points correspond to the indicated values of $x_B = 0$, 0.5, 1.0, and 2.0. Those above the original point ($x_B = 1.5$) imply borrowing, namely, $x_A < 0$ and $x_B > 1$, and those points below the original point imply lending.

5) Should an investor first short the risky asset, namely, $x_B < 0$, and then invest the short sale proceeds in the risk-free asset x_A, the analysis presented herein will still be applicable. However, in such cases, the variance and the standard deviation will be positive—since a negative number x_B squared is positive. The effects of negative x_B values is showed in Figure 3.11.

6) In the world of business, it is a common practice that when moneys are borrowed, they usually come at a higher rate (say, 8%) than the rate at which they are being lent (say, 6%). This is being illustrated in Figure 3.14.

This plot shows the loci of the $\mu_{AB}–\sigma_{AB}$ combinations plotted as two lines, where

$$\mu_{AB} = x_A\mu_A + x_B\mu_B$$

and

$$\sigma_{AB} = x_A\sigma_A + x_B\sigma_B$$

The first line is associated with the lower lending rate, and the second line is associated with the higher borrowing rate. Again, one may assume that the risky asset offers an Expected Return of 10% and a Risk of 15%. This efficient investment system is shown in Figure 3.14 in which the efficient frontier is

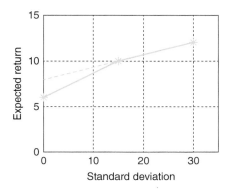

Figure 3.14 Effects between rate differences for "borrowed" and "lent" funds.

represented by solid lines, and the options that are available if the investor could lend at 8% are shown in the broken line.

Moreover, the rates charged for borrowing may inverse with the loan amounts, so the locus of the $\sigma_{AB}-\mu_{AB}$ combinations may increase at a decreasing rate as the risk, σ_{AB}, increases beyond the amount for a full unlevered portfolio containing the risky asset: $x_B = 1$. For such conditions, one may expect decreasing returns for taking risks!

Example 3.10 Portfolios Consisting of Two Risky Assets

If two risky assets, A and B, are included in a portfolio, one should take into account the expected returns, variances, and covariance (or correlation) between the returns of these two assets. Compared with the previous case, in which one of the two assets was risk-free, the differences occur in the formulation of the portfolio variance. The appropriate formulation of the variance, in terms of risks and correlations, takes the form:

$$
\begin{aligned}
\sigma_{AB}^2 &= \text{var}_{1-2} \\
&= x_A^2\sigma_A^2 + 2x_Ax_B\sigma_{AB}\sigma_A\sigma_B + x_B^2\sigma_B^2
\end{aligned}
\tag{3.54}
$$

for which σ_{AB} is the correlation coefficient between the returns of the two assets A and B.

To Combine Two Perfectly Positively Correlated Risky Assets If the two returns are perfectly positively correlated, then $\sigma_{AB} = 1$, and (3.54) becomes

$$
\begin{aligned}
\sigma_{AB}^2 &= x_A^2\sigma_A^2 + 2x_Ax_B(1)\sigma_A\sigma_B + x_B^2\sigma_B^2 \\
&= x_A^2\sigma_A^2 + 2x_Ax_B\sigma_A\sigma_B + x_B^2\sigma_B^2 \cdots \text{(which is a perfect square!)} \\
&= (x_A\sigma_A + x_B\sigma_B)^2
\end{aligned}
$$

so that

$$
\begin{aligned}
s_p &= \sigma_{AB} \\
&= |\sqrt{(\sigma_{AB}^2)}| \\
&= |\sqrt{(x_A\sigma_A + x_B\sigma_B)^2}| \\
&= |(x_A\sigma_A + x_B\sigma_B)|
\end{aligned}
\tag{3.55}
$$

Remarks:

1) The notation $|f(x)|$ denotes the absolute value of the enclosed expression $f(x)$.
2) Since neither σ_A nor σ_B is negative, when both x_A and x_B are nonnegative, the expression $(x_A\sigma_A + x_B\sigma_B)$ in (3.55) will be nonnegative, and the absolute value of s_p is understood implicitly. However, if one of the two x-values is sufficiently negative, then the absolute value of s_p should be used explicitly.

3) For long positions in the two assets, namely, x_A, $x_B \geq 0$, consider the following combinations:

$$s_p = x_A \sigma_A + x_B \sigma_B \qquad \text{(i)}$$

$$e_p = x_A \mu_A + x_B \mu_B \qquad \text{(ii)}$$

Since

$$x_A + x_B = 1 \quad \text{or} \quad x_A = 1 - x_B \qquad \text{(iii)}$$

By substituting for x_A in (i), using (iii), one obtains from (i):

$$
\begin{aligned}
s_p &= x_A \sigma_A + x_B \sigma_B, \quad \text{using (i)} \\
&= (1 - x_B)\sigma_A + x_B \sigma_B, \quad \text{substituting for } x_A \text{ from (iii)} \\
&= \sigma_A - \sigma_A x_B + x_B \sigma_B
\end{aligned}
$$

or

$$s_p = \sigma_A + x_B(\sigma_B - \sigma_A) \qquad (3.56)$$

Similarly, by substituting for x_A in (ii), using (iii), one obtains from (ii):

$$
\begin{aligned}
e_p &= x_A \mu_A + x_B \mu_B, \quad \text{using (ii)} \\
&= (1 - x_B)\mu_A + x_B \mu_B, \quad \text{substituting for } x_A \text{ from (iii)} \\
&= \mu_A - \mu_A x_B + x_B \mu_B
\end{aligned}
$$

or

$$e_p = \mu_A + x_B(\mu_B - \mu_A) \qquad (3.57)$$

4) All such relationships will lie on a straight line, which represents the two assets (see Figure 3.15).
5) Extension of the values of x_B to $x_B > 1$ or $x_B < 0$:

   Consider the *Minimum-Variance Portfolio*, namely, the combination that results in the least possible risk. To achieve a variance of 0, one may seek the value of x_B for which $s_p = 0$. And using (3.56):

$$s_p = \sigma_{AB} = \sigma_A + x_B(\sigma_B - \sigma_A) = 0$$

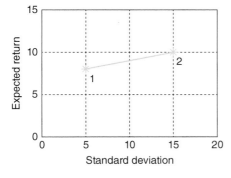

Figure 3.15 Two perfectly positively correlated risky assets: $\sigma_{AB} = 1$, Point 1: Asset A, for which $\mu_A = 8$, $\sigma_A = 5$; Point 2: Asset B, for which $\mu_B = 10$, $\sigma_B = 15$.

Solving for x_B:,

$$x_B = -\sigma_A/(\sigma_B - \sigma_A) \tag{3.58a}$$

and

$$
\begin{aligned}
x_A &= 1 - x_B \\
&= 1 - \{-\sigma_A/(\sigma_B - \sigma_A)\} \tag{3.58b} \\
&= \sigma_B/(\sigma_B - \sigma_A)
\end{aligned}
$$

Applying this result to the foregoing numerical example, it may be seen that a risk-free portfolio may be obtained by using (3.58a):

$$
\begin{aligned}
x_B &= -\sigma_A/(\sigma_B - \sigma_A) \\
&= -(5)/(15 - 5) \\
&= -\tfrac{1}{2} = -0.5
\end{aligned}
$$

and (3.58b):

$$
\begin{aligned}
x_A &= 1 - x_B \\
&= 1 - (-\tfrac{1}{2}) \\
&= 1\tfrac{1}{2} = 1.5
\end{aligned}
$$

6) This portfolio may be achieved by taking a short position in asset B equal to half of the total investment funds and investing this amount, together with the original amount of funds, in asset A. This action may need the pledging of some other collateral to form an adequate guarantee to the lender of asset B that the short position may be covered whenever required.

7) The need to create a risk-free portfolio by choosing offsetting positions in two perfectly positively correlated assets results in creating a configuration similar to the case for a risk-free asset combining with a risky asset. The following steps show how this goal may be achieved:
Let the expected return on the zero-variance portfolio be

$$\mu_0 = \mu_A + x_B(\mu_B - \mu_A)$$

Hence, for the foregoing numerical examples, $\mu_A = 8$, $\mu_B = 10$, and $x_B = -0.5$,

$$
\begin{aligned}
\mu_0 &= 8 + (-0.5)(10 - 8) \\
&= 8 - 1 \\
&= 7
\end{aligned}
$$

And the portfolio returns and risks may be obtained by considering any combinations of the risk-free asset A and either asset A or asset B. Either option provides the graph associated with a risk-free asset and a risky asset. This result is illustrated in Figure 3.16.

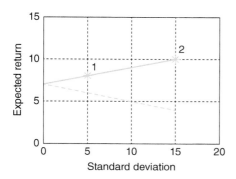

Figure 3.16 Creating a risk-free portfolio: $\mu_0 = 7$, by choosing off-setting positions in two perfectly positively correlated assets: Point 1: Asset A, for which $\mu_A = 8$, $\sigma_A = 5$; Point 2: Asset B, for which $\mu_B = 10$, $\sigma_B = 15$.

8) In this case, as in every case involving a risk-free and a risky asset, the relationship is linear. This is easily seen. Recall that e_p is always linear in x_B, as shown earlier. If asset A is risk-free, s_p will also be linear in x_B, since both σ_A^2 and covar_{A-B} will equal zero. In such a case,

$$\sigma_{AB}^2 = x_B^2 \sigma_B^2$$

namely,

$$x_A \sigma_A + x_B \sigma_B = \text{sqrt}(x_B^2 \sigma_B^2)$$
$$= |x_B| \sqrt{(\sigma_B^2)}$$
$$= |x_B| \sigma_B$$

where $|x_B|$ denotes the absolute value of x_B.

For cases in which it is possible to take short positions, this result is applicable: First assume one may either go *long* (viz., to lend) with the risk-free asset A or take a short position in it (viz., to borrow) at the same rate of interest μ_A. Then the foregoing formulas may be applied directly: This case has been illustrated in Figure 3.10 and Example 3.8.

9) Figure 3.17 shows the locus of mean standard deviation combinations for values of x_B between 0 and 1, when

$$\mu_A = 6, \quad \mu_B = 10, \quad \sigma_A = 0, \quad \sigma_B = 15, \quad \text{and} \quad \text{covar}_{A-B} = 0$$

Again, in this special case, as in cases involving a risk-free asset and a risky asset, the relationship is linear. This observation may be shown as follows: Since

$$e_p = x_A \mu_A + x_B \mu_B$$

e_p is linear in x_B; and when asset A is risk-free, s_p will also be linear in x_B. Since

$$s_p = x_A \sigma_A + x_B \sigma_B$$

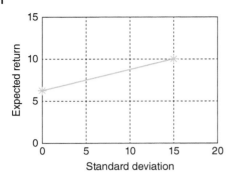

Figure 3.17 Combining a risk-free asset and a risky asset *when* $\mu_A = 6, \mu_B = 10, \sigma_A = 0$, $\sigma_B = 15$, and covar$_{A-B} = 0$.

both σ_A^2 and covar$_{A-B}$ will equal zero. Therefore,

$$\sigma_{AB}^2 = (x_B)^2 \sigma_B^2$$

and

$$
\begin{aligned}
s_p &= \sqrt{\{\sigma_{AB}^2\}} \\
&= \sqrt{\{(x_B)^2 \sigma_B^2\}} \\
&= |x_B| \sqrt{(\sigma_B^2)} \\
&= |x_B| \sigma_B
\end{aligned}
$$

where $|x_B|$ is the absolute value of x_B.

Example 3.11 Effects of Shorting the Risky Asset *B*

In this scenario, illustrated in Figure 3.18, the risky asset *B* is *shorted*, namely, $x_2 < 0$, with the proceeds from the short sale invested in the risk-free asset *A*. Again, starting from the relationship:

$$\sigma_p = \text{SD}(R_p) = \sqrt{\{\text{var}(R_p)\}} = \sqrt{(x_A^2 \sigma_A^2 + x_B^2 \sigma_B^2 + 2 x_A x_B \sigma_{AB})} \qquad (3.26)$$

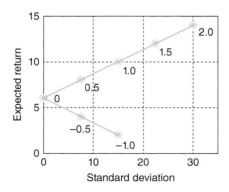

Figure 3.18 Effects of shorting the risky asset.

since a negative value of x_2, when squared, is positive, the resultant relationship indicates the effects of negative values of x_2 (shown in Figure 3.16).

Remark: It should not escape one's attention that when an investment involves *short* positions, and between ending asset and liability values, additional liquid capital must be pledged to cover these *short* positions, as well as possible liability values. Alternatively, a higher rate may have to be charged accordingly for all the short positions.

Example 3.12 Effects of Imperfectly Correlated Risky Assets

It is well known that, in many cases, correlation coefficients are less than 1.0, which means that recourse must be had to compensate this technical shortfall. This is usually achieved by the use of *diversification* in assets allocation.

Now consider an asset portfolio with *long* positions in two risky assets:

$$x_1 > 0, x_2 > 0$$

The variance of this portfolio will be given by (3.21) and (3.25):

$$\rho_{AB} = \text{cor}(R_A, R_B) = \sigma_{AB}/(\sigma_A \sigma_B) \tag{3.21}$$

$$\sigma_p^2 = \text{var}(R_p) = x_A^2 \sigma_A^2 + x_B^2 \sigma_B^2 + 2x_A x_B \sigma_{AB} \tag{3.25}$$

From (3.21), one may write

$$\sigma_{AB} = \rho_{AB}(\sigma_A \sigma_B) \tag{3.21a}$$

and substituting for σ_{AB} in (3.25) using (3.21a), the result is

$$\sigma_p^2 \equiv \text{var}(R_p) = x_A^2 \sigma_A^2 + x_B^2 \sigma_B^2 + 2x_A x_B \{\rho_{AB}(\sigma_A \sigma_B)\}$$

or

$$\sigma_p^2 = x_A^2 \sigma_A^2 + 2x_A x_B \rho_{AB} \sigma_A \sigma_B + x_B^2 \sigma_B^2 \tag{3.27}$$

Consider two portfolios that are similar in all the properties—the Proportions (x_A, x_B), the Expected Returns (μ_A, μ_B), the Standard Deviations (σ_A, σ_B), *except* the Correlation Coefficient ρ_{AB}.

Now, let the variance of one of the two portfolios for which $\rho_{AB} = 1$ and var (R_p) be the variance of the other portfolio for which $\rho_{AB} = r < 1$. Only the middle term in the equation for portfolio variance, namely, (3.27), will differ in the two computations. As all the components in that term, $2x_A x_B \rho_{AB} \sigma_A \sigma_B$, except ρ_{AB}, are positive, it is seen that $\sigma_p^2 \equiv \text{var}(R_p)$.

Now consider two cases—*similar* in every respect, *except* the Correlation Coefficient ρ_{AB}:

- Let var(1) be the variance of one portfolio, for which the Correlation Coefficient $\rho_{AB} = 1$
- Let var(r) be the variance of the other portfolio, for which $\rho_{AB} = r < 1$.

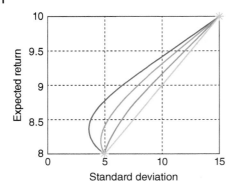

Figure 3.19 Effects of correlation coefficients for imperfectly correlated risky assets
in an investment portfolio.

In (3.27), only the second term, $2x_A x_B \rho_{AB} \sigma_A \ \sigma_B$, may differ in the two computations. Since all the components of that term, except ρ_{AB}, are positive, it follows that $\sigma_p^2 \equiv \text{var}(R_p)$ and ρ_{AB} *may have different signs.*

All other factors being equal, it is seen that the smaller the Correlation Coefficient between the two assets A and B, namely, ρ_{AB}, the smaller the risk of a portfolio of long positions in the two assets A and B. The combinations of risk and return for such portfolio when $\mu_1 = 8$, $\sigma_1 = 5$ and $\mu_2 = 10$, $\sigma_2 = 15$ are shown in Figure 3.19: Each curve is characteristic of a different correlation between the two assets' returns.

Remarks:

1) The cases are coincidental at the endpoints $(x_1 = 1, \ x_2 = 0)$ and $(x_1 = 0, x_2 = 1)$.
2) For all interior combinations, when the Correlation Coefficient $\rho_{AB} < 1.0$, risk is less than proportional to the risks of two assets; the greater the extent of risk reduction, the smaller the correlation coefficient:
   a) The yellow curve, $r_{12} = 1.0$, provides no risk reduction, only risk-averaging.
   b) The red curve, $r_{12} = 0.5$, provides some risk reduction.
   c) The green curve, $r_{12} = 0$, provides some more risk reduction.
   d) The blue curve, $r_{12} = 0.5$, provides even more.

Example 3.13 A Portfolio of Two Perfectly Positively Correlated Risky Assets

If the two returns are perfectly positively correlated, then $\rho_{AB} = 1$. Now, for a two-asset portfolio:

$$\sigma_p^2 = x_A^2 \sigma_A^2 + 2x_A x_B \rho_{AB} \sigma_A \sigma_B + x_B^2 \sigma_B^2$$
$$= x_A^2 \sigma_A^2 + 2x_A x_B (1) \sigma_A \sigma_B + x_B^2 \sigma_B^2, \quad \text{since } \rho_{AB} = 1$$
$$= x_A^2 \sigma_A^2 + 2x_A x_B \sigma_A \sigma_B + x_B^2 \sigma_B^2, \quad \text{namely, a binomial perfect square}$$
$$= (x_A \sigma_A + x_B \sigma_B)^2 \tag{3.27}$$

and

$$\sigma_p = |(x_A \sigma_A + x_B \sigma_B)| \tag{3.55}$$
$$R_p = x_A R_A + x_B R_B \tag{3.23}$$
$$\mu_p = E[R_p] = x_A \mu_A + x_B \mu_B \tag{3.24}$$

Remarks:

1) $|f(x_A, x_B)|$ represents the absolute value of the expression enclosed.
2) If both x_A and x_B are nonnegative, namely, zero or positive, then the expression $(x_A \sigma_A + x_B \sigma_B)$ will be nonnegative, since neither σ_A nor σ_B is negative.
3) If either x_A or x_B is sufficiently negative, the expression $(x_A \sigma_A + x_B \sigma_B)$ may become negative, and the absolute value should apply.
4) *Risks:* From the combinations of long positions in these two assets, namely,

$$x_A \geq 0, \quad x_B \geq 0$$

and in any such combinations: $\sigma_p = x_A \sigma_A + x_B \sigma_B$, from (3.23),

$$\sigma_p = x_A \sigma_A + x_B \sigma_B, \quad \text{from (3.23)}$$
$$= (1 - x_B) \sigma_A + x_B \sigma_B$$
$$= \sigma_A - x_B \sigma_A + x_B \sigma_B$$

namely,

$$\sigma_p = \sigma_A + x_B (\sigma_B - \sigma_A) \tag{3.56}$$

5) Similarly, for returns, for such combinations:

$$\mu_p = E[R_p] = x_A \mu_A + x_B \mu_B, \quad \text{from (3.24)}$$
$$= (1 - x_B) \mu_A + x_B \mu_B$$
$$= \mu_A - x_B \mu_A + x_B \mu_B$$

namely,

$$\mu_p = \mu_A - x_B (\mu_A - \mu_B) \tag{3.57}$$

6) In Figure 3.20, the σ–μ plot represents the following system:

$$\mu_A = 8, \quad \sigma_A = 5, \quad \mu_B = 10, \quad \sigma_B = 15, \quad \text{and} \quad \sigma_{AB} = \text{var}_{1-2} = 1$$

Both the risk and the return will be proportional to x_B, such portfolios will be on a straight line—connecting the points representing the two assets, namely, point 1: $(\sigma_A = 5, \mu_A = 8)$, and point 2: $(\sigma_B = 15, \mu_B = 10)$.

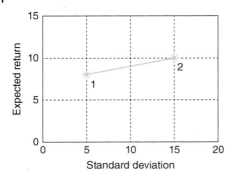

Figure 3.20 A portfolio of two perfectly positively correlated risky assets.

Example 3.14 The Minimum-Variance Portfolio

The strongest case of diversification occurs when $\rho_{AB} = -1.0$.
For this case,

$$\sigma_p^2 = x_A^2 \sigma_A^2 + 2x_A x_B \rho_{AB} \sigma_A \sigma_B + x_B^2 \sigma_B^2 \qquad (3.27)$$

$$\begin{aligned} &= x_A^2 \sigma_A^2 + 2x_A x_B (-1) \sigma_A \sigma_B + x_B^2 \sigma_B^2 \\ &= x_A^2 \sigma_A^2 - 2x_A x_B \sigma_A \sigma_B + x_B^2 \sigma_B^2 \end{aligned} \qquad (3.27a)$$

namely,

$$\sigma_p^2 = (x_A \sigma_A - x_B \sigma_B)^2$$

so that

$$\sigma_p = |(x_A \sigma_A - x_B \sigma_B)|$$

Thus, for this case, the minimum-variance portfolio will be assigned to the risk-free asset A that may be determined as follows:
From $|(x_A \sigma_A - x_B \sigma_B)| = 0$, it follows that

$$x_A \sigma_A - x_B \sigma_B = 0 \qquad \text{(i)}$$

and given that

$$x_A + x_B = 1 \qquad \text{(ii)}$$

or

$$x_B = 1 - x_A \qquad \text{(iii)}$$

Substituting $x_B = 1 - x_A$ in (i), one obtains

$$x_A \sigma_A - (1 - x_A) \sigma_B = 0$$

or

$$x_A (\sigma_A + \sigma_B) = \sigma_B$$

namely,

$$x_A = \sigma_B/(\sigma_A + \sigma_B)$$

and

$$x_B = 1 - x_A = 1 - \{\sigma_B/(\sigma_A + \sigma_B)\} = \sigma_A/(\sigma_A + \sigma_B)$$

When $\sigma_A = 5$ and $\sigma_B = 15$,

$$x_A = \sigma_B/(\sigma_A + \sigma_B) = 15/(5 + 15) = 15/20 = 3/4 = 0.75$$

and

$$x_B = 1 - x_A = 1/4 = 0.25$$

so that

$$\sigma_p = |(x_A\sigma_A - x_B\sigma_B)| = |(0.75)(5) - (0.25)(15)| = |3.75 - 3.75| = 0$$

namely, the two assets are *perfectly negatively correlated*!

And its Expected Return will be given by

$$\begin{aligned}
\mu_p = E[R_p] &= x_A\mu_A + x_B\mu_B \\
&= (0.75)(8\%) + (0.25)(10\%) \\
&= 6.0\% + 2.5\% \\
&= 8.5\%
\end{aligned} \tag{3.24}$$

The above results are graphically presented, in white, in Figure 3.21, which also repeats the results illustrated in Figures 3.10–3.12.

Remarks:

1) In Figure 3.21, the minimum-variance portfolio (in white) is a familiar and distinctive diagram—see Figures 3.9–3.12, 3.15, and 3.17—for portfolios that has one risk-free asset.

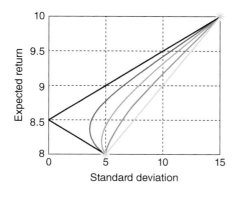

Figure 3.21 The minimum-variance portfolio (in white).

2) Note that *long* positions in two perfectly negatively correlated assets are similar to the following:

a) A *long* position in one of two perfectly positively correlated assets.

b) A *short* position in the other asset.

c) In many cases, asset correlations have values between −1 and +1.

d) A general formula for the minimum-variance portfolio may be derived as follows: Starting with a reduced-form equation where the variance var_p of a two-asset portfolio is being expressed as a function of x_B, starting with

$$\text{var}_{A-B} = \text{var}_A + 2x_B(\text{covar}_{A-B} - \text{var}_A) + x_B^2(\text{var}_A - 2\text{covar}_{A-B} + \text{var}_B) \tag{3.54}$$

for the two-asset portfolio, assets A and B.

Differentiating both sides of (3.54), with respect to x_B, namely, undertaking the $\partial/\partial x_B$ operation on both sides of (3.54), the result is

$$d(\text{var}_{A-B})/dx_B = 0 + 2(\text{covar}_{A-B} - \text{var}_A) + 2x_B(\text{var}_A - 2\,\text{cover}_{A-B} + \text{var}_B) \tag{3.55}$$

To obtain a stationary point, one sets (3.55) to zero, and solve for the value of x_B that will provide the minimum-variance portfolio, namely, solving for $x_{B\text{-min}}$. The result is

$$2(\text{covar}_{A-B} - \text{var}_A) + 2x_{B-\text{min}}(\text{var}_{A-2}\text{covar}_{A-B} + \text{var}_B) = 0$$

from which

$$x_{B-\text{min}} = (\text{var}_A - \text{covar}_{A-B})/(\text{var}_A - 2\text{covar}_{A-B} + \text{var}_B) \tag{3.56}$$

This minimum-variance portfolio may have a lower risk than either of the two-component assets, and may also have a higher return!

3) Considering the point at which $x_B = 0$, one has

$$d(e_p)/d(x_B) = \mu_B - \mu_A$$
$$d(v_p)/d(x_B) = \text{covar}_{A-B} - \sigma_A^2$$

and

$$d(e_p)/d(v_p) = \{d(e_p)/d(x_B)\}/\{d(v_p)/d(x_B)\}$$
$$= (\mu_B - \mu_A)/(\text{covar}_{A-B} - \sigma_A^2) \tag{3.57}$$

Assuming that $\mu_B > \mu_A$, if the slope $d(e_p)/d(v_p)$ is to be negative, then (3.57) shows that

$$(\text{covar}_{A-B} - \sigma_A^2) < 0$$

or

$$\text{covar}_{A-B} < \sigma_A^2 \tag{3.58}$$

For example, if $\sigma_A = 5$, $\sigma_B = 15$, $\mu_B > \mu_A$, and $\sigma_{AB} < 5/15$, the minimum-variance portfolio will dominate asset A, resulting in both higher expected return and lower risk—a double bonus!

3.2.5.2 The Tangency Portfolio

Next, one may proceed to expand the foregoing analysis by considering portfolios consisting of asset A, asset B, and risk-free asset (say, T-Bills). For this case, the efficient set will also be given by (3.51):

$$\mu_B = \{(\mu_B - r_f)/\sigma_B\}\sigma_p + r_f \tag{3.51}$$

namely, a straight line in the (μ_p, σ_p) space with intercept r_f. For this efficient set, its slope is the maximum Sharpe Ratio, namely, it is tangential to the efficient set consisting of only the two risky assets: A and B.

Tangent Portfolios are portfolios of stocks and bonds designed for long-term investors. To find the Tangent Portfolios, one may start by asking how much one would be prepared to lose in a worst-case scenario without dropping out of the market: 20%? 25%? or 33%? Once the maximum loss level is chosen, the Tangent Portfolios will try to deliver a high rate of return for that level of risk.

There is a human temptation to invest on the basis of the *BLASH* (**B**uy-**H**igh-**A**nd-**S**ell-**L**ow) route! Thus, most investors tend to take on large amounts of risk during good times (Buy High), and then sell out during bad times (Sell Low)—ruining their returns in the process. The Tangent Portfolios are designed to let the investor do well enough during *both* good *and* bad times to keep one in the markets throughout. This allows the investor reap the long-term benefits from investing in stocks and bonds with a simple, low-maintenance solution.

Figure 3.22, same as Figure 3.1, illustrates the Efficient Frontier. The hyperbola is sometimes referred to as the *Markowitz Bullet*, and is the efficient frontier if no risk-free asset is available. With a risk-free asset, the straight line is the efficient frontier.

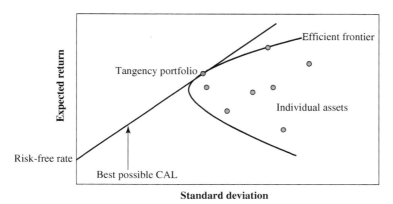

Figure 3.22 The tangency portfolio (same as Figure 3.1).

Efficient Frontiers   Different combinations of assets may produce different levels of return. The *Efficient Frontier* represents the *best* of these combinations, that is, those that produce the *maximum expected return for a given level of risk.* The efficient frontier is the *basis* for modern portfolio theory.

Example of Efficient Frontiers   Markowitz, in 1952 (see Figure 3.23), published a formal portfolio selection model in *The Journal of Finance*. He continued to develop and publish research on the subject over the next 20 years, eventually winning the 1990 Nobel Memorial Prize in Economic Science for his work on the *efficient frontier* and other contributions to modern portfolio theory. According to Markowitz, for every point on the efficient frontier, there is *at least one* portfolio that can be constructed from all available investments that has the expected risk and return corresponding to that point. An example is given here: Notice that the efficient frontier allows investors to understand how an expected returns of a portfolio vary with the amount of risk taken.

An important part of the efficient frontier is the relationship that the invested assets have with one another. Some assets' prices move in the *same* direction under similar circumstances, while others move in *opposite* directions. The more out of step that the assets in the portfolio are (i.e., the lower their *covariance*), the smaller the risk (*standard deviation*) of the portfolio that combines them. The efficient frontier is curved because there is a diminishing marginal return to risk. Each unit of risk added to a portfolio *gains* a smaller and

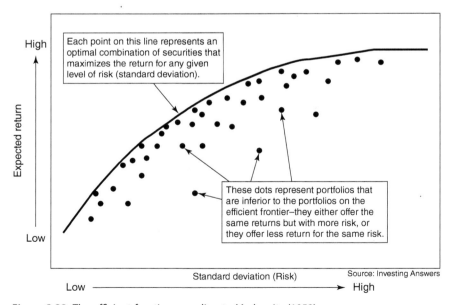

Figure 3.23 The efficient frontier according to Markowitz (1952).

smaller amount of return. When Markowitz introduced the *efficient frontier*, it was a seminal contribution to financial engineering science: One of its greatest contributions was its clear demonstration of the power of *diversification*.

Markowitz's theory relies on the claim that investors tend to choose, either purposely or inadvertently, portfolios that generate the largest possible returns with the least amount of risk. In other words, they seek out portfolios on the efficient frontier!

It should not be unaware that there is no one efficient frontier because individual investors as well as portfolio managers can and do edit the number and characteristics of the assets in the investing universe to conform to their own personal specific needs. For example, one individual investor may require the portfolio to have a minimum dividend yield, or another client may rule out investments in ideologically (e.g., politically, ethically, ethnically, or religiously) nonpreferred industries. Thus, only the remaining assets are included in the efficient frontier calculations.

Recent Historical Performance of the Stock Market   *Maximum Losses*: For all the rolling 12-month periods from 2010 going back to 1926:

- The Tangent 20 portfolio had a maximum 1 year inflation-adjusted loss of 20%
- The Tangent 25 portfolio had a maximum 1 year inflation-adjusted loss of 25%
- The Tangent 33 portfolio had a maximum 1 year inflation-adjusted loss of 33%

The period 1926–2010 includes the following extraordinary events:

i) The stock market crash of 1929
ii) The Great Depression
iii) The World War II
iv) The Cold War
v) Sputnik
vi) Assassination of a U.S. President
vii) Race riots
viii) The Vietnam War
ix) Inflation
x) The stock market crash of 1987
xi) 9/11 (2001)
xii) Bubbles and collapse of the .com bubbles
xiii) The panic of 2008
xiv) and so on.

These estimates of losses form historical benchmarks of the bad times that the stock market might face!

The following table shows that the performance of various portfolios respond after correcting for inflation over rolling 12-month periods from 1926 to 2008, before adjusting for applicable taxes and other relevant expenses.

Remarks:

1) The Tangent 20 portfolio delivered almost twice the average returns of a portfolio of Treasury Bills, with only slightly more risk of 1-year loss.
2) The Tangent 33 portfolio delivered most of the returns of the U.S. stock market, with substantially less risk.
3) These are hypothetical and historical notations for reference.
4) There is no guarantee that these levels of risk or returns will be maintained in the future.
5) Individual investment losses or gains may vary, depending on the levels of risk tolerance, investment objectives, and so on (Table 3.3).

3.2.5.3 Computing for Tangency Portfolios

To estimate the Tangency Portfolios, one may begin by determining the values of x_A and x_B that maximize the Sharpe Ratio of the portfolio that is on the envelope of the parabola.

For a given set of two available assets A and B to formally solve for the Tangency Portfolio, the task consists of finding the values of x_A and x_B that *maximize* the Sharpe Ratio (SR) of a portfolio that is on the envelope of the parabola: This calls for solving the following constrained maximization problem:

$$\max_{x_A, x_B} \mathrm{SR}_p = (\mu_p - r_f)/\sigma_p \tag{3.59a}$$

such that

$$
\left.
\begin{aligned}
x_A + x_B &= 1 \\[4pt]
\mu_p &= x_A \mu_A + x_B \mu_B \\[4pt]
\sigma_p^2 &= x_A^2 \sigma_A^2 + x_B^2 \sigma_B^2 + 2 x_A x_A \sigma_{AB}
\end{aligned}
\right\}
\tag{3.59b}
$$

Table 3.3 Estimated total stock market returns, 1926–2008.

Portfolios	Average year	Worst year
U.S. Stocks	7.9%	−65%
Tangent 33	7.6%	−33%
Tangent 25	6.0%	−25%
Tangent 20	4.6%	−20%
T-Bills	2.4%	−16%

This problem, as stated in (3.59a) and (3.59b), may be reduced to

$$\max_{x_A}[\{x_A(\mu_A - r_f) + (1 - x_A)(\mu_B - r_f)\} / \{(x_A^2\sigma_A^2 + (1 - x_A)^2\sigma_B^2 + 2_A^2(1 - x_A)\sigma_{AB}\}^{1/2}]$$

for which the solutions are as follows:

$$x_T^A = \frac{(\mu_A - r_f)\sigma_B^2 - (\mu_B - r_f)\sigma_{AB}}{(\mu_A - r_f)\sigma_B^2 + (\mu_B - r_f)\sigma_A^2 - (\mu_A - r_f + \mu_B - r_f)\sigma_{AB}} \qquad (3.60a)$$

$$x_T^B = 1 - x_T^A \qquad (3.60b)$$

A Numerical Example for the Tangency Portfolio for the Sample Data For the example data in Table 3.2 and using (3.60a) and (3.60b), one obtains

$$
\begin{aligned}
x_T^A &= \frac{(\mu_A - r_f)\sigma_B^2 - (\mu_B - r_f)\sigma_{AB}}{(\mu_A - r_f)\sigma_B^2 + (\mu_B - r_f)\sigma_A^2 - (\mu_A - r_f + \mu_B - r_f)\sigma_{AB}} \\
&= \frac{(0.2 - 0.03)(0.01) - (0.04 - 0.03)(-0.0038).}{(0.2 - 0.03)(0.01) + (0.04 - 0.03)(0.05) - (0.2 - 0.03 + 0.04 - 0.03)(-0.0038)} \\
&= \frac{0.00170 - (-0.00004)}{(0.00170) + 0.00050 - (-0.00068)} = 0.00174/0.00288 = 0.60417
\end{aligned}
$$

$$(3.61a)$$

and

$$x_T^B = 1 - x_T^A = 1 - 0.60417 = 0.39583$$

The expected return, variance, and standard deviation on this tangency portfolio are

$$
\begin{aligned}
\mu_T &= x_T^A\mu_A + x_T^B\mu_B, \quad \text{from (3.59b)} \\
&= (0.60417)(0.2) + (0.39583)(0.04) \\
&= 0.12083 + 0.01583 \\
&= 0.13667
\end{aligned}
$$

$$
\begin{aligned}
\sigma_T^2 &= (x_T^A)^2\sigma_A^2 + (x_T^B)^2\sigma_B^2 + 2x_T^Ax_T^B\sigma_{AB} \qquad \text{from (3.59b)} \\
&= (0.60417)^2(0.05) + (0.39583)^2(0.01) + 2(0.60417)(0.39583)(-0.0038) \\
&= 0.01825 + 0.00157 + (-0.00182) \\
&= 0.01800
\end{aligned}
$$

$$
\begin{aligned}
\sigma_T &= |\sqrt{(\sigma_T^2)}| \\
&= \sqrt{(0.01800)} \\
&= 0.13410
\end{aligned}
$$

Cleary, if repeated computations are required, a simple R code may be used to undertake the numerical calculations involving (3.59) and (3.60)

3.2.6 The Mutual Fund Separation Theorem

Until this point, it has been shown that the efficient portfolios are combinations of two classes of assets:

i) Tangency portfolios
ii) Risk-free assets, such as T-Bill

Hence, applying (3.56) and (3.57b):

$$\mu_p = r_f + x_A(\mu_A - r_f) \tag{3.56}$$

$$\sigma_p = x_A \sigma_A \tag{3.57a}$$

one may write the expected return and standard deviation of any *efficient portfolios*:

$$\mu_p^e = r_f + x_T(\mu_T - r_f) \tag{3.62}$$

$$\sigma_p^e = x_T \sigma_T \tag{3.57b}$$

where

x_T represents the fraction of investments in the tangency portfolio,
$(1 - x_T)$ represents the fraction of wealth invested in risk-free assets (e.g., T-Bills), and
μ_T and σ_T represent, respectively, the expected return and standard deviation of the tangency portfolio.

This result is called the *Mutual Fund Separation Theorem.*

Remarks:

1) The Tangency Portfolio may be considered as a mutual fund of two risky assets—in which the shares of the two risky assets are determined by the tangency portfolio weights: x_T^A and x_T^B determined from (3.60a) and (3.60b), and the T-Bills may be considered as a mutual fund of risk-free assets.
2) The exact combination of the tangency portfolio and the T-Bills will be dependent on the risk preference of the investor: If the investor is highly risk-adverse, then this investor may choose a portfolio with low volatility, namely, a portfolio with very small weight in the tangency portfolio together with a very large weight in the T-Bills! Clearly, this option will produce a portfolio with an expected return close to the risk-free rate, and a variance that is nearly zero! On the other hand, if the investor can tolerate a large amount risk, then the preferred portfolio will have high expected return regardless of the volatility.

This portfolio may consist of borrowing at the risk-free rate (known as "leveraging") and investing the proceeds in the tangency portfolio to achieve an overall high expected returns.

3.2.7 Analyses and Interpretation of Efficient Portfolios

For a given risk level, efficient portfolios have high expected returns—as measured by portfolio standard deviations. Thus, for those portfolios that yield expected returns *above* the T-Bill rates, the *efficient portfolios* may also be characterized as those that have minimum risks for a given target expected return—as measured by the portfolio standard deviation.

Example 3.15 Assets Allocation for Efficient Portfolios

An investor has the choice of investing $1,000,000 in a risk-free asset *A* or a risky asset *B*:

- The risky asset *B* will either halve or double, with equal probability:

$2,000,000 (DOUBLED !)
/
$1,000,000
\
$50,000   (HALVED !)

- The risk-free asset *A* will yield a certain return of 3% p.a., namely, $1,030,000.
- How should one decide which, and how much, of these assets to build the portfolio?

To analyze the portfolio allocation challenge, the following steps (known as Modern Portfolio Theory—this approach has its origin in Mean-Variance Portfolio Analysis—developed by Harry Markowitz in the 1960s, and is considered as the first step in the development of modern financial engineering) may be taken:

- The critical issue in this asset allocation problem is to answer the question: "How much of one's wealth should be invested in each asset so as to maximize the overall gain?"
- Prior to the time that the mean-variance analysis of Markowitz became known, an investment advisor would be expected to provide advice like "If one is young, one should be putting money into a couple of good growth stocks, maybe even into a few small stocks. For now it is the time to take risks.

However, if one is close to retirement, one should be putting all of one's money into bonds and safe stocks, and nothing into the risky stocks—do not take risks with one's portfolio at this stage in life.
- This advice was intuitively compelling, but it was just wrong!
- It is now known that the optimal portfolio of risky assets is *exactly* the same for everyone, *no matter what their tolerance for risk.*
  1) Investors should control the risk of their portfolio not by reallocating among risky assets, but through the split between risk-free and risky assets.
  2) The portfolio of risky assets should contain a large number of assets—it should be a *well-diversified portfolio.*

Note: The results are derived under the following assumptions:

a) 1) All returns are normally distributed.
   2) Investors care only about mean return and variance.
b) All assets are tradable.
c) There are no transaction costs.

(Any relaxing of these assumptions will warrant further qualifications and discussions.)

The analysis of this asset allocation problem may be considered in two steps:

Step 1: What risky assets should be held in the portfolio?
Step 2: How should one distribute the investment between the optimal risky assets and the risk-free assets?

Each step will now be examined on its own, before combining them. In the approach, a theoretical framework will be used in examining the effects of risks and concomitant returns. The following step-by-step approach follows:

1) First compute the Expected Return for each possible asset in the investment:
   a) For the risk-free asset A, the Expected Return is

   $$ER_A = (1,030,000/1,000,000) - 1 = 1.03 - 1 = 0.03 = 3\%$$

   b) For the risky asset B, the Expected Return, for a 50/50 split in the risks is

   $$ER_B = 1/2\{(100/50) - 1\} + 1/2\{(25/50) - 1\}$$
   $$= 1/2(2 - 1) + 1/2(1/2 - 1) = 1/2(1 - 1/2)$$
   $$= 1/2(1/2) = 1/4 = 0.25 = 25\%$$

2) Next compute the Risk Premium P_r on the risky asset B.
   The Excess Return R_E is the return net of the risk-free rate:

   $$R_E = R_r - R_f$$

The Risk Premium P_r is the expected excess return $E(R_E)$ given by

$$E(R_E) = E(R_r - R_f) = E(R_r) - R_f$$
$$= 1/2(100\% - 3\%) - 1/2(50\% + 3\%)$$
$$= 1/2(97\%) - 1/2(53\%) = 48.5\% - 26.5\% = 22\%$$

3) Then, compute the relative amount of risks for the two assets A and B: A reasonable *estimate* for the risk may be the Variance σ^2, or the Standard Deviation σ, of the return.
   - For the risk-free asset A, the return variance $\sigma_A^2 = 0$.
   - For the risky asset B, return variance σ_B^2 is

   $$= 1/2\{(1.00 - 0.25)^2 + (-0.50 - 0.25)^2\}$$
   $$= 1/2\{(0.75)^2 + (-0.75)^2\}$$
   $$= (0.75)^2$$
   $$= 0.5625$$

   And, the Return Standard Deviation is the positive square root of σ_B^2:

   $$\sigma_B = |(\sqrt{\sigma_B^2})| = \sqrt{(0.5625)} = 0.7500 = 75\%$$

4) Assess the foregoing results if this is a reasonable level of risk for the additional expected return.

   Assume that

   a) most investors prefer high expected returns and dislike large variances,
   b) investors are generally risk averse,
   c) their Utility or Satisfaction index is of the following form:

   $$U(r) = E(r) - 1/2A\sigma^2(F) \tag{3.63}$$

where

- $U(r)$ is the Certainty Equivalent Rate of Return for a risky asset, namely, the return such that one is indifferent between that asset and earning an anticipated rate of return.
- $E(r)$ is the *Expected Returns (People like high expected returns!)*
- A measures the investor's index of risk-aversion.
- The higher the value of A, the more an investor *dislike* risks.
- $\sigma^2(F)$ is the variance. *(People also dislike variances!)*

This approach is illustrated in Figure 3.15, where $E(r) = 0.25$ and Figure 3.24 graphically demonstrates the following:

i) Investor choices may be represented as the indifference curves: in red, green, blue, and yellow.

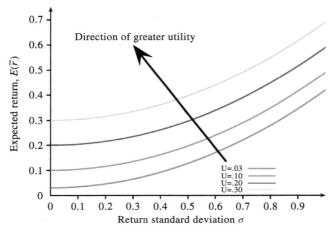

Figure 3.24 The tangency portfolio (indexed by the utility U).

ii) Each curve represents different application levels for fixed risk aversion.
iii) Each curve plots the combinations of $E(r)$, $\sigma^{\circledR}$ yielding the same level of utility $U(r)$, as defined by (3.63).

Figure 3.25 graphically demonstrates that:

i) Each curve (red/green/blue/yellow) represents the same Utility level U for different values of Risk Aversion A.

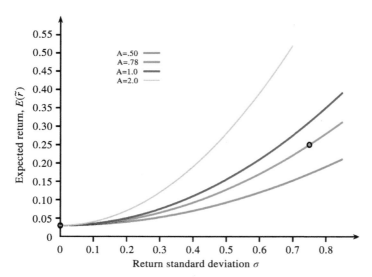

Figure 3.25 Assets allocation for efficient portfolios.

ii) For a given value of Return Standard Deviation σ, most investors prefer higher mean return to achieve the same level of Utility U.

Thus, in Example 3.15, $ER_B = 0.25$ and $\sigma_B^2 = 0.5625$, one may ascertain $U(r)$ for several levels of risk aversion:

A	U(r) %
0.04	24
0.50	11
0.78	3
1.00	−3

Questions:

1) If $A = 0.50$, would/should one hold the risk-free asset or the risky asset?
2) To be indifferent between risk-free and risky assets, what level of risk-aversion is to be expected?
3) If one is more risk-averse, will $U(r)$ be higher or lower?

3.3 The Black–Litterman Model

With the foregoing discussions of the development of the study of asset allocation in terms of Markowitz's Modern Portfolio Theory of the mean-variance approach to asset allocation and portfolio optimization, this may well be the suitable point of departure to go from the Markowitz model on to the Black–Litterman model.

Asset allocation is the continuing decision facing an investor who must decide on the optimum allocation of the assets in the portfolio across a few (say up to 20) asset classes. For example, a globally invested mutual fund must select the proportions of the total investment for allocation to each major country or global financial region.

It is true that the Modern Portfolio Theory (the mean-variance approach of Markowitz) may provide a plausible solution to this problem once the expected returns and covariances of the assets are available. Thus, while Modern Portfolio Theory is an important theoretical approach, its application does encounter a serious problem: Although the covariances of a few assets may be adequately estimated, *it is difficult to ascertain (with reasonable estimates) of the Expected Returns!*

The Black–Litterman approach resolves this problem by the following:

• *Not* requiring the use of input estimates of expected return.

- Instead, it assumes that the initial expected returns are whatever is required so that the equilibrium asset allocation is equal to what one observes in the markets.
- The user is only required to state how one's assumptions about expected returns differ from the expected returns in the market, and to state one's degree of confidence in the alternative assumptions.
- From this, the Black–Litterman method computes the *preferred* Mean-Variance Efficient asset allocation.

In general, to overcome portfolio constraints—for example, when short sales are not allowed— the way to find the optimal portfolio is to use the Black–Litterman model to generate the expected returns for the assets, and then use a mean-variance optimization procedure to solve the resultant optimization problem.

Efficient Frontier

What It Is Different combinations of securities produce different levels of return. The *efficient frontier* represents the best of these securities combinations—those that produce the maximum expected return for a given level of risk. The efficient frontier is the basis for modern portfolio theory.

How It Works (Example) In 1952, Harry Markowitz published a formal portfolio selection model in *The Journal of Finance*. He continued to develop and publish research on the subject over the next 20 years, eventually winning the 1990 Nobel Prize in Economic Science for his work on the efficient frontier and other contributions to modern portfolio theory.

According to Markowitz, see Figure 3.26, for every point on the efficient frontier, there is at least one portfolio that can be constructed from all available investments that have the expected risk and return corresponding to that point.

An example appears below. Note how the efficient frontier allows investors to understand how a portfolio's expected returns vary with the amount of risk taken.

To have relationship securities with each other is an important part of the efficient frontier. Some securities' prices move in the same direction under similar circumstances, while others move in opposite directions. The more the out of sync securities in the portfolio (i.e., the lower their covariance), the smaller the risk (*standard deviation*) of the portfolio that combines them. The efficient frontier is curved because there is a diminishing marginal return to risk. Each unit of risk added to a portfolio gains a smaller and smaller amount of return.

Why It Matters When Markowitz introduced the *efficient frontier*, it was groundbreaking in many respects. One of its largest contributions was its clear demonstration of the power of *diversification*.

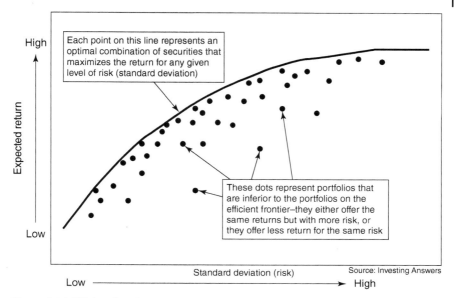

Figure 3.26 Efficient frontier.

Markowitz's theory relies on the claim that investors tend to choose, either purposefully or inadvertently, portfolios that generate the largest possible returns with the least amount of risk. In other words, they seek out portfolios on the efficient frontier.

However, there is no one efficient frontier because portfolio managers and investors can edit the number and characteristics of the securities in the investing universe to conform to their specific needs. For example, a client may require the portfolio to have a minimum dividend yield, or the client may rule out investments in ethically or politically undesirable industries. Only the remaining securities are included in the efficient frontier calculations.

3.4 The Black–Scholes Option Pricing Model

$x1 = x_{A=}$
$x2 = x_B$
$e1 = \mu_A$
$e2 = \mu_B$
$ep = x1^*e1 + x2^*e2 = x_A\,\mu_A + x_B\,\mu_B$
$r12 = \sigma_{AB} = var_{1\text{-}2}$
$s1 = \sigma_A$
$s2 = \sigma_B$
$sp = x1^*s1 + x2^*s2 = x_A\,\sigma_A + x_B\,\sigma_B$

$v1 = \text{var}\,A = \sigma^2_A$

$v2 = \text{var}\,B = \sigma^2_B$

$vp = var_{1\text{-}2} = \sigma^2_{AB}$

$c12 = covar_{A\text{-}B}$

https://web.stanford.edu/~wfsharpe/mia/rr/mia_rr5.htm

(6)

p. 22 of http://faculty.washington.edu/ezivot/econ424/introductionPortfolioTheory.pdf

Package:　　　　　Sim.DiffProc

3.4.1 Keep on Modeling!

Modern Portfolio Theory (MPT) has also been criticized for its assumptions that returns follow a Gaussian distribution. In the 1960s, several workers had shown the inadequacy of this assumption and proposed the use of stable distributions as well strategies for deriving optimal portfolios in such settings. Some notable suggestions were as follows:

1) Since the introduction of *MPT* in 1952, a number of attempts have been advanced with the sole purpose to improve this model by including more Realistic assumptions. These include the following:

    a) The *Post-Modern Portfolio Theory* (PMPT), which extends *MPT* by using nonnormally distributed asymmetric measures of risk. This approach has been found useful in some, but not all, cases.

    b) In the 1970s, Conroy, using concepts from MPT, applied the economic analysis in the field of regional science, modeled the labor force in the economy by a portfolio–theoretic approach to analyze growth and variability in the labor force. This work led to the analysis of the relationship between volatility and economic growth.

    Thus, while *MPT* is a useful introductory framework, more work is needed to render it useful and reliable in practice. Since the introduction of *MPT* in 1952, notable attempts made to improve the model, especially by using more realistic assumptions, *Post-modern Portfolio Theory* extends MPT by adopting nonnormally distributed, asymmetric measures of risk— This adjustment helps with some of these problems, but not all!

2) **Other Formulations**

    In the 1970s, concepts from MPT found their way into the field of Regional Science. In a series of seminal works, Conroy modeled the labor force in the economy using portfolio–theoretic methods to examine growth and variability in the labor force. This was followed by an extensive study on the relationship between economic growth and volatility.

    Recently, modern portfolio theory has been applied to the following:

    a) *Social Psychology*—For Modeling the Self-Concept: When the self attributes comprising the self-concept constitute a well-diversified portfolio, then psychological outcomes at the level of the individual such as mood and self-esteem should be more stable than when the self-concept is undiversified. This prediction has been confirmed in studies involving human subjects.

b) *Information Science*: For modeling the uncertainty and correlation between documents in information retrieval. Given an open question, one may maximize the relevance of a ranked list of documents and at the same time minimize the overall uncertainty of the ranked list.

c) *Other Nonfinancial Assets*: MPT has also been applied to portfolios of assets besides financial instruments. When MPT is applied to non-financial portfolios, the characteristics among the different types of portfolios should be considered, for example:

   A) The assets in financial portfolios are, for practical purposes, contin-uously divisible while portfolios of projects are "lumpy." Thus, for example, while one may compute that the optimal portfolio position for four stocks is, say, 45, 26, 20, and 9%, the optimal position for a project portfolio may not allow us to simply change the amount spent on a project. Projects may be all or nothing or, at least, have logical units that cannot be separated. Thus, a portfolio optimization method would have to take the discrete nature of projects into account.

   B) Assets of financial portfolios are liquid: They may be assessed or reassessed at any point in time. However, opportunities for starting new projects may be limited and may occur in limited time windows. Projects that have already been initiated may not be abandoned without the loss of the *sunk costs* (i.e., there may be little or no salvage value of a partly complete project).

Neither of these factors necessarily eliminate the possibility of using MPT for such portfolios. However, they require the need to run the optimization with an additional set of mathematically expressed constraints that would not normally apply to financial portfolios. Furthermore, some of the simplest elements of MPT are applicable to virtually any kind of portfolio. The concept of ascertaining the risk tolerance of an investor by documenting the quantity of risk that is acceptable for a given return may be applied to a variety of decision analysis problems. MPT uses historical variance as a measure of risk, but portfolios of assets like major projects generally do not have a well-defined "historical variance." In such cases, the MPT investment boundary may be expressed in more general terms like "the chance of an ROI (Return-On-Investment) less than cost of capital," or "the chance of losing more than half of the investment." Thus, when the risk is expressed in terms of uncertainty about forecasts, and possible losses, then the concept of MPT may be transferable to various types of investment.

3) **Black–Litterman Model (BL)**, introduced in 1990, advanced that the optimization should be an extension of unconstrained Markowitz

optimization that incorporates relative and absolute *views* on inputs of risks and returns. This will be fully discussed in Section 3.5.

3.5 The Black–Litterman Model

The goals of the Black–Litterman model were as follows:

- To create a systematic method of specifying a portfolio
- To incorporate the views of the analyst/portfolio manager views into the estimation of market parameters.

Let

$$A = \{a_1, a_2, a_3, \ldots, a_n\} \tag{3.64}$$

be a set of random variables representing the returns of n assets. In the BL model, the joint distribution of A is taken to be multivariate normal, that is,

$$A \sim N(\mu, \Sigma). \tag{3.65}$$

The model then considers incorporating an analyst's views into the estimation of the market mean μ. If one considers

- μ to be a random variable that is itself normally distributed and that
- its dispersion is proportional to that of the market,

then

$$\mu \sim N(\pi, \tau\Sigma) \tag{3.66}$$

where π is some parameter that may be determined by the analyst by some established procedure—*as will be seen in the remainder of this section*: On this point, Black and Litterman proposed (based on equilibrium considerations) that this should be obtainable from the intercepts of the capital-assert pricing model.

Next, upon the consideration that the analyst has certain *subjective* views on the actual mean of the return for the holding period, this part of the BL model may allow the analyst to include *personal* views. BL suggested that such views should best be made as linear combinations, namely, *portfolios*, of the asset return variable mean μ:

- Each such personal view may be allocated a certain mean-and-error, (μ_i, ε_i), so that a typical view would take the form

$$p_{i1}\mu_1 + p_{i2}\mu_2 + p_{i3}\mu_3 + \cdots + p_{ij}\mu_j + \cdots + p_{in}\mu_n = q_i + \varepsilon_i \tag{3.67}$$

where $\varepsilon_i \sim N(0, \sigma_i^2)$.

The standard deviations σ_i^2 of each view may be assumed to control the confidence in each. Expressing these views in the form of a matrix, call the "pick" matrix, one obtains the "general" view specification:

$$P\mu \leftarrow \sim \leftarrow N(\mu, \Omega) \tag{3.68}$$

in which Ω is the diagonal matrix $\text{diag}(\sigma_1^2, \sigma_2^2, \sigma_3^2, \ldots, \sigma_n^2)$. It may be shown, using Bayes' law, that the posterior distribution of the market mean conditional on these views is

$$\mu|_{q,\Omega} \sim N(\mu_{\text{BL}}, \Sigma_{\text{BL}}^{\mu}) \tag{3.69}$$

where

$$\mu_{\text{BL}} = \{(\tau\Sigma)^{-1} + P^T\Omega^{-1}P\}^{-1}\{(\tau\Sigma)^{-1}\pi + P^T\Omega^{-1}q\} \tag{3.70}$$

$$\Sigma_{\text{BL}}^{\mu} = \{(\tau\Sigma)^{-1} + P^T\Omega^{-1}P\}^{-1} \tag{3.71}$$

One may then obtain the posterior distribution of the market by taking

$$A|_{q,\Omega} = \mu|_{q,\Omega} + Z \quad \text{and} \quad Z \sim N(0, \Sigma) \tag{3.72}$$

which is independent of μ.

One may then obtain

$$E[A] = \mu_{\text{BL}} \tag{3.73}$$

and

$$\Sigma_{\text{BL}} = \Sigma + \Sigma_{\text{BL}}^{\mu} \tag{3.74}$$

The Canonical Black–Litterman Reference Model   The remainder of the Black–Litterman model is built on the reference model for returns. It assumes which variables are random, and which are not. It also defines which parameters are modeled, and which are not. Other reference models have been Alternative Reference Model or Beyond Black–Litterman, which do not have the same theoretical basis as the canonical one that was initially specified in Black and Litterman (1992).

Starting with normally distributed expected returns,

$$r \sim N(\mu, \Sigma) \tag{3.75}$$

The fundamental objective of the Black–Litterman model is to model these expected returns, which are *assumed* to be normally distributed with mean μ and variance $\sum$. Note that one will need at least these values, the expected returns, and covariance matrix later as inputs into a portfolio selection model.

Define μ, the unknown mean return, as a random variable itself distributed as

$$\mu \sim N(\pi, \Sigma_\pi) \tag{3.76}$$

π is the estimate of the mean and $\sum_\pi$ is the variance of the unknown mean, μ, about the estimate. Another way to view this linear relationship is shown in the following formula:

$$r \sim N(\mu, \Sigma) \tag{3.75}$$

$$\mu = \pi + \varepsilon \tag{3.77}$$

These prior returns are normally distributed around π with a disturbance value ε. Figure 3.27 shows the distribution of the actual mean about the estimated mean of 5% with a standard deviation of 2% (or a variance of 0.0004). Clearly, this is not a precise estimate from the width of the peak. One may complete the reference model by defining $\sum_r$ as the variance of the returns about the initial estimate π.

From (3.77) and the assumption above that ε and μ are not correlated, the formula for calculating $\sum_\pi$ is

$$\Sigma_r = \Sigma + \Sigma_\pi \tag{3.78}$$

(3.78) shows that the proper relationship between the variances is ($\sum_r \geq \sum$, $\sum_\pi$).

One may check the reference model at the boundary conditions to ensure that it is correct. In the absence of estimation error, that is, $\varepsilon = 0$, $\sum_r = \sum$. As the estimate gets worse, if $\sum_\pi$ increases, then $\sum_r$ also increases.

For the Black–Litterman model expected return, the canonical reference model is

$$r \sim N(\pi, \Sigma_r) \tag{3.79}$$

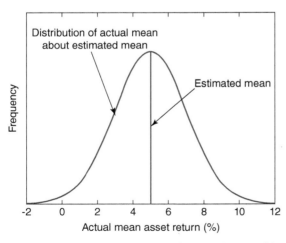

Figure 3.27 The estimated mean and the distribution of the actual mean about the estimated mean ε, which is normally distributed with mean 0 and variance $\sum_\pi$ and is assumed to be uncorrelated with μ.

It should be emphasized that (3.79) is the canonical Black–Litterman reference model, and *not* (3.76).

Remarks:

1) A common misconception about the Black–Litterman model is that (3.76) is the reference model, and also
2) μ is a point estimate in the model.

Computing the Equilibrium Returns   The Black–Litterman model begins with a neutral equilibrium portfolio for the prior estimate of returns. The model relies on General Equilibrium theory, namely, that *"if the aggregate portfolio is at equilibrium, then each sub-portfolio must also be at equilibrium."* It may be used with any utility function, making it very flexible. In fact, most practitioners use the Quadratic Utility function and assume a risk-free asset, and thus the equilibrium model simplifies to the *Capital Asset Pricing Model* (CAPM). In fact, the neutral portfolio in this case is the CAPM Market portfolio.

Some workers have used other utility functions, and others consider other measures of portfolio risk without applying the same theoretical basis. In order to preserve the symmetry of the model, the practitioner should use the same utility function to identify both the neutral portfolio and the portfolio selection area. Here the approach uses the Quadratic Utility function, CAPM, and unconstrained mean-variance because it is a well-understood model. Given these assumptions, the prior distribution for the Black–Litterman model is the estimated mean excess return from the CAPM market portfolio. The process of computing the CAPM equilibrium excess returns is straightforward. CAPM is based on the concept that there is a linear relationship between risk (as measured by standard deviation of returns) and return. Furthermore, it requires returns to be normally distributed. This model is of the form

$$E(r) = r_f + \beta r_m + \alpha \tag{3.80}$$

where

r_f is the risk-free rate,
r_m is the excess return of the market portfolio,
β is the a regression coefficient computed as $\beta = \rho \, (\sigma_p/\sigma_m)$, and
α is the residual or asset-specific excess return.

Under CAPM, any irregular risks associated with an asset is uncorrelated with that from other assets, and this issue may be reduced via diversification. And the investor may be rewarded for taking systematic risk measured by β, but is not compensated for taking irregular risks associated with α.

Under CAPM, all investors should hold the same risky CAPM market portfolio. Since all investors hold risky assets only in the market portfolio, at equilibrium their weights in the market portfolio will be determined by the market capitalization of the various assets. On the efficient frontier, the CAPM market portfolio has the maximum Sharpe Ratio of any portfolio. Here, the Sharpe Ratio is the excess return divided by the excess risk, or $(r - r_f)/\sigma$.

The investor may also invest in a risk-free asset. This risk-free asset has essentially a fixed positive return for the time period over which the investor is concerned. It is generally similar to the sovereign bond yield curve for the investors' local currency—such as the U.S. Government Treasury Bills. Depending on how the asset allocation decision will be framed, this risk-free asset can range from a 4-week Treasury Bill (1-month horizon) to a 20-year inflation protected bond.

The CAPM market portfolio contains all investable assets that makes it difficult to specify. Since the system is in equilibrium, all submarkets must also be in equilibrium and any submarket chosen is part of the global equilibrium. While this permits one to reverse optimize the excess returns from the market capitalization and the covariance matrix, forward optimization from this point to identify the investors optimal portfolio within CAPM is difficult since one does not have information for the entire market portfolio. In general, this is not actually the question investors are asking, usually most investors select an investable universe and search for the optimal asset allocation within the universe: Thus, the theoretical problem with the market portfolio may be initially ignored.

The Capital Market Line   The Capital Market Line is a line through the risk-free rate and the CAPM market portfolio. The *Two Fund Separation Theorem*, closely related to the CAPM, states that *all investors should hold portfolios on the Capital Market Line*. Any portfolio on the Capital Market Line dominates all portfolios on the Efficient Frontier, the CAPM market portfolio being the only point on the Efficient Frontier and on the Capital Market Line. Depending on one's risk aversion, an investor will hold arbitrary fractions of their investment in the risk-free asset and/or the CAPM market portfolio. Figure 3.20 shows the relationship between the Efficient Frontier and the Capital Market Line.

To start the market portfolio, one may choose to start with a set of weights that are all greater than zero and naturally sum to 1. The market portfolio only includes risky assets, because by definition investors are rewarded only for taking on systematic risk. Thus, in the CAPM, the risk-free assets with $\beta = 0$ will *not* be in the market portfolio. At a later stage, one will see that the Bayesian investor may invest in the risk-free asset based on their confidence in their return estimates. The problem may be constrained by asserting that the covariance matrix of the returns, $\sum$, is known. In fact, this covariance matrix may be estimated from historical return data. It is often calculated from higher frequency data and then scaled up to the time frame required for the asset allocation problem.

By calculating it from actual historical data, one may ensure that the covariance matrix is positive definite. Without basing the estimation process on actual data, there may be significant issues involved in ensuring the covariance matrix is positive definite. One may apply shrinkage or random matrix theory filters to the covariance matrix in an effort to make it robust.

In this approach, one may use a common notation, similar to that used in He and Litterman (1999). Note that this notation is unusually different.

First, one derives the equations for 'reverse optimization' starting from the quadratic utility function:

$$U = w^T \Pi - (\delta/2)w^T \Sigma w \tag{3.81}$$

where

U is the investors' utility, the objective function during Mean-Variance Optimization,
w is the vector of weights invested in each asset,
Π is the vector of equilibrium excess returns for each asset,
δ is the risk aversion parameter, and
Σ is the covariance matrix of the excess returns for the assets.

Now, U, being a convex function, will have a single global maxima. If one maximizes the utility with no constraints, then there is a closed-form solution. The exact solution may be found by taking the first derivative of (3.81) with respect to the weight w and setting it to zero:

$$dU/dw = \Pi - \delta \Sigma w = 0 \tag{3.82}$$

Solving (3.82) for Π (the vector of excess returns) yields:

$$\Pi = \delta \Sigma w \tag{3.83}$$

In order to use (3.83) to solve for the CAPM market portfolio, one needs to have a value for δ, the risk aversion coefficient of the market. One way to find δ is by multiplying both sides of (3.83) by w^T and replacing vector terms with scalar terms:

$$(r - r_f) = \delta \sigma^2 \tag{3.84}$$

Here the expression at equilibrium is that the excess return to the portfolio is equal to the risk aversion parameter multiplied by the variance of the portfolio. From (3.84),

$$\delta = (r - r_f)/\sigma^2 \tag{3.85}$$

where

r is the total return on the market portfolio $(r = w^T \Pi + r_f)$,

r_f is the risk-free rate, and

σ^2 is the variance of the market portfolio ($\sigma^2 = w^T \sum w$)

Many specify the value of δ used. For global fixed income, some use a Sharpe Ratio of 1.0. Black and Litterman (1992) use a Sharpe Ratio closer to 0.5. Given the Sharpe Ratio (SR), one rewrite (3.85) for δ in terms of SR as

$$\delta = SR/\sigma_m \tag{3.86a}$$

One may now calibrate the returns in terms of formulas (3.85) or (3.86). As part of the analysis, one should arrive at the terms on the right-hand side of which formula one chooses to use. For (3.85), this is r, r_f, and σ^2 in order to calculate a value for δ. For (3.86), this is the Sharpe Ratio SR and σ.

To use formula (3.85), one needs to have an implied return for the market portfolio that may be more difficult to estimate than the SR of the market portfolio.

With this value of δ, substitute the values for w, δ, and $\sum$ into (3.83) to obtain the set of equilibrium asset returns. Equation (3.83) is therefore the closed-form solution to the reverse optimization problem for calculating asset returns given an optimal mean-variance portfolio in the absence of constraints. One may rearrange Equation (3.83) to yield the formula for the closed-form computation of the optimal portfolio weights in the absence of constraints.

$$w = (\sigma \Sigma)^{-1} \Pi \tag{3.86b}$$

Herold (2005) provides insights into how implied returns can be calculated in the presence of simple equality constraints such as the budget or full investment ($\sum w = 1$) constraint. It was shown how errors may be introduced during a reverse optimization process if constraints are assumed to be nonbinding when they are, in fact, binding for a given portfolio. Note that because one is dealing with the market portfolio that has only positive weights summing to 1, one may assume that there are no binding constraints on the reverse optimization.

The only missing item is the variance of the estimate of the mean. Considering the reference model, $\sum_\pi$ is needed:

Black and Litterman made the simplifying assumption that the structure of the covariance matrix of the estimate is proportional to the covariance of the returns $\sum$. Thus, a parameter τ is created as the constant of proportionality. Given the assumption $\sum_\pi = \tau \sum$, the prior distribution $P(A)$ is

$$P(A) \sim N(\Pi, \tau\Sigma) \tag{3.87a}$$

and

$$r_A \sim N(P(A), \Sigma) \tag{3.87b}$$

This becomes the prior distribution for the Black–Litterman model. It represents the *estimate* of the mean, which is expressed as a distribution of the actual unknown mean *about* the estimate.

Using (3.79), one may rewrite (3.87a) and (3.87b) in terms of $\prod$ as

$$r_A \sim N\{\Pi, (1 + \tau)\Sigma\} \tag{3.88}$$

Investors with no views and using an unconstrained mean-variance portfolio selection model may often invest 100% in the neutral portfolio, but this is only true if one will apply a budget constraint. Because of their uncertainty in the estimates, they will invest $\tau/(1 + \tau)$ in the risk-free asset and $1/(1 + \tau)$ in the neutral portfolio! This may be seen in the case as follows, starting from (3.83):

$$\Pi = \delta\Sigma w \tag{3.83}$$

$$\Rightarrow w = (\delta\Sigma)^{-1}\Pi$$
$$\Rightarrow \underline{w} = \{(1 + \tau)\delta\Sigma)^{-1}\Pi$$
$$\Rightarrow \underline{w} = \{1/(1 + \tau)\}\delta\Sigma)^{-1}\Pi$$
$$\Rightarrow \underline{w} = \{1/(1 + \tau)\}w, \quad by\ (3.83)\ \text{again}$$

Figure 3.28 demonstrates this concept graphically.

Also, one may view the Bayesian efficient frontier as a shift to the right if one plots the efficient frontier generated with the increased covariance matrix and a budget constraint. In such a case, the uncertainty adjusts each point further to the right in the risk/return space. Figure 3.29 demonstrates the Risk-Adjusted Bayesian Efficient Frontier.

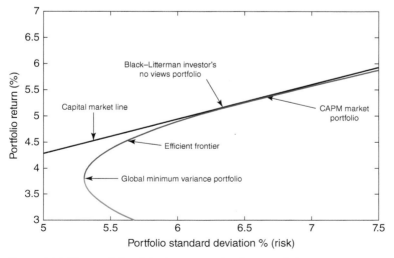

Figure 3.28 The portfolio of the investor in the absence of views.

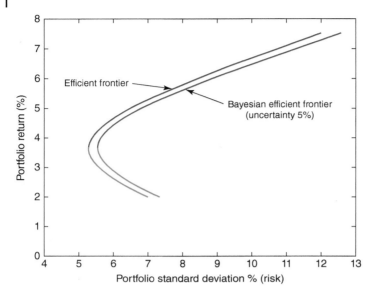

Figure 3.29 Risk-adjusted Bayesian efficient frontier.

Illustrating and Specifying the Views The investors' views on the estimated mean excess returns may be described as follows. First, one defines the combination of the investors' views as the conditional distribution as follows:

1) By construction, one requires each view to be unique and uncorrelated with the other views. This gives the conditional distribution the property that the covariance matrix will be diagonal, with all off-diagonal entries equal to 0. By constraining the problem this way in order to improve the stability of results, one simplifies the problem. Estimating the covariances between views would be even more complicated and error-prone than estimating the view variances.

2) Second, one requires views to be fully invested, either the sum of weights in a view is 0 (relative view) or is 1 (an absolute view). One does not require a view on any or all assets. In addition, it is actually possible for the views to conflict, the mixing process will merge the views based on the confidence in the views and the confidence in the prior.

3) Then, one will represent the investors' k views on n assets using the following matrices:
   - *P, a $k \times n$ matrix of the asset weights within each view. For a relative view, the sum of the weights will be 0 and for an absolute view the sum of the weights will be 1. Different analysts compute the various weights within the view differently. For example,
     a) He and Litterman (1999) and Idzorek (2005) use a market capitalization-weighted scheme.

b) Satchell and Scowcroft (2000) use an equal weighted scheme in their examples.

In practice, weights will be a mixture depending on the process used to estimate the view returns.

- Q, a $k \times 1$ vector of the returns for each view.

Ω, a $k \times k$ matrix of the covariance of the views. Ω is diagonal as the views are required to be independent and uncorrelated. Ω^{-1} is known as the confidence in the investor's views. The ith diagonal element of Ω is represented as ω_i.

One does not require P to be invertible. Meucci (2006) describes a method of augmenting the matrices to make the P matrix invertible while not changing the net results.

Ω is symmetric and zero on all nondiagonal elements, but may also be zero on the diagonal if the investor is certain of a view. This means that Ω may or may not be invertible. At a practical level, one may require that $\omega > 0$ so that Ω is invertible, but one should reformulate the problem so that Ω is not required to be inverted.

Example 3.16 Matrices for the Black–Litterman Model

As an example of how these matrices would be populated, one may examine some investors' views. For an example consisting of four assets and two views:

- First, a relative view in which the investor believes that asset A will outperform asset C by 2% with confidence ω_A.
- Second, there exists an absolute view in which the investor believes that asset B will return 3% with confidence ω_2.
- The investor has no view on asset D, and thus its return should not be directly adjusted. These views are specified as follows:

$$P = \begin{bmatrix} 1 & 0 & -1 & 0 \\ 0 & 1 & 0 & 0 \end{bmatrix}; \quad Q = \begin{bmatrix} 2 \\ 3 \end{bmatrix}; \quad \Omega = \begin{bmatrix} \omega_{11} & 0 \\ 0 & \omega_{22} \end{bmatrix}$$

Given this specification of the views, one may formulate the conditional distribution mean and variance in view space as follows:

$$P(B|A) \sim N(Q, \Omega) \tag{3.84}$$

Generally, one cannot convert this into a useful expression into asset space because of the mixture of relative and absolute views, and because the P matrix is not required to be of full rank. If one did express the views in asset space, the formula is shown below:

$$P(B|A) \sim N(P^{-1}Q, [P^T \Omega^{-1} P]^{-1}) \tag{3.85}$$

This representation is *not* of any practical use. Incomplete views and relative views make the variance noninvertible, and relative views also impact the mean term. P may not be invertible, and even if P is invertible, $[P^T\Omega^{-1}P]$ is probably not invertible, making this expression impossible to evaluate in practice. Fortunately, to work with the Black–Litterman model, one does not need to evaluate (3.85). It is however, interesting to see how the views are projected into the asset space.

Specifying Ω Ω, the variance of the views, is inversely related to the investors' confidence in the views; however, the basic Black–Litterman model does not provide an intuitive way to quantify this relationship. It is up to the investor to compute the variance of the views Ω.

The following are ways to calculate Ω:

- Proportional to the variance of the prior
- Use a confidence interval
- Use the variance of residuals in a factor model
- Use Idzorek's method to specify the confidence along the weight dimension

Proportional to the Variance of the Prior   Assume that the variance of the views will be proportional to the variance of the asset returns, just as the variance of the prior distribution is. Both He and Litterman (1999) and Meucci (2006) use this method, although they use it differently. He and Litterman (1999) expressed the variance of the views as follows:

$$\omega_{ij} = p(\tau\Sigma)p^T, \quad \forall\, i = j \tag{3.86a}$$
$$= 0, \quad \forall i \neq j$$

Or

$$\Omega = \mathrm{diag}\{P(\tau\Sigma)P^T\} \tag{3.86b}$$

This specification of the variance, or uncertainty, of the views essentially equally weights the investor's views and the market equilibrium weights. By including τ in the expression, the posterior estimate of the returns becomes independent of τ as well.

Meucci (2006) did not bother with the diagonalization at all, and just sets

$$\Omega = (1/c)P\Sigma P^t \tag{3.87}$$

who also sets $c > 1$, with one obvious choice for c being τ^{-1}. This form of the variance of the views lends itself to some simplifications of the Black–Litterman formulas.

Use a Confidence Interval   The investor may specify the variance using a confidence interval around the estimated mean return, for example, asset B has an estimated 3.0% mean return with the expectation that it is 68% likely to be within the interval (2.0%, 4.0%). Knowing that 68% of the normal distribution falls within 1 standard deviation of the mean allows one to translate this into a variance for the view of 1%.

Now, Ω is the uncertainty in the estimate of the mean, one is not specifying the variance of returns about the mean. This formulation of the variance of the view is consistent with the canonical reference model.

Using the Variance of Residuals from a Factor Model   If the investor is using a factor model to compute the views, one may use the variance of the residuals from the model to drive the variance of the return estimates. The general expression for a factor model of returns is

$$r = \sum_{i=1}^{n} \beta_i f_i + \varepsilon \tag{3.88}$$

where

r is the return of the asset,
β_i is the factor loading for factor (i).
f_i is the return owing to factor (i), and
ε is an independent normally distributed residual.

The general expression for the variance of the return from a factor model is

$$V(r) = \mathrm{BV}(F)B^T + V(\varepsilon) \tag{3.89}$$

where

B is the factor loading matrix, and
F is the vector of returns owing to the various factors.

Given Equation (3.88), and the assumption that ε is independent and normally distributed, one may calculate the variance of ε directly as part of the regression. While the regression might yield a full covariance matrix, the mixing model will be more robust if only the diagonal elements are used.

Beach and Orlov (2006) describe their work using GARCH style factor models to generate their views for use with the Black–Litterman model. They generated the precision of the views using the GARCH models.

Using the Idzorek Method   Idzorek (2005) describes a method for specifying the confidence in the view in terms of a percentage move of the weights on the interval from 0% to 100% confidence. We will look at Idzorek's algorithm in the section on extensions.

The Estimation Model   The original Black–Litterman paper references Theil's Mixed Estimation model rather than a Bayesian estimation model, although one may obtain results from both methodologies. Let us start with Theil's model because of its relative simplicity. Also, for completeness, the Bayesian version of the derivations will be reviewed. For either approach, the canonical Black–Litterman Reference Model will be used. For this reference model, the estimation model is used to compute the distribution of the estimated returns about the mean return, and then estimate the distribution of returns about the mean return. This distinction is important in understanding the values used for τ and Ω, and for the computations of the variance of the prior and posterior distributions of returns. The posterior estimate of the mean generated by the estimation model is more precise than either the prior estimate or the investor's views. Note that one should not expect large changes in the variance of the distribution of returns about the mean because the estimate of the mean is more precise. The prototypical example of this would be to blend the distributions:

$$P(A) \sim N(10\%, \ 20\%)$$

and

$$P(B|A) \sim N(12\%, \ 20\%)$$

If one applies estimation model in a straightforward fashion,

$$P(A|B) \sim N(11\%, \ 10\%)$$

Clearly, with financial data, one does not really cut the variance of the return distribution about the mean in half just because one has a slightly better estimate of the mean. In this case, the mean is the random variable, thus the variance of the posterior corresponds to the variance of the estimated mean around the mean return; not the variance of the distribution of returns about the mean return. In this case, the posterior result of

$$P(A|B) \sim N(11\%, \ 10\%)$$

makes sense. By blending these two estimates of the mean, one has an estimate of the mean with much less uncertainty (less variance) than either of the estimates, even though one does not have a better estimate of the actual distribution of returns around the mean.

The Theil Mixed Estimation Model   Theil's mixed estimation model was created for the purpose of estimating parameters from a mixture of complete prior data and partial conditional data. This is a good fit with the present problem as it allows one to express views on only a subset of the asset returns, there is no requirement to express views on all of them. The views can also be expressed on a single asset, or on arbitrary combinations of the assets.

The views do not even need to be consistent, the estimation model will take each into account based on the investors' confidence.

Theil's Mixed Estimation model starts from a linear model for the parameters to be estimated. One may use Equation (3.77) from the reference model as a starting point. A simple linear model is shown below:

$$x\beta = \pi + u \tag{3.90}$$

where

π is the $n \times 1$ vector of equilibrium returns for the assets,
x is the $n \times n$ matrix in which are the factor loadings for the model,
β is the $n \times 1$ vector of unknown means for the asset return process, and
u is an $n \times n$ matrix of residuals from the regression where

$$E(u) = 0, V(u) = E(u'u) = \Phi$$

The Black–Litterman model uses a very simple linear model, the expected return for each asset is modeled by a single factor that has a coefficient of 1. Thus, x is the identity matrix. Since β and u are independent and x is constant, one may model the variance of π as follows:

$$V(\pi) = x'V(\beta)x + V(u) \tag{3.91}$$

which may be simplified to

$$V(\pi) = \Sigma + \Phi \tag{3.92}$$

where

$\sum$ is the historical covariance matrix of asset returns as used earlier.
Φ is the covariance of residuals or of the estimate about the actual mean.

This approach connects with (3.78) in the reference model:

$$\Sigma_r = \Sigma + \Sigma_\pi \tag{3.78}$$

which shows that the proper relationship between the variances is ($\sum_r \geq \sum$, $\sum_\pi$). The total variance of the estimated return is the sum of the variance of the actual return process plus the variance of the estimate of the mean. This relationship will be revisited at a later stage.

Next, consider some additional information that one would like to combine with the prior. This information may be considered as a subjective view or may be derived from statistical data. It may also consider it to be incomplete, meaning that one might not have an estimate for each asset return.

$$P\beta = q + v \tag{3.93}$$

where

q is the $k \times 1$ vector of returns for the views,
p is the $k \times n$ matrix mapping the views onto the assets,

β is the $n \times 1$ vector of unknown means for the asset return process, ν is a $k \times 1$ vector of residuals from the regression, where $E(\nu) = 0$, $V(\nu) = E(\nu'\nu) = \Omega$, and Ω is nonsingular.

One may combine the prior and conditional information by writing:

$$[x] \quad \underline{\beta} = [\pi] + [u]$$
$$[p] \qquad [q] \quad [v] \tag{3.94}$$

where the expected value of the residual is 0, and the expected value of the variance of the residual is

$$V([u]) = E([u][u'v']) = [\Phi \ 0]$$
$$([v]) \quad ([v]) \qquad) \quad [0 \ \Omega] \tag{3.95}$$

One can then apply the generalized least-squares procedure, which leads to the estimating of $\underline{\beta}$ as

$$\underline{\beta} = (x'p')[\Phi \ 0]^{-1}[x])^{-1}[x'p'][\Phi \ 0]^{-1}[\pi]$$
$$= (\quad [0 \ \Omega] \ [p]) \qquad [0 \ \Omega] \quad [q] \tag{3.96}$$

This may be rewritten without the matrix notation as

$$\underline{\beta} = [x'\Phi^{-1}x + p'\Omega^{-1}p]^{-1}[\Omega^{-1}\pi + p'\Omega^{-1}q] \tag{3.97}$$

For the Black–Litterman model that is a single factor per asset, one may drop the variable x as it is the identity matrix. If one preferred using a multifactor model for the equilibrium, then x would be the equilibrium factor loading matrix.

$$\underline{\beta} = [\Phi^{-1} + p'\Omega^{-1}p]^{-1}[\Phi^{-1}\pi + p'\Omega^{-1}q] \tag{3.98}$$

This new $\underline{\beta}$ is the weighted average of the estimates, where the weighting factor is the precision of the estimates. The precision is the inverse of the variance. The posterior estimate $\underline{\beta}$ is also the best linear unbiased estimate given the data, and has the property that it minimizes the variance of the residual.

Given a new $\underline{\beta}$, one should also have an updated expectation for the variance of the residual.

If one were using a factor model for the prior, then one should keep x, the factor weightings, in the equation. This would result in a multifactor model, where all the factors will be priced into the equilibrium.

One may reformulate the combined relationship in terms of the estimate of $\underline{\beta}$ and a new residual $\underline{u}$ as

$$[x] \quad \underline{\beta} = [\pi] + \underline{u}$$
$$[p] \qquad [q] \tag{3.99}$$

Once again $E(u) = 0$, so one may derive the expression for the variance of the new residual as

$$V(\underline{u}) = E(\underline{u}'\underline{u}) = [\Phi^{-1} + p'\Omega^{-1}p]^{-1} \tag{3.100}$$

and the total variance is

$$V([y\pi]) = V(\underline{\beta}) + V(\underline{u}) \tag{3.101}$$

One began this section by asserting that the variance of the return process is a known quantity. Improved estimation of the quantity $\underline{\beta}$ does not change the estimate of the variance of the return distribution, $\sum$. Because of the improved estimate, one does expect that the variance of the estimate (residual) has decreased, thus the total variance has changed. One may simplify the variance formula (3.92):

$$V(\pi) = \Sigma + \Phi \tag{3.92}$$

to

$$V([y\pi]) = \Sigma + V(\underline{u}) \tag{3.102}$$

This is an intuitive result, consistent with the realities of financial time series. One has combined two estimates of the mean of a distribution to arrive at a better estimate of the mean. The variance of this estimate has been reduced, but the actual variance of the underlying process remains unchanged. Given the uncertain estimate of the process, the total variance of the estimated process has also improved incrementally, but it has the asymptotic limit that it cannot be less than the variance of the actual underlying process.

This is the convention for computing the covariance of the posterior distribution of the canonical reference model as shown in He and Litterman (1999).

In the absence of views, (3.102) simplifies to

$$V([y \pi]) = \Sigma + \Phi \tag{3.103}$$

which is the variance of the prior distribution of returns.

Bayes' Theorem for the Estimation Model   In the Black–Litterman model, the prior distribution is based on the equilibrium-implied excess returns. One of the major assumptions made by the Black–Litterman model is that *the covariance of the prior estimate is proportional to the covariance of the actual returns, but the two quantities are independent.* The parameter τ serves as the constant of proportionality. The prior distribution for the Black–Litterman model was specified in 3.87a and 3.87b:

$$P(A) \sim N(\Pi, \tau\Sigma) \tag{3.87a}$$

and

$$r_A \sim N(P(A), \Sigma) \tag{3.87b}$$

The conditional distribution is based on the investor's views. The investor's views are specified as returns to portfolios of assets, and each view has an uncertainty that will impact the overall mixing process. The conditional distribution from the investor's views was specified in Equation (3.85):

$$P(B|A) \sim N(P^{-1}Q, [P^T\Omega^{-1}P]^{-1}) \tag{3.85}$$

The posterior distribution from Bayes' theorem is the precision-weighted average of the prior estimate and the conditional estimate. One may now apply Bayes' theory to the problem of blending the prior and conditional distributions to create a new posterior distribution of the asset returns. Given Equation (3.85) and Equations (3.87a) and (3.87b), for the prior and conditional distribution, respectively, one may apply Bayes' theorem to derive the following equation for the posterior distribution of the asset returns:

$$P(A|B) \sim N\{[(\tau\Sigma)^{-1}\Pi + P^T\Omega^{-1}Q][(\tau\Sigma)^{-1} + P^T\Omega^{-1}P]^{-1}, \ [(\tau\Sigma)^{-1} + P^T\Omega^{-1}P]^{-1}\} \tag{3.104}$$

This is the *Black–Litterman Master Formula*.

An alternative representation of the same formula for the mean returns Π and covariance M takes the form:

$$\underline{\Pi} = \Pi + \tau\Sigma P^T[(P\tau\Sigma P^T) + \Omega]^{-1}[Q - P\Pi] \tag{3.105}$$

$$M = \{(\tau\Sigma)^{-1} + P^T\Omega^{-1}P\}^{-1} \tag{3.106}$$

Note that M, the posterior variance, is the variance of the posterior mean estimate about the actual mean. It is the uncertainty in the posterior mean estimate, and is *not* the variance of the returns.

Calculating the posterior covariance of returns requires adding the variance of the estimate about the mean to the variance of the distribution about the estimate the same as in (3.102). This is indicated in He and Litterman (1999).

Now,

$$\Sigma_p = \Sigma + M \tag{3.107}$$

Substituting the posterior variance from (3.106), one obtains

$$\Sigma_p = \Sigma + \{(\tau\Sigma)^{-1} + P^T\Omega^{-1}P\}^{-1} \tag{3.108}$$

In the absence of views, this reduces to

$$\begin{aligned} \Sigma_p &= \Sigma + (\tau\Sigma) \\ &= (1 + \tau)\Sigma \end{aligned} \tag{3.109}$$

Thus, when applying the Black–Litterman model in the absence of views, the variance of the estimated returns will be, according to (3.109), *greater* than the prior distribution variance. The impact of this equation was highlighted in the results shown in He and Litterman (1999), in which the investor's weights sum to less than 1 if they have no views. Idzorek (2005) and most other authors do not compute a new posterior variance, but instead use the known input variance of the returns about the mean.

If the investor has only partial views, that is, views on a subset of the assets, then by using a posterior estimate of the variance, one will tilt the posterior weights toward assets with lower variance (higher precision of the estimated mean) and away from assets with higher variance (lower precision of the estimated mean). Thus, the existence of the views and the updated covariance will tilt the optimizer toward using or not using those assets. This tilt will not be very large if one is working with a small value of τ, but it will be measurable.

Since one often builds the known covariance matrix of returns, $\sum$, from historical data one may use methods from basic statistics to compute τ, as $\tau\sum$ is analogous to the standard error. One may also estimate τ based on one's confidence in the prior distribution. Note that both of these techniques provide some intuition for selecting a value of τ that is closer to 0 than to 1. Black and Litterman (1992), He and Litterman (1999), and Idzorek (2005) all indicate that in their calculations, they used small values of τ on the order of 0.025–0.050. Satchell and Scowcroft (2000) state that many investors use a τ, around 1, which has no intuitive connection to the data, and in fact shows that their paper uses the *Alternative Reference Model*.

One may check the results by testing if the results match one's intuition at the boundary conditions.

Given (3.105), let $\Omega \to 0$ and show that the return under 100% certainty of the views is

$$\underline{\Pi} = \Pi + \Sigma P^T [P\Sigma P^T]^{-1} [Q - P\Pi] \tag{3.110}$$

Hence, under 100% certainty of the views, the estimated return is insensitive to the value of τ used.

Also, if P is invertible, which means that one has offered a view on every asset, then

$$\underline{\Pi} = P^{-1}Q \tag{3.111}$$

If the investor is not sure about the views, $\Omega \to \infty$, then (3.105) reduces to

$$\underline{\Pi} = \Pi \tag{3.112}$$

Thus, seeking an analytical path to calculate the posterior variance under 100% certainty of the views is still a challenge!

(This is shown in Table 4 and mentioned on page 11 of He and Litterman (1999).)

Next, the problem remains to obtain an analytically tractable way to express and calculate the posterior variance under 100% certainty.

The alternative formula for the posterior variance derived from (3.106) using the Woodbury Matrix Identity is

$$M = \tau\Sigma - \tau\Sigma P^T (P\tau\Sigma P^T + \Omega)^{-1} P\tau\Sigma \tag{3.113}$$

If $\Omega \to 0$ (total confidence in views, and every asset is in at least one view), then (3.113) may be simplified to

$$
\begin{aligned}
M &= \tau\Sigma - \tau\Sigma P^T (P\tau\Sigma P^T + 0)^{-1} P\tau\Sigma \\
&= \tau\Sigma - \tau\Sigma P^T (P\tau\Sigma P^T)^{-1} P\tau\Sigma \\
&= \tau\Sigma - \tau\Sigma P^T (P^T)^{-1} \\
&= \tau\Sigma - \tau\Sigma
\end{aligned}
$$

namely,

$$M = 0 \tag{3.114}$$

And if the investor is not confident in the views, $\Omega \to \infty$, then (3.113) can be reduced to

$$
\begin{aligned}
M &= \to \tau\Sigma - \tau\Sigma P^T (P\ \tau\Sigma P^T + \infty)^{-1} P\tau\Sigma \\
&= \to \tau\Sigma - 0
\end{aligned}
$$

namely,

$$M = \tau\Sigma \tag{3.115}$$

Alternative Reference Model Consider now the most common alternative reference model used with the Black–Litterman estimation model: It is the one used in Satchell and Scowcroft (2000), and in the work of Meucci prior to his introduction of "Beyond Black–Litterman":

$$E(r) \sim N(\mu, \Sigma) \tag{3.116}$$

In this reference model, μ is normally distributed with variance $\sum$. One may estimate μ, but μ is not considered a random variable. This is commonly described as having a $\tau = 1$, but more precisely one is making a point estimate and thus have eliminated τ as a parameter. In this model, Ω becomes the covariance of the returns to the views around the unknown mean return,, just as $\sum$ is the covariance of the prior return about its mean. Given that one is now using point estimates, the posterior is now a point estimate and one no longer needs to be concerned about posterior covariance of the estimate. In this model, one does not have a posterior precision to use downstream in one's portfolio selection model.

Rewriting Equation (3.105):

$$\underline{\Pi} = \Pi + \tau \Sigma P^T [(P \tau \Sigma P^T) + \Omega]^{-1}[Q - P\Pi] \tag{3.105}$$

noting that one may move around the τ term:

$$\underline{\Pi} = \Pi + \tau \Sigma P^T [(P \tau \Sigma P^T) + (\Omega / \tau]^{-1}[Q - P\Pi] \tag{3.106}$$

Now it is seen that τ only appears in one term in this formula. Because the Alternative Reference Model does not include updating the covariance of the estimates, this is the only formula. Given that the investor is selecting both Ω and τ to control the blending of the prior and their views, one may eliminate one of the terms. Since τ is a single term for all views and Ω has a separate element for each view, one shall keep Ω. One may then rewrite the posterior estimate of the mean as follows:

$$\underline{\Pi} = \Pi + \Sigma P^T [(P \Sigma P^T) + \Omega]^{-1}[Q - P\Pi] \tag{3.107}$$

The primary artifacts of this new reference model are as follows:

i) τ is gone.
ii) The investor's portfolio weights in the absence of views equal the equilibrium portfolio weights.
iii) Finally, at implementation time, there is no need or use of (3.106) or (3.107).

Remark:

1) None of the authors prior to Meucci (2008), except for Black and Litterman (1992) and He and Litterman (1999), make any mention of the details of the Canonical Reference Model or of the fact that different authors actually use quite different reference models.
2) In the Canonical Reference Model, the updated posterior covariance of the unknown mean about the estimate will be smaller than the covariance of either the prior or conditional estimates, indicating that the addition of more information will reduce the uncertainty of the model. The variance of the returns from (3.107) will never be less than the prior variance of returns. This matches the intuition as adding more information should reduce the uncertainty of the estimates.
3) Since there is some uncertainty in this value (M), (3.107) provides a better estimator of the variance of returns than the prior variance of returns.

Impact of τ   For users of the Black–Litterman model, the meaning and impact of the parameter τ may cause confusion. Investors using the Canonical Reference Model use τ and that it does have a very *precise* meaning in the model. An author who selects an essentially random value for τ probably does not using the Canonical Reference Model, but instead uses the Alternative Reference Model.

Given the Canonical Reference Model, one may still understand the impact of τ on the results. One may start with the expression for Ω similar to the one used by He and Litterman (1999). Rather than using only the diagonal, one may retain the entire structure of the covariance matrix to simplify the methodology:

$$\Omega = P(\tau\Sigma)P^T \qquad (3.108)$$

One may substitute (3.108) into (3.105):

$$\underline{\Pi} = \Pi + \tau\Sigma P^T[(P\tau\Sigma P^T) + \Omega]^{-1}[Q - P\Pi] \qquad (3.105)$$

to obtain

$$
\begin{aligned}
\underline{\Pi} &= \Pi + \tau\Sigma P^T[(P\tau\Sigma P^T) + \Omega]^{-1}[Q - P\Pi^T] \\
&= \Pi + \tau\Sigma P^T[(P\tau\Sigma P^T) + (P\tau\Sigma P^T)^{-1}[Q - P\Pi^T] \\
&= \Pi + \tau\Sigma P^T[2(P\tau\Sigma P^T)]^{-1}[Q - P\Pi^T] \\
&= \Pi + (1/2)\tau\Sigma P^T(P^T)^{-1}[P\tau\Sigma]^{-1}[Q - P\Pi^T] \\
&= \Pi + (1/2)\tau\Sigma(\tau\Sigma)^{-1}P^{-1}[P\tau\Sigma]^{-1}[Q - P\Pi^T] \\
&= \Pi + (1/2)P^{-1}[Q - P\Pi^T]
\end{aligned}
$$

namely,

$$\underline{\Pi} = \Pi + (1/2)[P^{-1}Q - \Pi^T] \qquad (3.109)$$

Thus, using (3.108) is just a simplification and does not do justice to investors' views, but one may still see that setting Ω proportional to τ will eliminate τ from the final formula for Π.

In the Canonical Reference Model, it does not eliminate τ from the equations for posterior covariance given by (3.107):

$$\Sigma_p = \Sigma + M \qquad (3.107)$$

In the general form, if one formulates Ω as

$$\Omega = P(\alpha\tau\Sigma)P^T \qquad (3.110)$$

then one may rewrite (3.109) as

$$\underline{\Pi} = \Pi + \{1/(1+\alpha)\}[P^{-1}Q - \Pi] \qquad (3.111)$$

One may see a similar result if one substitutes Equation (3.108) into Equation (3.113):

$$\Omega = P(\tau\Sigma)P^T \qquad (3.108)$$

$$M = \tau\Sigma - \tau\Sigma P^T(P\,\tau\Sigma P^T + \Omega)^{-1}P\tau\Sigma \qquad (3.113)$$

Resulting in:

$$
\begin{aligned}
M &= \tau\Sigma - \tau\Sigma P^T (P\tau\Sigma P^T + \Omega)^{-1} P\tau\Sigma \\
&= \tau\Sigma - \tau\Sigma P^T (P\tau\Sigma P^T + P(\tau\Sigma)P^T)^{-1} P\tau\Sigma \\
&= \tau\Sigma - \tau\Sigma P^T (2(P\tau\Sigma P^T)^{-1} P\tau\Sigma \\
&= \tau\Sigma - (1/2)(\tau\Sigma)(P^T)(P^T)^{-1}(\tau\Sigma)^{-1}(P)^{-1}(P)(\tau\Sigma) \\
&= \tau\Sigma - (1/2)\tau\Sigma
\end{aligned}
\tag{3.113}
$$

namely,

$$
M = (1/2)\tau\Sigma \tag{3.117}
$$

Note that τ is not eliminated from (3.117). One may also observe that if τ is on the order of 1, and if one were to use the equation

$$
\Sigma_p = \Sigma + M \tag{3.107}
$$

then the uncertainty in the estimate of the mean would be a significant portion of the variance of the returns. With the Alternate Reference Model, no posterior variance computations are preformed and the mixing is weighted by the variance of returns.

In both cases, the choice for Ω has evenly weighted the prior and conditional distributions in the estimation of the posterior distribution. This matches the intuition when one considers one has blended two inputs, for both of which one has the same level of uncertainty. The posterior distribution will be the average of the two distributions. If instead one solves for the more useful general case of

$$
\Omega = \alpha P(\tau\Sigma)P^T \tag{3.118}
$$

where $\alpha \geq 0$, substituting into (3.105) and following the same logic as used to derive (3.117) one obtains

$$
\underline{\Pi} = \Pi + \{1/(1+\alpha)\}[P^{-1}Q - \Pi] \tag{3.119}
$$

This parameterization of the uncertainty is specified in Meucci (2005) and it allows us an option between using the same uncertainty for the prior and views, and having to specify a separate and unique uncertainty for each view. Given that one is essentially multiplying the prior covariance matrix by a constant, this parameterization of the uncertainty of the views does not have a negative impact on the stability of the results.

Note that this specification of the uncertainty in the views changes the assumption from the views being uncorrelated to the views having the same correlations as the prior returns.

In summary, if the investor uses the Alternative Reference Model and makes Ω proportional to $\sum$, then it is necessary only to calibrate the constant of proportionality, α, which indicates their relative confidence in their views versus the equilibrium. If the Canonical Reference Model is used and set Ω proportional to $\tau\sum$, then the return estimate will not depend on the value of τ, but the posterior covariance of returns will depend on the proper calibration of τ.

Calibration of τ   Some empirical ways to select and calibrate the value of τ will be considered:

The first method to calibrate τ relies on basic statistics: When estimating the mean of a distribution, the uncertainty (variance) of the mean estimate will be proportional to the inverse of the sample sizes. Given that one is estimating the covariance matrix from historical data, then

$\tau = 1/T$ The maximum-likelihood estimator

$\tau = 1/(T-k)$ The best quadratic unbiased estimator

where $T=$ the number of samples and $k=$ the number of assets.

While there are other estimators, usually the first definition above is used. Given that one usually aims for a number of samples around 60 (viz., 5 years of 12 monthly samples), τ is on the order of 0.02 ($1/60 = 0.01666\ldots$). This is consistent with several of the papers that indicate they used values of τ on the range of (0.025, 0.05).

The most intuitively easy way to calibrate τ is as part of a confidence interval for the prior mean estimates. One may illustrate this concept with a simple example: Consider the scenario where $\tau = 0.05$, and one considers only a single asset with a prior estimate of 7% as the excess return and 15% as the known standard deviation of returns about the mean. If one uses a confidence interval of (1%, 5%) with 68% confidence, one keeps the ratio of View Precision to Prior Precision fixed between the two scenarios, one with $\tau = 0.05$ and one with $\tau = 1$. In the first scenario, even though one uses a seemingly small $\tau = 0.05$, the prior estimate has relatively low precision based on the width of the confidence interval, and thus the posterior estimate will be heavily weighted toward the view. In the second scenario with $\tau = 1$, the prior confidence interval is so wide as to make the prior estimate close to worthless. In order to keep the posterior estimate the same across scenarios, the view estimate also has a wide confidence interval indicating the investor is really not confident in any of their estimates!

Example 3.17 Asset Allocation and Confidence Levels

Consider the following preliminary asset allocation (Table 3.4):

Given such wide intervals for the $\tau = 1$ scenario, 16% confidence that one's asset has a mean return less than -8%, it is reasonable to doubt the final asset allocation. Understanding the interplay between the selection of τ and the specification of the variance of the views is essential and critical. This example illustrates the difference between the parameters for the Canonical Reference

Table 3.4   Assets allocation.

τ	Prior @ 68% confidence	Prior precision	View σ	View precision	View @ 68% confidence	View/prior precision
0.05	(4.6%, 9.4%)	888	2.00%	2500	(1%, 5%)	2.81
1.00	(−8%, 22%)	44.4	8.90%	125	(−5.9%, 11.9%)	2.81

Model and the Alternative Reference Model. Specifying $\tau = 1$ is the Alternative Reference Model, but it just generates garbage outputs from the Canonical Reference Model.

Remarks:

1) One could instead calibrate τ to the amount invested in the risk-free asset given the prior distribution. Here one sees that the portfolio invested in risky assets given the prior views will be

$$w = \Pi[\delta(1 + \tau)\Sigma]^{-1} \tag{3.120}$$

2) Thus, the weights allocated to the assets are smaller by $[1/(1+\tau)]$ than the CAPM market weights. This is on account of the Bayesian investor is uncertain in the estimate of the prior and does not want to be 100% invested in risky assets.

Results

It is of interest and considerable enlightenment to review and compare the results of the various authors on this subject. The Java programs used to compute these results are all available as part of the akutan open source finance project at Π sourceforge.net. All of the mathematical functions were built using the Colt open-source numerics library for Java. Selected formulas are also available as MATLAB and/or SciLab scripts on the website black-litterman.org. Any small differences between all the authors' reported results are most likely the result of rounding of inputs and/or results. When reporting results, most authors have just reported the portfolio weights from an unconstrained optimization using the posterior mean and variance. Given that the vector Π is the excess return vector, one does not need a budget constraint ($\sum w_i = 1$) as one may safely assume any "missing" weight is invested in the risk-free asset that has expected return 0 and variance 0. This calculation comes from (3.86b):

$$w = (\delta\Sigma)^{-1}\Pi \tag{3.86b}$$

or

$$w = \Pi(\delta\Sigma)^{-1} \tag{3.86c}$$

As a first test of the algorithm, one verifies that when the investor has no views that the weights are correct, substituting (3.109) into (3.86b)

$$\Sigma_p = (1 + \tau)\Sigma \tag{3.109}$$

one obtains

$$
\begin{aligned}
w_{nv} &= \Pi\{\delta(1 + \tau)\Sigma\}^{-1} \\
&= \Pi(\delta\Sigma)^{-1}/(1 + \tau)
\end{aligned}
$$

or

$$w_{nv} = w/(1 + \tau) \tag{3.121}$$

Hence, it is clear that the output weights with no views will be impacted by the choice of τ when the Black–Litterman reference model is used. He and Litterman (1999) indicate that if the investor is a Bayesian, then one shall not be certain of the prior distribution and thus would not be fully invested in the risky portfolio in the beginning. This is consistent with (3.121).

Matching the Results of He and Litterman   First one shall consider the results shown in He and Litterman (1999). These results are easy to reproduce as they clearly implement the Canonical Reference Model and they provide all the data required to reproduce their results in the paper.

He and Litterman (1999) set

$$\Omega = \text{diag}\{P^T(\tau\Sigma)P\} \tag{3.122}$$

This makes the uncertainty of the views equivalent to the uncertainty of the Equilibrium estimates. A small value for τ, 0.05, was selected, and the Canonical Reference Model was used. The updated posterior variance of returns is calculated in (3.113) and (3.107).

The results of Table 3.5 correspond to Table 7 in (He and Litterman, 1999).

Figures 3.30 and 3.31 show the pdf for the prior, view, and posterior for each view defined in the problem. The y-axis uses the same scale in each graph. Note how in Figure 3.27 the conditional distribution of the estimated mean is much more diffuse because the variance of the estimate is *larger* (viz., precision of the estimate is *smaller*). Note how the precision of the prior and views impacts the precision (width of the peak) on the pdf. In Figure 3.27 with the less precise view, the posterior is also less precise.

Matching the Results of Idzorek   This section reproduces the results of Idzorek (2005). In trying to match Idzorek's results, it is found that Idzorek used the

Table 3.5 Results computed using the akutan implementation of Black–Litterman and the input data for the equilibrium case and the investor's views from He and Litterman (1999).

Asset	P_0	P_1	μ	weq/$(1+\tau)$	w^*	w^* - weq/$(1+\tau)$
Australia	0.0	0.0	4.3	16.4%	1.5%	0.0%
Canada	0.0	1.0	8.9	2.1%	53.9%	51.8%
France	−0.295	0.0	9.3	5.0%	−0.5%	−5.4%
Germany	1.0	0.0	10.6	5.2%	23.6%	18.4%
Japan	0.0	0.0	4.6	11.0%	11.0%	0.0%
UK	−0.705	0.0	6.9	11.8%	−1.1%	−13.0%
USA	0.0	−1.0	7.1	58.6%	6.8%	−51.8%
Q	5.0	4.0				
ω/τ	0.043	0.017				
λ	0.193	0.544				

The values shown for w^* exactly match the values shown in their paper: Figure 5—Distributions of actual means about estimates means.

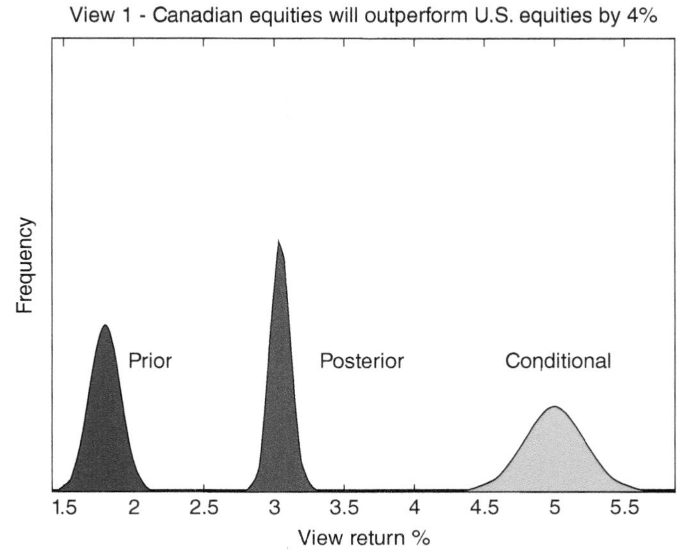

Figure 3.30 Canadian outperforming U.S. equities by 4%.

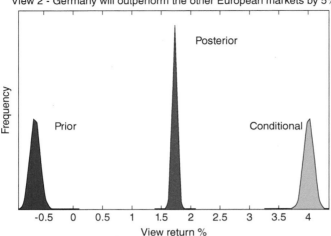

View 2 - Germany will outperform the other European markets by 5%

Figure 3.31 German outperforming other European equities by 5%.

Alternative Reference Model. which leaves $\sum$, the known variance of the returns from the prior distribution, as the variance of the posterior returns. This is a significant difference from the Canonical Reference Model, but in the end the differences amounted to only 50 basis points per asset. Tables 3.6 and 3.7 illustrate calculated results with the data from Idzorek's paper and how the results differ between the two versions of the model.

Table 3.6 contains results generated using the data from Idzorek (2005) and the Canonical Reference Model. Table 3.7 shows the same results as generated by the Alternative Reference Model.

Table 3.6 Canonical Reference model with Idzorek data (corresponds with data in Idzorek's Table 6).

Asset class	μ	w_{eq}	w	Black–Litterman reference model	Idzorek's results
U.S. Bonds	0.07	18.87%	28.96%	10.09%	10.54%
Intl Bonds	0.50	25.49%	15.41%	−10.09%	−10.54%
US LG	6.50	11.80%	9.27%	−2.52%	−2.73%
US LV	4.33	11.80%	14.32%	2.52%	−2.73%
US SG	7.55	1.31%	1.03%	−0.28%	−0.30%
US SV	3.94	1.31%	1.59%	0.28%	0.30%
Intl Dev	4.94	24.18%	27.74%	4.15%	3.63%
Intl Eng	6.84	3.40%	3.40%	0%	0%

Table 3.7 Alternative Reference model with Idzorek data (corresponds with data in Idzorek's Table 6).

Country	μ	w_{eq}	w	Alternative reference model	Idzorek's results
US Bonds	0.07	19.34%	29.89%	10.55%	10.54%
Intl Bonds	0.50	26.13%	15.58%	−10.55%	−10.54%
US LG	6.50	12.09%	9.37%	−2.72%	−2.73%
US LV	4.33	12.09%	14.81%	2.72%	−2.73%
US SG	7.55	1.34%	1.04%	−0.30%	−0.30%
US SV	3.94	1.34%	1.64%	0.30%	0.30%
Intl Dev	4.94	24.18%	27.77%	3.59%	3.63%
Intl Eng	6.84	3.49%	3.49%	0.00%	0.00%

Note that the results in Table 3.6 are close, but for several of the assets the difference is about 50 basis points. The values shown in Table 3.7 are within four basis points, essentially matching the results reported by Idzorek.

Additional Work   Consider efforts to reproduce results from some of the major research papers on the Black–Litterman model.

Of the major papers on the Black–Litterman model, there are two that would be very useful to reproduce:

1) Satchell and Scowcroft (2000)
2) Black and Litterman (1992)

Satchell and Scowcroft (2000) did not provide enough data in their paper to reproduce their results. They have several examples, one with 11 countries equity returns plus currency returns, and one with 15 countries. They did not provide the covariance matrix for either example, and so their analysis cannot be reproduced. It would be interesting to confirm that they use the Alternative Reference Model by reproducing their results.

Black and Litterman (1992) did provide what seems to be all the inputs to their analysis, although they chose a nontrivial example, including partially hedged equity and bond returns. This requires the application of some constraints to the reverse optimization process that have been formulated. It should be useful to continue this work with the goal of verifying the details of the Black–Litterman implementation used in Black and Litterman (1992).

Extensions to the Black–Litterman Model   This section covers the extensions to the Black–Litterman model proposed in Idzorek (2005), Fusai and Meucci (2003), Krishnan and Mains (2006), and Qian and Gorman (2001).

Idzorek (2005) presents a means to calibrate the confidence or variance of the investors views in a simple and straightforward method.

Next is a section on measures of extremity or quality of views. Fusai and Meucci (2003) proposed a way to measure how consistent a posterior estimate of the mean is with regard to the prior, or some other estimate. Braga and Natale (2007) described how to use Tracking Error to measure the distance from the equilibrium to the posterior portfolio. Also included are additional original work on using relative entropy to measure quality of the views.

Finally, larger extensions to the model such as Krishnan and Mains (2006) present a method to incorporate additional factors into the model. Qian and Gorman (2001) presented a method to integrate views on the covariance matrix as well as views on the returns.

Idzorek's Extension Idzorek's apparent goal was to reduce the complexity of the Black–Litterman model for nonquantitative investors. He achieved this by allowing the investor to specify the investors' confidence in the views as a percentage (0–100%) where the confidence measures the change in weight of the posterior from the prior estimate (0%) to the conditional estimate (100%). This linear relation is shown in Equation (3.123):

$$\text{confidence} = (\underline{w} - w_{\text{mkt}})/(w_{100} - w_{\text{mkt}}) \tag{3.123}$$

where

w_{100} is the weight of the asset under 100% certainty in the view,
w_{mkt} is the weight of the asset under no views, and
w is the weight of the asset under the specified view.

Also provided was a method to back out the value of ω required to generate the proper tilt (change in weights from prior to posterior) for each view. These values may then be combined to form Ω, and the model is used to compute posterior estimates.

Idzorek includes τ in the formulas, but because of the use of the Alternative Reference Model and his formula (3.123), there is no need to use τ with the Idzorek method.

The paper discussed solving for ω using a least-squares method. One may actually solve this analytically. The next section will provide a derivation of the formulas required for this solution.

First one may use the following form of the uncertainty of the views. Idzorek includes τ in this formula, but he uses the Alternative Reference Model so that one may drop τ from the formulation of his method:

$$\Omega = \alpha P \Sigma P^T \tag{3.124}$$

where α, the coefficient of uncertainty, is a scalar quantity in the interval $[0, \infty]$.

- When the investor is 100% confident in their views, then α will be 0.
- When they are totally uncertain, then α will be ∞.

Note that (3.124) is exact, it is identical to Equation (3.122) and the Ω used by He and Litterman (1999) because it is a 1×1 matrix. This allows one to find a closed-form solution to the problem of Ω for Idzorek's confidence.

First one substitutes (3.119)

$$\underline{\Pi} = \Pi + \{1/(1+\alpha)\}[P^{-1}Q - \Pi] \tag{3.119}$$

into

$$\underline{\Pi} = \delta\Sigma\underline{w}, \tag{3.83}$$

or

$$\underline{w} = \Pi(\delta\Sigma)^{-1}$$

yielding

$$\underline{w} = \{\Pi + \{1/(1+\alpha)\}[P^{-1}Q - \Pi]\}(\delta\Sigma)^{-1} \tag{3.125}$$

Now one may solve (3.125) at the boundary conditions for α:

$$\lim_{\alpha \to \infty} w_{\text{mkt}} = \Pi(\delta\Sigma)^{-1} \tag{3.126}$$

$$\lim_{\alpha \to 0} w_{100} = P^{-1}Q(\delta\Sigma)^{-1} \tag{3.127}$$

And recombining some of the terms in (3.125), one arrives at

$$\underline{w} = \Pi(\delta\Sigma)^{-1} + [1/(1+\alpha)][P^{-1}Q(\delta\Sigma)^{-1} - \Pi(\delta\Sigma)^{-1}] \tag{3.128}$$

Substituting w_{mkt} and w_{100}, from (3.126) and (3.127) respectively, into (3.128), the result is

$$\underline{w} = w_{\text{mkt}} + [1/(1+\alpha)][w_{100} - w_{\text{mkt}}] \tag{3.129}$$

And comparing the above with (3.123):

$$\text{confidence} = (\underline{w} - w_{\text{mkt}})/(w_{100} - w_{\text{mkt}}) \tag{3.123}$$

It is seen that

$$\text{confidence} = 1/(1+\alpha) \tag{3.130}$$

However, if one solves for α from (3.130), one obtains

$$\alpha = (1 - \text{confidence})/\text{confidence} \tag{3.131}$$

Using (3.131) and (3.124), the investor can easily calculate the value of ω for each view, and then roll them up into a single Ω matrix. To check the results for each view, one then solve for the posterior estimated returns using (3.105) and plug them back into (3.123). When the investors apply all their views at once, the interaction among the views may pull the posterior estimate for individual assets away from the results generated when the views were taken one at a time.

Idzorek's method greatly simplified the investor's process of specifying the uncertainty in the views when the investor does not have a quantitative model driving the process. In addition, this model does not add meaningful complexity to the process.

An Example of Idzorek's Extension   Idzorek described the steps required to implement his extension in his paper, but did not provide a complete worked example. Here, one will work through the example from where Idzorek left off! Idzorek's example includes three views:

- International Development Equity will have absolute excess return of 5.25%, Confidence 25.0%.
- International Bonds will outperform U.S. bonds by 25 bps, Confidence 50.0%.
- U.S. Growth Equity will outperform U.S. Value Equity by 2%, Confidence 65.0%.

Idzorek defined the steps that include calculations of w_{100} and then the calculation of ω for each view given the desired change in the weights. From the previous section, one noted that one only need to take the investor's confidence for each view, plug it into (3.131), and calculate the value of alpha. Then one plugs α, P, and $\sum$ into Equation (3.124) and compute the value of ω for each view. At this point, one may assemble the Ω matrix and proceed to solve for the posterior returns using (3.104) or (3.105).

In presenting this example, the results for each view will be shown including w_{mkt} and w_{100} in order to make the working of the extension more transparent. Tables 3.8–3.10 show the results for a single view.

Then one may use the freshly computed values for the Ω matrix with all views specified together and arrive at the final result shown in Table 3.11 blending all three views together.

Table 3.8   Calibrated results for View 1.

Asset	ω	w_{mkt}	w^*	w 100%	Implied confidence
Intl Dev Equity	0.002126625	24.18%	25.46%	29.28%	25.00%

Table 3.9   Calibrated results for View 2.

Asset	ω	w_{mkt}	w^*	w 100%	Implied confidence
US Bonds	0.000140650	19.34%	29.06%	38.78%	50.00%
Intl Bonds	0.000140650	26.13%	16.41%	6.69%	50.00%

Table 3.10   Calibrated results for View 3.

Asset	ω	w_{mkt}	w^*	w 100%	Implied confidence
US LG	0.000466108	12.09%	9.49%	8.09%	65.00%
US LV	0.000466108	12.09%	14.69%	16.09%	65.00%
US SG	0.000466108	1.34%	1.05%	0.90%	65.00%
US SV	0.000466108	1.34%	1.63%	1.78%	65.00%

Table 3.11   Final results for Idzorek's confidence extension example.

Asset	View 1	View 2	View 3	μ	σ	w_{mkt}	Posterior weight	Change
US Bonds	0.0	−1.0	0.0	0.1	3.2	19.3%	29.6%	10.3%
Intl Bonds	0.0	1.0	0.0	0.5	8.5	26.1%	15.8%	10.3%
US LG	0.0	0.0	0.9	6.3	24.5	12.1%	8.9%	3.2%
US LV	0.0	0.0	−0.9	4.2	17.2	12.1%	15.2%	3.2%
US SG	0.0	0.0	0.1	7.3	32.0	1.3%	1.0%	−0.4%
US SV	0.0	0.0	−0.1	3.8	17.9	1.3%	1.7%	0.4%
Intl Dev	1.0	0.0	0.0	4.8	16.8	24.2%	26.0%	1.8%
Intl Eng	0.0	0.0	0.0	6.6	28.3	3.5%	3.5%	−0%
Total							101.8%	
Return	5.2	2	2.0					
Omega/Tau	0.08507	0.00563	0.01864					
Lambda	0.002	−0.006	−0.002					

Impact of the Several Views   Consider the methods used in the literature to measure the impact of the views on the posterior distribution. In general, one may divide these measures into two groups:

1) The first group allows one to test the hypothesis that the views or posterior contradict the prior.
2) The second group allows one to measure a distance or information content between the prior and posterior.

Theil (1971) and Fusai and Meucci (2003) described measures that are designed to allow a hypothesis test to ensure the views or the posterior does not contradict the prior estimates. Theil (1971) described a method of performing a hypothesis test to verify that the views are compatible with the prior. That work is being extended to measure compatibility of the posterior and the prior. Fusai and Meucci (2003) described a method for testing the compatibility of the posterior and prior when using the alternative reference model. He and Litterman (1999) and Braga and Natale (2007) described measures that may be used to measure the distance between two distributions, or the amount of tilt between the prior and the posterior. These measures did not lend themselves to hypothesis testing, but they may be used as constraints on the optimization process. He and Litterman (1999) define a metric, Λ, which measured the tilt induced in the posterior by each view. Braga and Natale (2007) used Tracking Error Volatility (TEV) to measure the distance from the prior to the posterior.

Theil's Measure of Compatibility Between the Views and the Prior Theil (1971) described this as testing the compatibility of the views with the prior information. Given the linear mixed estimation model, one has the prior (3.132) and the conditional (3.133):

$$x\beta = \pi + u \tag{3.132}$$

$$p\beta = q + v \tag{3.133}$$

The mixed estimation model defines u as a random vector with mean 0 and covariance $\tau\sum$, and v as a random vector with mean 0 and covariance Ω.

The approach being taken here is very similar to the approach taken when analyzing a prediction from a linear regression. One has two estimates of the views, the conditional estimate of the view and the posterior estimate of the returns.

One may define the posterior estimate as $\underline{\beta}$. One may measure the estimation error between the prior and the views as

$$\zeta = (x\underline{\beta} - \mu) = x(\underline{\beta} - \beta) + u \tag{3.134}$$

The vector ζ has mean 0 and variance $V(\zeta)$. One shall form one's own hypothesis test using the formulation

$$\xi = E(\zeta)V(\zeta)^{-1}E(\zeta) \tag{3.135}$$

The quantity ξ is known as the Mahalanobis distance (multidimensional analog of the z-score) and is distributed as $\chi^2(n)$. In order to use this form, one needs to solve for the $E(\zeta)$ and $V(\zeta)$.

If one considers only the information in the views, the estimator of β is

$$\underline{\beta} = (P^T\Omega^{-1}P)^{-1}P^T\Omega^{-1}Q \tag{3.136}$$

Note that since P is not required to be a matrix of full rank, one might not be able to evaluate this formula as written. One works in return space here (as opposed to view space) as it seems more natural. Later on one will transform the formula into view space to make it computable.

One then substitutes the new estimator into the formula (3.134) and eliminate x as it is the identity matrix in the Black–Litterman application of mixed estimation:

$$\zeta = (P^T\Omega^{-1}P)^{-1}P^T\Omega^{-1}Q - \beta + u \tag{3.137}$$

Next one substitutes formula (3.133) for Q.

$$\zeta = (P^T\Omega^{-1}P)^{-1}P^T\Omega^{-1}(P\beta + v) - \beta + u \tag{3.138}$$

$$\zeta = (P^T\Omega^{-1}P)^{-1}(P^T\Omega^{-1}P)\beta + (P^T\Omega^{-1}P)^{-1}P^T\Omega^{-1}v - \beta + u \tag{3.139}$$

$$\zeta = (P^T\Omega^{-1}P)^{-1}P^T\Omega^{-1}v + u \tag{3.140}$$

Given the estimator, one needs to find the variance of the estimator, as follows:

$$V(\zeta) = E(\zeta\,\zeta^T) \tag{3.141}$$

$$V(\zeta) = E[(P^T\Omega^{-1}P)^{-1}P^T\Omega^{-1}v\,v^T\Omega^{-1}P(P^T\Omega^{-1}P)^{-1}$$
$$-2(P^T\Omega^{-1}P)^{-1}P^T\Omega^{-1}v\,u^T + uu^T] \tag{3.142}$$

Now, $E(vu) = 0$, so one may eliminate the cross-term, and simplify the formula:

$$V(\zeta) = E[(P^T\Omega^{-1}P)^{-1}P^T\Omega^{-1}vv^T\Omega^{-1}P(P^T\Omega^{-1}P)^{-1} + uu^T] \tag{3.143}$$

$$V(\zeta) = E[(P)^{-1}vv^T(P^T)^{-1} + uu^T] \tag{3.144}$$

$$V(\zeta) = (P^T\Omega^{-1}P)^{-1} + \tau\Sigma \tag{3.145}$$

The last step is to take the expectation of (3.137). At the same time, one will substitute the posterior estimate (μ) for β.

$$E(\zeta) = (P^T\Omega^{-1}P)^{-1}P^T\Omega^{-1}Q - \Pi \tag{3.146}$$

Now substitute the various values into (3.135) as follows:

$$\xi = \{(P^T\Omega^{-1}P)^{-1}P^T\Omega^{-1}Q - \Pi\}[(P^T\Omega^{-1}P)^{-1}$$
$$+\tau\Sigma]^{-1}\{(P^T\Omega^{-1}P)^{-1}P^T\Omega^{-1}Q - \Pi\}^T \tag{3.147}$$

Unfortunately, under the usual circumstances, one cannot compute ξ. Because P does not need to contain a view on every asset, several of the terms are not always computable as written. However, one may easily convert it to view space by multiplying by P and P^T.

$$\xi = (Q - P\Pi)[\Omega + P\tau\Sigma P^T]^{-1}(Q - P\Pi)^T \tag{3.148}$$

This new test statistic ξ in (3.148) is distributed as $\chi^2(q)$, where q is the number of views. It is the square of the Mahalanobis distance of the posterior return estimate versus the posterior covariance of the estimate. One may use this test statistic to determine if the views are consistent with the prior by means of a standard confidence test:

$$P(q) = 1 - F\{\xi(q)\} \tag{3.149}$$

where $F(\xi)$ is the CDF of the $\chi^2(q)$ distribution.

One may also calculate the sensitivities of this measure to the views using the chain rule:

$$\partial P/\partial q = (\partial P/\partial\xi)(\partial\xi/\partial q) \tag{3.150}$$

Substituting the various terms,

$$\partial P/\partial q = -f(\xi)[2((\Omega + P\tau\Sigma P^T)^{-1}(Q - P\Pi)^T)] \tag{3.151}$$

where $f(\xi)$ is the PDF of the $\chi^2(q)$ distribution.

The Theil Measure of the Source of Posterior Information   Theil (1963) describes a measure that may be used to determine the contribution to the posterior precision of the prior and the views. This measure was called θ sums to 1 across all sources, and conveniently also sums across the views if one measures the contribution of each view.

The measure for the contribution to posterior precision from the prior is

$$\theta_{\text{prior}} = (1/n)\text{tr}\{(\tau\Sigma)^{-1}[(\tau\Sigma)^{-1} + P^T\Omega^{-1}P]^{-1}\} \tag{3.152}$$

where n is the number of assets. Equation (3.153) can be used for all views by using the full matrices P and Ω. For a single view i, use the relevant slices of P and Ω:

$$\theta_i = (1/n)\text{tr}((P_i^T O_{i,j}^{-1} P_i)[(\tau\Sigma)^{-1} + P^T\Omega^{-1}P]^{-1} \tag{3.153}$$

Equations (3.152) and (3.153) provide the equations one may use to compute the contribution toward the posterior precision from both the prior and the views. One may use this diagnostic to identify if the relative contributions match the intuitive view on this proportion.

Fusai and Meucci's Measure of Consistency   Next consider the work of Fusai and Meucci (2003). In their paper they present a way to quantify the statistical difference between the posterior return estimates and the prior estimates. This provided a way to calibrate the uncertainty of the views and ensured that the posterior estimates are not extreme when viewed in the context of the prior equilibrium estimates.

In this paper they used the Alternative Reference Model. The measure is analogous to Theil's Measure of Compatibility, but because the alternative reference model used the prior variance of returns for the posterior, they did not need any derivation of the variance. One may apply a variant of their measure to the Canonical Reference Model as well. They proposed the use of the squared Mahalanobis distance of the posterior returns from the prior returns. This includes τ to match the Canonical Reference Model, but their work did not include τ as they used the Alternative Reference Model:

$$M(q) = (\mu_{BL} - \mu)(\tau\Sigma)^{-1}(\mu_{BL} - \mu) \tag{3.154}$$

It is essentially measuring the distance from the prior, μ, to the estimated returns, μ_{BL}, normalized by the uncertainty in the estimate. One may use the covariance matrix of the prior distribution as the uncertainty. The squared Mahalanobis distance is distributed as $\chi^2(q)$ where q is the number of assets. This may easily be used in a hypothesis test. Thus, the probability of this event occurring can be computed as

$$P(q) = 1 - F(M(q)) \tag{3.155}$$

where $F(M(q))$ is the CDF of the chi-square distribution of $M(q)$ with n degrees of freedom.

Finally, to identify which views contribute most highly to the distance away from the equilibrium, one may also compute sensitivities of the probability to each view. Use the chain rule to calculate the partial derivatives

$$\partial P(q)/\partial q = (\partial P/\partial M)(\partial M/\partial \mu_{BL})(\partial \mu_{BL}/\partial q) \tag{3.156}$$

$$\partial P(q)/\partial q = -f(M)[2(\mu_{BL} - \mu)][(P(\tau\Sigma)P + \Omega)^{-1}P] \tag{3.157}$$

where $f(M)$ is the PDF of the chi-square distribution with n degrees of freedom for $M(q)$. Note that this measure is very similar to (3.151) Theil's measure of compatibility between the prior and the views.

An example in their paper resulted in an initial probability of 94% that the posterior is consistent with the prior. They specified that their investor desires this probability to be no less than 95% (a commonly used confidence level in hypothesis testing), and thus they would adjust their views to bring the probability in line. Given that they also computed sensitivities, their investor can identify that views are providing the largest marginal increase in their measure and the investor may then adjust these views. These sensitivities are especially useful since some views may actually be pulling the posterior toward the prior, and the investor could strengthen these views or weaken views that pull the posterior away from the prior. The last point may seem nonintuitive. Given that the views are indirectly coupled by the covariance matrix, one would expect that the views only push the posterior distribution away from the prior. However, because the views can be conflicting, either directly or via the correlations, any individual view may have a net impact pushing the posterior closer to the prior, or pushing it further away.

They proposed to use their measure in an iterative method to ensure that the posterior is consistent with the prior to the specified confidence level. With the Canonical Reference Model, one could rewrite (3.154) using the posterior variance of the return instead of $\tau\Sigma$ yielding

$$M(q) = (\mu_{BL} - \mu)((\tau\Sigma)^{-1} + P^T\Omega^{-1}P)^{-1}(\mu_{BL} - \mu) \tag{3.158}$$

Otherwise, their Consistency Measure and its use is the same for both reference models.

He and Litterman Lambda He and Litterman (1999) used a measure, Λ, to measure the impact of each view on the posterior. They defined the Black–Litterman unconstrained posterior portfolio as a blend of the equilibrium portfolio (prior) and a contribution from each view; that contribution is measured by Λ.

Deriving the formula for Λ, one will start with (3.86b) and substitute in the various values from the posterior distribution

$$w = (\delta\Sigma)^{-1}\Pi \tag{3.86b}$$

and also substitute the return for $\prod$:

$$\underline{w} = (1/\delta)\ \underline{\Sigma}^{-1}M^{-1}[(\tau\Sigma)^{-1}\Pi + P^T\Omega^{-1}Q] \tag{3.159}$$

The covariance term will first be simplified:

$$\underline{\Sigma}^{-1}M^{-1} = (\Sigma + M^{-1})^{-1}M^{-1} \tag{3.160a}$$

$$\underline{\Sigma}^{-1}M^{-1} = (\Sigma M + I)^{-1} \tag{3.160b}$$

$$\underline{\Sigma}^{-1}M^{-1} = \Sigma^{-1}(\Sigma^{-1} + M)^{-1} \tag{3.160c}$$

$$\underline{\Sigma}^{-1}M^{-1} = \Sigma^{-1}[\Sigma^{-1} + (\tau\Sigma)^{-1} + P\Omega^{-1}P^T]^{-1} \tag{3.160d}$$

$$\underline{\Sigma}^{-1}M^{-1} = \Sigma^{-1}[\{(1+\tau)/\tau\}\Sigma^{-1} + P\Omega^{-1}P^T]^{-1} \tag{3.160e}$$

$$\underline{\Sigma}^{-1}M^{-1} = (P^T\tau\Sigma P)^{-1}[\{(1+\tau)(P^T\Sigma P)^{-1} + \tau\Omega^{-1}]^{-1} \tag{3.160f}$$

$$\underline{\Sigma}^{-1}M^{-1}(P^T\tau\Sigma P)^{-1}[\{(P^T\Sigma P)/(1+\tau)\} - \{(P^T\Sigma P)/(1+\tau)\}$$
$$\{(P^T\Sigma P)/(1+\tau)\}\{(\Omega/r) + (P^T\Sigma P)/(1+\tau)\}] \tag{3.160g}$$

$$\underline{\Sigma}^{-1}M^{-1} = \{\tau/(1+\tau)\}[I - \{(P^T\Sigma P)/(1+\tau)\}\{(\Omega/\tau)$$
$$+(P^T\Sigma P)/(1+\tau)\}^{-1}] \tag{3.160h}$$

Then, one may define

$$A = [(\Omega/\tau) + \{(P^T\Sigma P)/(1+\tau)\}] \tag{3.161}$$

And finally rewrite as

$$\underline{\Sigma}^{-1}M^{-1} = \{\tau/(1+\tau)\}[I - [(P^T A^{-1} P)\{\Sigma/(1+\tau)\}]] \qquad (3.162)$$

One uses the multiplier, $\{\tau/(1+\tau)\}$, because in the Black–Litterman Reference Model the investor is not fully invested in the prior (equilibrium) portfolio.

To find Λ, one may substitute (3.162) into (3.159), and then gather terms:

$$\underline{w} = (1/\delta)\{\tau/(1+\tau)[I - (P^T A^{-1} P)\{\Sigma/(1+\tau)\}][(\tau\Sigma)^{-1}\Pi + P^T \Omega^{-1} Q] \qquad (3.163)$$

$$\underline{w} = (1/\delta)\{\tau/(1+\tau)[(\tau\Sigma)^{-1}\Pi - (P^T A^{-1} P)\{\tau/(1+\tau)\Pi + P^T \Omega^{-1} Q$$
$$- (P^T A^{-1} P)\{\Sigma/(1+\tau)\}P^T \Omega^{-1} Q\}] \qquad (3.164)$$

$$\underline{w} = \{1/(1+\tau)\}\{w_{eq} + P^T[-A^{-1}P\Pi\{1/[\delta(1+\tau)]\} + (\tau\Omega^{-1}Q/\delta)$$
$$- (A^{-1}P)\{\Sigma/\delta(1+\tau)\}\}P^T \Omega^{-1} Q\}] \qquad (3.165)$$

$$\underline{w} = (\{1/(1+\tau)\}\{w_{eq} + P^T[(\tau/\delta)\Omega^{-1}Q - \{A^{-1}P\Sigma w_{eq}/(1+\tau)\}$$
$$- A^{-1}\{\tau(1+\tau)\}(P^T \Sigma P)(\Omega^{-1}Q/\delta)]\} \qquad (3.166)$$

Thus, along with (3.161), the following equation defines Λ:

$$\Lambda = (\tau/\delta)\Omega^{-1}Q\{A^{-1}P\Sigma w_{eq}/(1+\tau)\}$$
$$- A^{-1}\{\tau/(1+\tau)\}(P\Sigma P^T)(\Omega^{-1}Q/\delta) \qquad (3.167)$$

Λ, of He and Litterman, represents the weight on each of the view portfolios on the final posterior weight. As a result, one may use Λ as a measure of the impact of one's views.

One may also derive the He and Litterman Λ for the Alternative Reference Model: Call this Λ_A. One may start from the same reverse optimization equation3.86b:

$$w = (\delta\Sigma)^{-1}\Pi \qquad (3.86b)$$

and substitute the return from (3.104) for $\prod$,

$$P(A|B) \sim N\{[(\tau\Sigma)^{-1}\Pi + P^T \Omega^{-1} Q][(\tau\Sigma)^{-1} + P^T \Omega^{-1} P]^{-1}, \quad [(\tau\Sigma)^{-1} + P^T \Omega^{-1} P]^{-1}\} \qquad (3.104)$$

but using the prior covariance matrix as 1 is using the Alternative Reference Model.

$$\underline{w} = (1/\delta)\Sigma^{-1}[(\Sigma)^{-1} + P^T \Omega^{-1} P]^{-1}[\Sigma^{-1}\Pi + P^T \Omega^{-1} Q] \qquad (3.168)$$

First simplify the covariance term using the Woodbury Matrix Identity.

$$\Sigma^{-1}[(\Sigma)^{-1} + P^T \Omega^{-1} P]^{-1} = I^{-1} - I^{-1}P^T(\Omega + P\Sigma I^{-1}P^T)^{-1}P\Sigma I^{-1} \qquad (3.168a)$$
$$= I - P^T(\Omega + P\Sigma P^T)^{-1}P\Sigma \qquad (3.168b)$$

Then, one may define

$$A_A = \Omega + P^T \Sigma P \tag{3.169}$$

and one may substitute the above result back into (3.128) and expand the terms:

$$\underline{w} = \Pi(\delta\Sigma)^{-1} + [1/(1+\alpha)][P^{-1}Q(\delta\Sigma)^{-1} - \Pi(\delta\Sigma)^{-1}] \tag{3.128}$$

$$\underline{w} = (\delta)^{-1}(I - P^T A_A^{-1} P\Sigma)[\Sigma^{-1}\Pi + P^T \Omega^{-1} Q] \tag{3.170a}$$

$$\underline{w} = (\delta)^{-1}(\Sigma^{-1}\Pi + P^T \Omega^{-1} Q - P^T A_A^{-1} P\Pi - P^T A_A^{-1} P\Sigma P^T \Omega^{-1} Q) \tag{3.170b}$$

$$\underline{w} = \Sigma^{-1}\Pi(\delta)^{-1} + (1/\delta)P^T[(\Omega^{-1}Q - A_A^{-1}P\Pi - A_A^{-1}P\Sigma P^T \Omega^{-1} Q)] \tag{3.170c}$$

$$\underline{w} = \Sigma^{-1}\Pi(\delta)^{-1} + (1/\delta)P^T[(\Omega^{-1}Q - A_A^{-1}P\delta\Sigma w_{eq} - A_A^{-1}P\Sigma P^T \Omega^{-1} Q)] \tag{3.170d}$$

In He and Litterman, Λ took the similar form in the Alternative Reference Model as shown here:

$$\underline{w} = w_{eq} + P^T \Lambda_A \tag{3.171}$$

Note that in the Alternative Reference Model, the investor's prior portfolio has the exact same weights as the equilibrium portfolio.

One may note that the following formula defines Λ_A:

$$\Lambda_A = (\Omega^{-1}Q/\delta) - A_A^{-1}P\Sigma w_{eq} - A_A^{-1}P\Sigma P^T(\Omega^{-1}Q/\delta) \tag{3.172}$$

Braga and Natale and Tracking Error Volatility (*TEV*) Braga and Natale (2007) proposed the use of tracking error between the posterior and prior portfolios as a measure of distance from the prior. Tracking error is commonly used by investors to measure risk versus a benchmark, and may be used as an investment constraint. Since it is commonly used, most investors have an intuitive understanding and a level of comfort with *TEV*.

Tracking Error Volatility (TEV) is here defined as

$$\text{TEV} = \sqrt{(w_{actv}^T \Sigma w_{actv})}, \quad \text{where } w_{actv} = \underline{w} - w_r \tag{3.173}$$

and where

w_{actv} is the active weights or active portfolio,
$\underline{w}$ is the weight in the investor's portfolio,
w_r is the weight in the reference portfolio, and
$\sum$ is the covariance matrix of returns.

Moreover, the formula derived for tracking error sensitivities is as follows: Given that

$$\text{TEV} = f(w_{actv}) \tag{3.174}$$

and one may further refine

$$w_{actv} = g(q) \tag{3.175}$$

where q represents the views.

Then one may use the chain rule to decompose the sensitivity of TEV to the views:

$$\partial \text{TEV}/\partial q = (\partial \text{TEV}/\partial w_{\text{actv}})(\partial w_{\text{actv}}/\partial q) \tag{3.176}$$

Solving for the *first* term of (3.176) directly,

$$\partial \text{TEV}/\partial w_{\text{actv}} = \partial[\sqrt{(w_{\text{actv}}^T \Sigma w_{\text{actv}})}]/\partial w_{\text{actv}} \tag{3.177}$$

Let $x = w_{\text{actv}}^T \Sigma w_{\text{actv}}$, then applying the chain rule to solve for the *first* term of (3.176):

$$
\begin{aligned}
(\partial \text{TEV}/\partial w_{\text{actv}}) &= (\partial \text{TEV}/\partial x)(\partial x/\partial w_{\text{actv}}) \\
&= [1/\{2\sqrt{(w_{\text{actv}}^T \Sigma w_{\text{actv}})}\}][2\Sigma w_{\text{actv}}] \\
&= \Sigma w_{\text{actv}}/\sqrt{(w_{\text{actv}}^T \Sigma w_{\text{actv}})}
\end{aligned} \tag{3.178}
$$

And similarly solve for the *second* term of (3.176):

$$
\begin{aligned}
(\partial w_{\text{actv}}/\partial q) &= \partial\{\underline{w} - w_{\text{ref}}\}/\partial q \\
&= \partial\{(\delta\Sigma)^{-1}E(r) - (\delta\Sigma)^{-1}\Pi\}/\partial q \\
&= (\delta\Sigma)^{-1}[\partial\{E(r) - \Pi\}/\partial q] \\
&= (\delta\Sigma)^{-1}[\partial\{\tau\Sigma P^T\{(P\tau\Sigma P^T) + \Omega\}^{-1}[Q - P\Pi]/\partial Q] \\
&= (\delta\Sigma)^{-1}\tau\Sigma P^T[(P\tau\Sigma P^T) + \Omega]^{-1} \\
&= (1/\delta)P^T[(P\Sigma P^T) + (\Omega/\tau)]^{-1}
\end{aligned} \tag{3.179}
$$

This result differs somewhat from that found in the Braga and Natale (2007) paper because the form of the Black–Litterman model is used that requires less matrix inversions. The sensitivities equation is

$$\partial \text{TEV}/\partial q = \{\Sigma w_{\text{actv}}/(\sqrt{(w_{\text{actv}}^T \Sigma w_{\text{actv}})})\}(1/\delta)P^T[(P\Sigma P^T) + (\Omega/\tau)]^{-1} \tag{3.180}$$

One may come up with the equivalent metric for the Canonical Reference Model. Notice that in a tracking error scenario, the covariance matrix should be the most accurate, which would be the posterior covariance matrix:

$$\partial \text{TEV}/\partial q = [\underline{\Sigma} w_{\text{actv}}/\{\sqrt{(w_{\text{actv}}^T \underline{\Sigma} w_{\text{actv}})}\}](\partial w_{\text{actv}}/\partial q) \tag{3.181}$$

$$\partial \text{TEV}/\partial q = [\underline{\Sigma} w_{\text{actv}}/\{\sqrt{(w_{\text{actv}}^T \underline{\Sigma} w_{\text{actv}})}\}]\{P^T/(1 + \tau)\}(\partial \Lambda/\partial q) \tag{3.182}$$

yielding the TEV sensitivities for the Canonical Reference Model:

$$
\begin{aligned}
\partial \text{TEV}/\partial q = [\underline{\Sigma} w_{\text{actv}}/\{\sqrt{(w_{\text{actv}}^T \underline{\Sigma} w_{\text{actv}})}\}]\{P^T/(1 + \tau)\} \\
[(\tau/\delta)\Omega^{-1}] - A^{-1}\{\tau/(1 + \tau)\}(P\Sigma P^T)(\Omega^{-1}/\delta)]
\end{aligned} \tag{3.183}
$$

Braga and Natale provided an example in their paper, but they did not provide all the raw data required to reproduce their results. Given their posterior distribution as presented in the paper, one should easily reproduce their TEV results. One advantage of the TEV is that most investors are familiar with it, and so they will have some intuition as to what it represents. The consistency metric introduced by Fusai and Meucci (2003) will not be as familiar to investors.

Metrics Introduced in Herold (2003) This diagnostic concept may be applied to portfolio models in order to validate the outputs. One of these diagnostics is the correlation between views as one diagnostic that may be used to determine how the updated portfolio may perform. By examining the correlation matrix of the views, $P \sum P^t$, one may determine how correlated the views are. If the views are highly correlated, then one may expect all views to contribute to the performance if they are correct, and underperformance if they are incorrect. If the views are not highly correlated, then one may expect a diversified contribution to performance.

As an example, if one uses the He and Litterman's data and their two views, one calculated this measure as

$$P\Sigma P^t = [0.0213 \quad 0.0020]$$
$$= [0.0020 \quad 0.0170]$$

The off-diagonal elements (0.0020 and 0.0020) are one order of magnitude *smaller* than the on-diagonal elements (0.0213 and 0.0170). This difference indicates that the views are *not* strongly correlated. This is consistent with He and Litterman's specification of the two views on mutually exclusive sets of assets that are loosely correlated.

Herold (2003) also discussed the marginal contribution to tracking error by view. This is the same metric computed by Braga and Natale, although he proposed deriving the formula for its calculation from the Alternative Reference Model version of He and Litterman's Λ, Λ_A.

Herold (2003) considers the case of Active Management, which meant that the Black–Litterman model was applied to an active management overlay on some portfolio. The prior distribution in this case corresponds to $w_{\text{eq}} = 0$. The posterior weights were the overlay weights. He uses the Alternative Reference Model and computed a quantity Φ, which is the active management version of Λ_A.

$$\Lambda_A = (\Omega^{-1}Q/\delta) - A_A^{-1}P\Sigma w_{\text{eq}} - A_A^{-1}P\Sigma P^T(\Omega^{-1}Q/\delta) \tag{3.184}$$

but $w_{\text{eq}} = 0$, so

$$\Phi = (\Omega^{-1}Q/\delta) - A_A^{-1}P\Sigma P^T(\Omega^{-1}Q/\delta) \tag{3.185}$$

where

$$w_{\text{actv}} = P^T\Phi \tag{3.186}$$

Herold starts from Equation (3.173) just as Braga and Natale, but then uses an alternative formula for tracking error:

$$\text{TEV} = \sqrt{(w_{\text{actv}}^T \Sigma w_{\text{actv}})}, \quad \text{where } w_{\text{actv}} = \underline{w} - w_r \tag{3.173}$$

and also

$$w_{\text{actv}} = \underline{w} - w_{\text{eq}} = P^T \Phi \tag{3.187}$$

$$\text{TEV} = \sqrt{(\Phi^T P \Sigma P^T \Phi)} \tag{3.188}$$

Then one may take $\delta\text{TEV}/\delta Q$ to find the marginal contribution to tracking error by views. By the chain rule:

$$\begin{aligned}
&\delta\text{TEV}/\delta Q \\
&= (\delta\text{TEV}/\delta\Phi) \times (\delta\Phi/\delta Q), \quad \text{by the chain rule} \\
&= \{(P\Sigma P^T \Phi)/\sqrt{(\Phi P \Sigma P^T \Phi)}\} \times \{(1/\delta)[\Omega^{-1} - (P\Sigma P^T + \Omega)^{-1} P\Sigma P^T \Omega^{-1}\}
\end{aligned} \tag{3.189}$$

Note that one could perform the same calculation for the Canonical Reference Model using He and Litterman's Λ, but one would also need to use the posterior covariance of the distribution rather than just Σ in the calculations.

A Demonstration of the Measures   Consider a sample problem to illustrate all of the metrics, and to provide some comparison of their features. One may start with the equilibrium from He and Litterman (1999) and for Example 1 use the views from their paper; Germany will outperform other European markets by 5% and Canada will outperform the United States by 4%.

Table 3.12 illustrates the results of applying the views and Table 3.13 displays the various impact measures.

Table 3.12   Example 1 returns and weights: equilibrium from He and Litterman (1999).

Asset	P_0	P_1	μ	μ_{eq}	$w_{\text{eq}}/(1+\tau)$	w^*	$w^* - w_{\text{eq}}/(1+\tau)$
Australia	0.0	0.0	4.45	3.9	1.50%	1.5%	0.0%
Canada	0.0	1.0	9.06	6.9	2.1%	53.3%	51.2%
France	−0.295	0.0	9.53	8.4	5.0%	−3.3%	−8.3%
Germany	1.0	0.0	11.3	9	5.2%	33.1%	27.9%
Japan	0.0	0.0	4.65	4.3	11.0%	11.0%	0.0%
UK	−0.705	0.0	6.98	6.8	11.8%	−7.8%	−19.6%
USA	0.0	−1.0	7.31	7.6	58.6%	7.3%	−51.3%
q	5.0	4.0					
ω/τ	0.02	0.017					

Table 3.13   Impact measures for Example 1.

Measure	Value (confidence level)	Sensitivity (V1)	Sensitivity (V2)
Theil's measure	1.672 (43.3%)	−5.988	−11.46
Theil's θ	0.858 (prior)	0.0712	0.0712
Fusai and Meucci's measure	0.8728 (99.7%)	−0.1838	−0.3327
Λ		0.2920	0.5380
TEV	8.28%	0.688	1.294

If one examines the change in the estimated returns versus the equilibrium, one sees where the U.S. returns decreased by only 29 bps, but the allocation decreased 51.25% caused by the optimizer favoring Canada whose returns increased by 216 bps and whose allocation increased by 51.25%. This shows that what appear to be moderate changes in return forecasts may cause very large swings in the weights of the assets, a common issue with unconstrained mean variance optimization. Here one uses unconstrained mean variance optimization not because one has to, but because it is well understood and transparent.

Looking at the impact measures, Theil's measure indicates that one may be confident only at the 43% level that the views are consistent with the prior. If one examines the diagram, one may see that indeed these views are significantly different from the prior. Fusai and Meucci's measure of compatibility of the prior and posterior, on the other hand, comes with a confidence level of 99.8%, so they are much more confident that the posterior is consistent with the prior. A major difference in their approaches is that Theil is working in view space, which for the example has order of 2, while Fusai and Meucci are working in asset space which has order of 7.

All of the metrics' sensitivities to the views indicate that the second view has a relative weight just about twice as large as the first view, so the second view contributes much more to the change in the weights.

The TEV of the posterior portfolio is 8.28%, which is significant in terms of how closely the posterior portfolio will track the equilibrium portfolio. This seems to be a very large TEV value given the scenario with which one is working. Next, one changes one's confidence in the views by multiplying the variance by 2, this will increase the change from the prior to the posterior and allow one to make some judgments based on the impact measures.

Examining the updated results in Table 3.14, one sees that the changes to the forecast returns have decreased—consistent with the increased uncertainty in the views. One now has a 20% increase in the allocation to Canada and a 20% decrease in the allocation to the United States. From Table 3.15, one may see that Theil's measure has increased, but only by a fraction and one continues to

Table 3.14 Example 2 returns and weights: equilibrium from He and Litterman, (1999).

Asset	P_0	P_1	μ	$w_{eq}/(1+\tau)$	w^*	$w^* - w_{eq}/(1+\tau)$
Australia	0.0	0.0	4.72	1.5%	1.5%	0.0%
Canada	0.0	1.0	10.3	2.1%	22.7%	20.6%
France	−0.295	0.0	10.2	5.0%	1.6%	−3.4%
Germany	1.0	0.0	12.4	5.2%	16.8%	11.6%
Japan	0.0	0.0	4.84	11.0%	11.0%	0.0%
UK	−0.705	0.0	7.09	11.8%	3.7%	−8.1%
USA	0.0	−1.0	7.14	58.6%	38.0%	−20.6%
q	5.0	4.0				
ω/τ	0.09	0.07				

have little confidence that the views are consistent with the prior. Fusai and Meucci's measure now is 99.8% confident that the posterior is consistent with the prior. It is unclear in practice what bound one would want to use, but their measure usually presents with a much higher confidence than Theil's measure. The TEV has decreased and is now 7.67%, which is not significantly smaller.

Once again all the sensitivities show the second view having more of an impact on the final weights.

Across both scenarios, one may draw some conclusions about these various measures. Theil's consistency measure test ranged from a high of 46% confident to a low of 44% confident that the views were consistent with the prior estimates.

This measure is very sensitive and it is unclear what would be a good confidence threshold. "50%" does seem intuitively appealing.

Table 3.15 Impact measures for Example 2.

Measure	Value (confidence level)	Sensitivity	Sensitivity
Theil's measure	1.537 (46.4%)	−4.27	−12.39
Theils θ	0.88	0.05	0.07
Fusai and Meucci's measure	0.7479 (99.8%)	−0.06	−0.24
Λ		0.193	0.544
TEV	7.67%	0.332	1.406

Fusai and Meucci's Consistency measure ranged from 98.38 to 99.98% confident, indicating that the posterior estimates was generally highly consistent with the prior. Fusai and Meucci present that an investor may have a requirement that the confidence level be 5%. In light of these results, that would seem to be a fairly large value. The sensitivities of the Consistency measure scale with the measure, and for low values of the measure the sensitivities are very low.

Theil's θ changed with the confidence of the views and generally indicated that much of the information in the posterior originated with the prior. It moved intuitively with the changes in confidence level.

Across the two scenarios, the TEV decreased by 61 bps, but against a starting point of 8.28%, it still indicates large active weights. It is not clear what a realistic threshold for the TEV is in this case, but these values are likely toward the upper limit that would be tolerated. Note that between the two scenarios, the sensitivity to the first view dropped by 50%, which is consistent with the change in confidence that we applied.

In analyzing these various measures of the tilt caused by the views, the TEV of the weights measures the impact of the views and the optimization process, which one may consider as the final outputs. If the investor is concerned about limits on TEV, they could be easily added as constraints on the optimization process.

He and Litterman's Lambda measures the weight of the view on the posterior weights, but only in the case of an unconstrained optimization. This makes it suitable for measuring impact and being a part of the process, but it cannot be used as a constraint in the optimization process. Theil's compatibility measure and Fusai and Meucci's consistency measure the posterior distribution, including the returns and the covariance matrix: the former in view space, the latter in asset space.

Active Management and the Black–Litterman Model   Active Management refers to the case when the investor is managing an overlay portfolio on top of a 100% invested passive benchmark portfolio. In this case, one is only interested in the overlay and not in the benchmark, so one starts with a prior distribution with 0 active weights and 0 expected excess returns over the benchmark. All returns to views input to the model are relative to the equilibrium benchmark returns rather than to the risk-free rate. The weights are for the active portfolio, so the weights should always sum to 0.

Herold (2003) discusses the application of the Black–Litterman model to the problem of Active Management. He introduces a measure Φ, which is He and Litterman's Λ_A modified for Active Management. When one uses the Black–Litterman model for Active Management versus a passive benchmark portfolio, then the equilibrium weights (w_{eq}) are 0, and thus the equilibrium returns ($\prod$) are also 0.

Because the value $\prod = 0$, the middle term in He and Litterman's Λ_A vanishes, and

$$\Phi = (1/\delta)[\Omega^{-1}Q - (P\Sigma P^T + \Omega)^{-1}P\Sigma P^T\Omega^{-1}Q] \tag{3.190}$$

Two-Factor Black–Litterman   Krishnan and Mains (2005) developed an extension to the alternative reference model that allows the inclusion of additional uncorrelated market factors. The main point involved was that the Black–Litterman model measures risk, like all MVO approaches, as the covariance of the assets. They advocated for a richer measure of risk. They specifically focus on a recession indicator, given the thesis that many investors want assets that perform well during recessions and thus there is a positive risk premium associated with holding assets that do poorly during recessions. Their approach is general and may be applied to one or more additional market factors given that the market has zero beta to the factor and the factor has a nonzero risk premium.

They started from the standard quadratic utility function, but add an additional term for the new market factors:

$$U = w^T\Pi - (\delta_0/2)w^T\Sigma w - \sum_{j=1}^{n} \delta_j w^T\beta_j \tag{3.191}$$

where

U is the investors utility, the objective function during portfolio optimization,
w is the vector of weights invested in each asset,
$\prod$ is the vector of equilibrium excess returns for each asset,
$\sum$ is the covariance matrix for the assets,
δ_0 is the risk aversion parameter of the market,
δj is the risk aversion parameter for the jth additional risk factor, and
β_j is the vector of exposures to the jth additional risk factor.

Given their utility function as shown in (3.191), one may take the first derivative with respect to w in order to solve for the equilibrium asset returns.

$$\prod_{j=1}^{n} = \delta_0\Sigma w + \Sigma\delta_j\beta j \tag{3.192}$$

Comparing this with (3.83)

$$\Pi = \delta\Sigma w \tag{3.83}$$

the simple reverse optimization formula, one sees that the equilibrium excess return vector ($\prod$) is a linear composition of (3.83) *and* a term linear in the β_j values. This matches the intuition as expect assets exposed to this extra factor to have additional return above the equilibrium return.

The following quantities will be further defined as follows:

r_m is the return of the market portfolio,
f_j is the time series of returns for the factor, and
r_j is the return of the replicating portfolio for risk factor j.

In order to compute the values of δ, one will need to perform a little more algebra. Given that the market has no exposure to the factor, one may find a weight vector v_j such that $v_j T \beta_j = 0$. In order to find v_j, one performs a least-squares fit of $\|f_j - v_j^T \Pi\|$ subject to the above constraint. v_0 will be the market portfolio and

$$v_0 \beta_j = 0, \quad \forall j \quad \text{by construction} \tag{3.193}$$

One may solve for the various values of δ by multiplying (3.192) by v and solving for δ_0:

$$v_0^T \Pi = \delta_0 v_0^T \Sigma v_0 + \sum_{j=1}^{n} \delta_j v_0^T \beta_j \tag{3.194}$$

By construction, $v_0 \beta_j = 0$ and $v_0 \prod = r_m$, so

$$\delta_0 = r_m / (v_0^T \Sigma v_0) \tag{3.195}$$

For any $j \geq 1$, one may multiply (3.192) by v_j and substitute δ_0 to obtain

$$v_j^T \Pi = \delta_0 v_j^T \Sigma v_j + \sum_{i=1}^{n} \delta_i v_j^T \beta_i \tag{3.196}$$

As these factors must all be independent and uncorrelated,

$$v_i \beta_j = 0, \quad \forall i \neq j \tag{3.197}$$

so that one may solve for each δ_j given by

$$\delta_j = (r_j - \delta_i v_j^T \Sigma_{vj}) / (v_i^T \beta_j) \tag{3.198}$$

It was indicated that this is only an approximation because the quantity $\|f_j - v_j^T \Pi\|$ may not be identical to 0. The assertion that $v_i \beta_j = 0 \ \forall \ i \neq j$ may also not be satisfied for all i and j. For the case of a single additional factor, one can ignore the latter issue.

In order to transform these formulas so that one may directly use the Black–Litterman model, Krishnan and Mains changed variables, letting

$$\underline{\Pi} = \Pi - \sum_{j=1}^{n} \delta_j \beta_j \tag{3.199}$$

Substituting back into (3.191), one is back to the standard utility function:

$$U = w^T \underline{\Pi} - (\delta_0/2) w^T \Sigma w \tag{3.200}$$

and from (3.85):

$$P(B|A) \sim N(P^{-1}Q, \quad [P^T\Omega^{-1}P]^{-1}) \tag{3.85}$$

one has

$$P\underline{\Pi} = P\left(\Pi - \sum_{j=1}^{n} \delta_i \beta_j\right) \tag{3.201}$$

$$P\underline{\Pi} = P\Pi - \sum_{j=1}^{n} \delta_i P\beta_j \tag{3.202}$$

and therefore

$$\underline{Q} = Q - \sum_{j=1}^{n} \delta_i P\beta_j \tag{3.203}$$

Given the additional factors, one may then directly substitute $\underline{\Pi}$ and $\underline{Q}$ into (3.106) for the posterior returns in the Black–Litterman model in order to calculate the returns:

$$\underline{\Pi} = \Pi + \tau\Sigma P^T[(P\tau\Sigma P^T) + (\Omega/\tau)]^{-1}[Q - P\Pi] \tag{3.106}$$

Note that the additional factors do not impact the posterior variance in any way.

Krishnan and Mains offered an example of their model for world equity models with an additional recession factor. This factor is comprised of the Altman Distressed Debt index and a short position in the S&P 500 index to ensure the market has a zero beta to the factor. They worked through the problem for the case of 100% certainty in the views. They provided all of the data needed to reproduce their results given the set of formulas in this section. In order to perform all the regressions, one would need to have access to the Altman Distressed Debt index along with the other indices used in their paper.

The Work of Qian and Gorman   Qian and Gorman (2001) discussed a method to provide both a conditional mean estimate and a conditional covariance estimate. They used a Bayesian framework referencing the Black–Litterman model, as well as using the alternative reference model as τ did not appear in their paper and they neglected the conditional (or posterior) covariance estimate.

In this section, one will compare the Qian and Gorman approach with the approach taken in the Black–Litterman Reference Model.

One may match the variance portion of Qian and Gorman's formula (3.204) with Equation (3.113):

$$M = \tau\Sigma - \tau\Sigma P^T(P\tau\Sigma P^T + \Omega)^{-1}P\tau\Sigma \tag{3.113}$$

if one sets $\Omega = 0$, and remove τ (this is the alternative reference model). This describes the scenario where the investor has 100% confidence in their views.

For those assets where the investor has absolute views, the variance of the posterior estimate will be zero. For all other assets, the posterior variance will be nonzero.

$$\varepsilon \sim N[0, \Sigma - (\Sigma P^T)(\Sigma_v)^{-1}(P\Sigma)] \tag{3.204}$$

where

$$\Sigma_v = P\Sigma P^T \tag{3.205}$$

$$\text{Var}(\varepsilon) = \Sigma - (\Sigma P^T)(P\Sigma P^T)^{-1}(P\Sigma)] \tag{3.206}$$

$$M = \tau\Sigma - \tau\Sigma P^T(P\tau\Sigma P^T + \Omega)^{-1}P\tau\Sigma \tag{3.113}$$

setting $\Omega = 0$ and removing τ:

$$M = \Sigma - \Sigma P^T(P\tau\Sigma P^T)^{-1}P\Sigma \tag{3.114}$$

In order to get Qian and Gorman's equation (3.207), one needs to reintroduce the covariance of the views, Ω, but rather than mixing the covariances as is done in the Black–Litterman model, one may rely on the lack of correlation and just add the two variance terms. One takes the variance of the conditional from (3.85).

$$P(B|A) \sim N(P^{-1}Q, [P^T\Omega^{-1}P]^{-1}) \tag{3.85}$$

$$\text{Var}(\varepsilon) = \Sigma - (\Sigma P^T)(P\Sigma P^T)^{-1}(\Sigma P^T) + (P^T\Omega^{-1}P)^{-1} \tag{3.115a}$$

$$\text{Var}(\varepsilon) = \Sigma + (\Sigma P^T)(P\Sigma P^T)^{-1}(\Sigma P^T) + (\Sigma P^T)(P^T\Omega^{-1}P)^{-1}(\Omega)(P\Sigma P^T)(\Sigma P^T) \tag{3.115b}$$

$$\text{Var}(\varepsilon) = \Sigma + (\Sigma P^T)(P\Sigma P^T)^{-1}(\Omega)(P\Sigma P^T)^{-1} - (P\Sigma P^T)^{-1}(\Sigma P^T) \tag{3.115c}$$

Note that (3.115c) exactly matches (3.207):

$$\underline{\Sigma} = \Sigma + (\Sigma P^T)(\Sigma_v^{-1}\underline{\Sigma}_v\Sigma_v^{-1} - \Sigma_v^{-1})(P\Sigma) \tag{3.207}$$

where

$$\Sigma_v = P\Sigma P^T \quad \text{and} \quad \underline{\Sigma}_V = \Omega$$

Upon substituting these into (3.207), the result is

$$\underline{\Sigma} = \Sigma + (\Sigma P^T)\{(P\Sigma P^T)^{-1}\Omega(P\Sigma P^T)^{-1} - (P\Sigma P^T)^{-1}\}(P\Sigma) \tag{3.116}$$

Qian and Gorman demonstrated a conditional covariance that is not derived from Theil's mixed estimation nor from Bayesian updating. Its behavior may increase the variance of the posterior versus the prior in the event that the view variance is larger than the prior variance. They described this as allowing the investor to have a view on the variance, and they do not suggest that Ω needs to be diagonal. In this model the conditional covariance is proportional to the investor's views on variance, but the blending process is not clear.

Compare this formula with (3.113) for the variance of the posterior mean estimate:

$$M = \tau\Sigma - \tau\Sigma P^T (P\tau\Sigma P^T + \Omega)^{-1} P\tau\Sigma \qquad (3.117)$$

Intuition indicates that the updated posterior estimate of the mean will be more precise (viz., lower variance) than the prior estimate, thus one would like the posterior (conditional) variance to always be less than the prior variance. The Black–Litterman posterior variance achieves this goal and arrives at the well-known result of Bayesian analysis in the case of unknown mean and known precision. As a result, it seems one should prefer these results over Qian and Gorman.

Directions for the Future Future directions for this research include reproducing the results from the original papers: Black and Litterman (1991) and Black and Litterman (1992). These results have the additional complication of including currency returns and partial hedging.

Future work should include more information on process and a synthesized model containing the best elements from the various authors. A full example from the CAPM equilibrium, through the views to the final optimized weights would be useful, and a worked example of the two-factor model from Krishnan and Mains (2005) would also be useful. Meucci (2006, 2008) provides further extensions to the Black–Litterman model for nonnormal views and views on parameters other than return. This allowed one to apply the Black–Litterman model to new areas such as alternative investments or derivatives pricing. His methods are based on simulation and do not provide a closed-form solution.

An Asset Allocation Process Using the Black–Litterman Model When used as part of an asset allocation process, the Black–Litterman model provided for estimates that lead to more stable and more diversified portfolios than estimates derived from historical returns when used with unconstrained mean-variance optimization. Owing to this significant property, an investor using mean-variance optimization is less likely to require artificial constraints to get a portfolio without extreme weights. *Unfortunately, this model requires a broad variety of data, some of which may not be readily available!*

- First, the investor needs to identify their investable universe and find the market capitalization of each asset. Then, they need to estimate a covariance matrix for the excess returns of the assets. This is most often done using historical data for an appropriate time window: Both Litterman (2003) and Bevan and Winkelmann (1998) provided details on the process used to compute covariance matrices at Goldman Sachs.
- In the literature, monthly covariance matrices are most commonly estimated from 60 months of historical excess returns.

- If the actual asset return itself cannot be used, then an appropriate proxy can be used, for example, S&P 500 Index for U.S. Domestic Large Cap equities. The return on a short-term sovereign bond, for example, U.S. 4- or 13-week treasury bill, would suffice for most U.S. investor's risk-free rate.
- When applied to the asset allocation problem, finding the market capitalization information for liquid asset classes might be a challenge for an individual investor, but likely presents little obstacle for an institutional investor because of their access to index information from the various providers.
- Given the limited availability of market capitalization data for illiquid asset classes, for example, real estate, private equity, commodities, even institutional investors might have a difficult time piecing together adequate market capitalization information.
- Return data for these same asset classes can also be complicated by delays, smoothing, and inconsistencies in reporting.
- Further complicating the problem is the question of how to deal with hedge funds or absolute return managers. The question of whether they should be considered a separate asset class calls for further research!
- Next, the investor needs to quantify their views so that they can be applied and new return estimates computed. The views can be derived from quantitative or qualitative processes, and can be complete or incomplete, or even conflicting.
- Finally, the outputs from the model need to be fed into a portfolio selection model to generate the efficient frontier, and an efficient portfolio selected.
- Bevan and Winkelmann (1999) provided a description of their asset allocation process (for international fixed income) and how they use the Black–Litterman model within that process. This includes their approaches to calibrating the model and information on how they compute the covariance matrices. The standard Black–Litterman model does not provide direct sensitivity of the prior to market factors besides the asset returns. It is fairly simple to extend the Black–Litterman model to use a multifactor model for the prior distribution. Krishnan and Mains (2005) have provided extensions to the model that allow adding additional cross-asset factors that are not priced in the market.
- Examples of such factors are a recession, or credit, market factor. Their approach is general and could be applied to other factors if desired. Most of the Black–Litterman literature reports results using the closed-form solution for unconstrained mean variance optimization. They also tend to use non-extreme views in their examples.
- As part of an investment process, it is reasonable to conclude that some constraints would be applied at least in terms of restricting short selling and limiting concentration in asset classes. Lack of a budget constraint is also consistent with a Bayesian investor who may not wish to be 100% invested in the market due to uncertainty about their beliefs in the market.

- Portfolio selection is normally considered as part of a two-step process: first compute the optimal portfolio, and then determine position along the Capital Market Line.

Remarks:

1) For the ensuing discussion, we will refer to the CAPM equilibrium distribution as the prior distribution, and the investors views as the conditional distribution.
2) This is consistent with the original Black and Litterman (1992) papers. It is also consistent with one's intuition about the outcome in the absence of a conditional distribution (no views in Black–Litterman terminology.) This is the opposite to the way most examples of Bayes' theorem are defined: they start with a nonstatistical prior distribution and then add a sampled (statistical) distribution of new data as the conditional distribution. The mixing model and the use of normal distributions will bring the investigation to the same outcome independent of these choices.

Final Remarks on an Asset Allocation Process   The Black–Litterman model is just one part of an asset allocation process. Bevan and Winkelmann (1998) documented the asset allocation process they used in the Fixed Income Group at Goldman Sachs. At a minimum. a Black–Litterman-oriented investment process would have the following steps:

- Determine which assets constitute the market.
- Compute the historical covariance matrix for the assets.
- Determine the market capitalization for each asset class.
- Use reverse optimization to compute the CAPM equilibrium returns for the assets.
- Specify views on the market.
- Blend the CAPM equilibrium returns with the views using the Black–Litterman Model.
- Feed the estimates (estimated returns, covariances) generated by the Black–Litterman model into a portfolio optimizer.
- Select the efficient portfolio that matches the investors' risk preferences.

A further discussion of each step is provided below:

1) The first step is to determine the scope of the market. For an asset allocation exercise, this would be identifying the individual asset classes to be considered.
2) For each asset class, the weight of the asset class in the market portfolio is required. Then a suitable proxy return series for the excess returns of the asset class is required.
3) Between these two requirements, it can be very difficult to integrate illiquid asset classes such as private equity or real estate into the model.

Furthermore, separating public real estate holdings from equity holdings (e.g., REITS in the S&P 500 index) may also be required.

4) Idzorek (2006) provides an example of the analysis required to include commodities as an asset class. Once the proxy return series have been identified, and returns in excess of the risk-free rate have been calculated, a covariance matrix can be calculated. Typically, the covariance matrix is calculated from the highest frequency data available, for example, daily, and then scaled up to the appropriate time frame.

5) Investor's often use an exponential weighting scheme to provide increased weights to more recent data and less to older data. Other filtering (Random Matrix Theory) or shrinkage methods could also be used in an attempt to impart additional stability to the process.

6) Now one may run a reverse optimization on the market portfolio to compute the equilibrium excess returns for each asset class. Part of this step includes computing a δ value for the market portfolio. This may be calculated from the return and standard deviation of the market portfolio. Bevan and Winkelmann (1998) discussed the use of an expected Sharpe Ratio target for the calibration of δ.

7) For their international fixed income investments, they used an expect Sharpe Ratio of 1.0 for the market. The investor then needs to calibrate τ in some manner. This value is usually on the order of 0.025–0.050. At this point, almost all of the machinery is in place. The investor needs to specify views on the market.

8) These views can impact one or more assets, in any combination. The views can be consistent, or they can conflict. An example of conflicting views would be merging opinions from multiple analysts, where they may not all agree. The investor needs to specify the assets involved in each view, the absolute or relative return of the view, and their uncertainty in the return for the view consistent with their reference model and measured by one of the methods discussed previously.

3.6 The Black–Litterman Model

3.6.1 Derivation of the Black–Litterman Model

3.6.1.1 Derivation Using Theil's Mixed Estimation
This discussion includes the derivation of the Black–Litterman master formula (3.58) using Theil's Mixed Estimation approach that is based on Generalized Least Squares.

Theil's Mixed Estimation Approach　This approach is from Theil (1971) and is similar to the reference in the original Black and Litterman (1992) papers. Koch (2005) also includes a derivation similar to this.

Start with a prior distribution for the returns. Assume a linear model such as

$$\pi = x\beta + u \tag{3.118}$$

where π is the mean of the prior return distribution, β is the expected return, and u is the normally distributed residual with mean 0 and variance Φ.

Next, let one consider some additional information, the conditional distribution:

$$q = p\beta + v \tag{3.119}$$

where q is the mean of the conditional distribution and v is the normally distributed residual with mean 0 and variance Ω.

Both Ω and $\sum$ are assumed to be nonsingular.

One may combine the prior and conditional information by writing

$$\begin{bmatrix} \pi \\ q \end{bmatrix} = \begin{bmatrix} x \\ p \end{bmatrix} \beta + \begin{bmatrix} u \\ v \end{bmatrix} \tag{3.120}$$

where the expected value of the residual is 0, and the expected value of the variance is given by

$$E([u][u'v']) = \begin{bmatrix} \Phi 0 \\ 0\Omega \end{bmatrix}) \tag{3.121}$$

One may then apply the generalized least-squares procedure, which leads to the estimating of β as

$$\beta = [[xp][\Phi 0]^{-1}[x']]^{-1}[x'p'][\Phi 0]^{-1}[\pi] \\ [[0\Omega][p']][0\Omega][q] \tag{3.122}$$

which may be rewritten *without the* matrix notation as

$$\beta = [[x\Phi^{-1}x' + p\Omega^{-1}p']^{-1}[x'\Phi^{-1}\pi + p'\Omega^{-1}q] \tag{3.123}$$

One may derive the expression for the variance using similar logic. Given that the variance is the expectation of $(\beta - \beta)^2$, one may start by substituting (3.122) into (3.123), obtaining

$$\beta = [x\Phi^{-1}x' + p\Omega^{-1}p']^{-1}[x'\Phi^{-1}(x\beta + u) + p'\Omega^{-1}(p\beta + v)] \\ = [x\Phi^{-1}x' + p\Omega^{-1}p']^{-1}[x\beta\Phi^{-1}x\beta + p'\Omega^{-1}p\beta] \tag{3.124a}$$

$$+ x\Phi^{-1}x' + p'\Omega p']^{-1}[x'\Phi^{-1}u + p'\Omega^{-1}v] \tag{3.124b}$$

$$= \beta + [x\Phi^{-1}x' + p\Omega^{-1}p']^{-1}[x'\Phi^{-1}u + p\Omega^{-1}v] \tag{3.124c}$$

namely,

$$\beta - \beta = [x\Phi^{-1}x' + p\Omega^{-1}p']^{-1}[x'\Phi^{-1}u + p\Omega^{-1}v] \tag{3.125}$$

The variance is the expectation of (3.125) squared.

$$
E[(\underline{\beta} - \beta)^2]
$$
$$
= ([x\Phi^{-1}x^T + p\Omega^1 p^T]^{-1}[x\Phi^{-1}u^T + p\Omega^{-1}v^T])^2 \tag{3.126}
$$
$$
= [x\Phi^{-1}x^T + p\Omega^1 p^r]^{-2}
$$
$$
[x\Phi^{-1}u^T u\Phi^{-1}x^T + p\Omega^{-1}v^T v\Omega^{-1}p^T + x\Phi^{-1}u^T v\Omega^{-1}p^T + p\Omega^{-1}v^T u\Phi^{-1}x^T] \tag{3.127}
$$

From the foregoing assumptions that $E[uu'] = \Phi$, $E[vv'] = \Omega$, and $E(uv') = 0$—because u and v are independent variables, taking the expectations one sees the cross-terms are 0, so that (3.126) becomes

$$
E[(\underline{\beta} - \beta)^2]
$$
$$
= [x\Phi^{-1}x^T + p\Omega^{-1}p^r]^{-2}
$$
$$
[[(x\Phi^{-1}\Phi\Phi^{-1}x^T) + p\Omega^{-1}p^T]^{-2}[x\Phi^{-1}\Phi\Phi^{-1}x^T + (p\Omega^{-1}\Omega\Omega^{-1}) + 0 + 0] \tag{3.128a}
$$
$$
= [x\Phi^{-1}x^T + p\Omega^{-1}p^r]^{-2} \tag{3.128b}
$$
$$
[x\Phi^{-1}\Phi^{-1}x^T + p\Omega^{-1}p^T]^{-2}[x\Phi^{-1}x^T + p\Omega^{-1}p^T]
$$

And for the Black–Litterman model, x is the identity matrix and $\Phi = \tau\sum$. Upon making these substitutions, the result is

$$
E[(\underline{\beta} - \beta)^2] = [(\tau\Sigma)^{-1} + p\Omega^{-1}p^T]^{-1} \tag{3.129}
$$

3.6.1.2 Derivation Using Bayes' Theory

This derivation provides an overview of the relevant portion of Bayes' theory in order to create a common vocabulary that may be used in analyzing the Black–Litterman model from a Bayesian viewpoint.

Introduction to Bayes' Theory Bayes' theory states

$$
P(A|B) = P(B|A)P(A)/P(B) \tag{3.130}
$$

in which

$P(A|B)$ is the conditional (or joint) probability of A, given B, it is also known as the *posterior distribution.*

$P(B|A)$ is the conditional probability of B given A, it is also known as the *sampling distribution*. Call this the conditional distribution.

$P(A)$ is the probability of A, also known as the *prior distribution*. Call this the prior distribution.

$P(B)$ is the probability of B, it is also known as the *normalizing constant.*

When applying this formula and solving for the posterior distribution, the normalizing constant will disappear into the constants of integration; so from this point on, it may be ignored.

A general problem in using Bayes' theory is to identify an intuitive and tractable prior distribution. One of the core assumptions of the Black–Litterman model (and Mean-Variance optimization) is that *asset returns are normally distributed*. For that reason, one confines to the case of normally distributed conditional and prior distributions. Given that the inputs are normal distributions, it follows that the posterior will also be normally distributed. When the prior distribution and the posterior have the same structure, the prior is known as a *conjugate prior*.

Given interest, there is nothing to keep us from building variants of the Black–Litterman model using different distributions; however, the normal distribution is generally the most straightforward.

Another core assumption of the Black–Litterman model is that the variance of the prior and the conditional distributions about the actual mean are known, but the actual mean is not known. This case, known as "Unknown Mean and Known Variance," is well documented in the Bayesian literature. This matches the model that Theil uses where one has an uncertain estimate of the mean, but know the variance.

Remarks: The significant distributions are defined here below:

1) The prior distribution

$$(P(A) \sim N(x, S/n) \tag{3.131}$$

where S is the sample variance of the distribution about the mean, with n samples, then S/n is the variance of the estimate of x about the mean.
    The conditional distribution

$$P(B|A) \sim N(\mu, \Omega) \tag{3.132}$$

where Ω is the uncertainty in the estimate μ of the mean, it is not the variance of the distribution about the mean.

2) Then the posterior distribution is specified by

$$P(A|B) - N([\Omega^{-1}\mu + nS^{-1}x]^T[\Omega^{-1} + nS^{-1}]^{-1}, (\Omega^{-1} + nS^{-1})^{-1}) \tag{3.133}$$

3) The variance term in (3.133) is the variance of the estimated mean about the actual mean. In Bayesian statistics, the inverse of the variance is known as the *precision*. One may describe the posterior mean as the weighted mean of the prior and conditional means, where the weighting factor is the respective precision.

4) Furthermore, the posterior precision is the sum of the prior and conditional precision. Equation (3.133) requires that the precision of the prior and conditional be noninfinite, and that the sum is nonzero. Infinite precision corresponds to a variance of 0, or absolute confidence.

5) Zero precision corresponds to infinite variance, or total uncertainty. As a first check on the formulas, one may test the boundary conditions to see if they agree with one's intuition. If one examines (3.133) in the absence of a conditional distribution, it should collapse into the prior distribution.

$$\sim N([nS^{-1}x][nS^{-1}]^{-1}, \quad (nS^{-1})^{-1}) \tag{3.134}$$

$$\sim N(x, S/n) \tag{3.135}$$

As one may see in Equation (3.133), it does indeed collapse to the prior distribution. Another important scenario is the case of 100% certainty of the conditional distribution, where S, or some portion of it is 0, and thus S is not invertible. One may transform the returns and variance from formula (3.133) into a form more easy to work within the 100% certainty case.

$$P(A \mid B) \sim N(x + (S/n)[\Omega + S/n]^{-1}[\mu - x], [(S/n) - (S/n)(\Omega + S/n)^{-1}(S/n)]) \tag{3.136}$$

This transformation relies on the result that

$$(A^{-1} + B^{-1})^{-1} = A - A(A + B)^{-1}A \tag{3.137}$$

It is easy to see that when S is 0 (viz., 100% confidence in the views), the posterior variance will be 0. If Ω is positive infinity (the confidence in the views is 0%), the posterior variance will be (S/n).

Later, one will revisit Equations (3.133) and (3.136) where one transforms these basic equations into the various parts of the Black–Litterman model. Equation (3.122) contains derivations of the alternative Black–Litterman formulas from the standard form, analogous to the transformation from (3.133) to (3.136).

3.6.2 Further Discussions on The Black–Litterman Model

This section contains a derivation of the Black–Litterman master equation using the standard Bayesian approach for modeling the posterior of two normal distributions. One additional derivation is in Mankert (2006) where the author derived the Black–Litterman "master formula" from Sampling theory, and also shows the detailed transformation between the two forms of this formula.

The PDF-Based Approach The PDF-based approach follows a Bayesian approach to obtain the PDF of the posterior distribution, when the prior and conditional distributions are both normally distributed. This section is based on the proof shown in DeGroot (1970). This is similar to the approach taken in Satchell and Scowcroft (2000).

This proof examines all the terms in the PDF of each distribution that depends on $E(r)$, neglecting the other terms as they have no dependence on $E(r)$ and thus are constant with respect to $E(r)$.

Starting with the prior distribution, one derives an expression proportional to the value of the PDF:

$$P(A) \propto N(x, S/n) \tag{3.138}$$

with n samples from the population.

So $\xi(x)$ from the PDF of $P(A)$ satisfies $\sum$:

$$\xi(x) \propto \exp[(S/n)^{-1}\{E(r) - x\}^2] \tag{3.139}$$

Next, consider the PDF for the conditional distribution:

$$P(B|A) \propto N(\mu, \Sigma) \tag{3.140}$$

So $\xi(\mu|x)$ from the PDF of $P(B|A)$ satisfies

$$\xi(\mu|x) \propto \exp[\Sigma^{-1}\{E(r) - \mu\}^2] \tag{3.141}$$

Substituting (3.139) and (3.141) into (3.75) from the text,

$$r \sim N(\mu, \Sigma) \tag{3.75}$$

one has an expression which the PDF of the posterior distribution will satisfy.

$$\xi(\mu|x) \propto \exp[-[\Sigma^{-1}\{E(r) - \mu\}^2 + (S/n)^{-1}\{E(r) - x\}^2]] \tag{3.142}$$

or

$$\xi(\mu|x) \propto \exp(-\Phi) \tag{3.143}$$

Considering only the quantity in the exponent and simplifying

$$\Phi = [\Sigma^{-1}\{E(r) - \mu\}^2 + (S/n)^{-1}\{E(r) - x\}^2] \tag{3.144a}$$

$$= [\Sigma^{-1}\{E(r)^2 - 2E(r)\mu + \mu^2\} + (S/n)^{-1}\{E(r)^2 - 2E(r)x + x^2\}] \tag{3.144b}$$

$$= E(r)^2[\Sigma^{-1} + (S/n)^{-1}] - 2E(r)[\mu\Sigma^{-1} + x(S/n)^{-1}] + \Sigma^{-1}\mu^2 + (S/n)^{-1}x^2 \tag{3.144c}$$

introduce a new term y, where

$$y = \{\mu\Sigma^{-1} + x(S/n)^{-1}\}/\{\Sigma^{-1} + (S/n)^{-1}\} \tag{3.145}$$

and then substitute in the second term:

$$\Phi = E(r)^2[\Sigma^{-1} + (S/n)^{-1}] - 2E(r)[\mu\Sigma^{-1} + x(S/n)^{-1}] + \Sigma^{-1}\mu 2 + (S/n)^{-1}x^2 \tag{3.146}$$

Then add

$$0 = y^2\{\Sigma^{-1} + (S/n)^{-1}\} - [\mu\Sigma^{-1} + x(S/n)^{-1}]^2 \ [\Sigma^{-1} + (S/n)^{-1}]^{-1} \tag{3.147}$$

giving

$$\Phi = E(r)^2[\Sigma^{-1} + (S/n)^{-1}] - 2E(r)y[\Sigma^{-1} + (S/n)^{-1}] + \Sigma^{-1}\mu^2 + (S/n)^{-1}x^2$$
$$+ y^2[\Sigma^{-1} + (S/n)^{-1}] - [\mu\Sigma^{-1} + x(S/n)^{-1}]^2[\Sigma^{-1} + (S/n)^{-1}]^{-1} \tag{3.148a}$$

$$= E(r)^2[\Sigma^{-1} + (S/n)^{-1}] - 2E(r)y[\Sigma^{-1} + (S/n)^{-1}] + y^2[\Sigma^{-1} + (S/n)^{-1}]$$
$$+ \Sigma^{-1}\mu^2 + (S/n)^{-1}x^2 - [\mu\Sigma^{-1} + x(S/n)^{-1}]^2[\Sigma^{-1} + (S/n)^{-1}]^{-1} \tag{3.148b}$$

$$= [\Sigma^{-1} + (S/n)^{-1}][E(r)^2 - 2E(r)y + y^2] + [\Sigma^{-1}\mu^2 + (S/n)^{-1}x^2]$$
$$- [\mu\Sigma^{-1} + x(S/n)^{-1}]^2[\Sigma^{-1} + (S/n)^{-1}]^{-1} \tag{3.148c}$$

$$= [\Sigma^{-1} + (S/n)^{-1}][E(r)^2 - 2E(r)y + y^2] - [\mu\Sigma^{-1} + x(S/n)^{-1}]^2[\Sigma^{-1} + (S/n)^{-1}]^{-1}$$
$$+ [\Sigma^{-1}\mu^2 + (S/n)^{-1}x^2][\Sigma^{-1} + (S/n)^{-1}][\Sigma^{-1} + (S/n)^{-1}]^{-1} \tag{3.148d}$$

$$= [\Sigma^{-1} + (S/n)^{-1}][E(r)^2 - 2E(r)y + y^2]$$
$$- [\mu^2\Sigma^{-2} + 2\mu x\Sigma^{-1}(S/n)^{-1} + x^2(S/n)^{-2}][\Sigma^{-1} + (S/n)^{-1}]^{-1}$$
$$+ [\Sigma^{-2}\mu^2 + (S/n)^{-1}\Sigma^{-1}x^2] + \mu^2\Sigma^{-1}(S/n)^{-1} + x^2(S/n)^{-2}][\Sigma^{-1} + (S/n)^{-1}]^{-1} \tag{3.148e}$$

$$= [\Sigma^{-1} + (S/n)^{-1}][E(r)^2 - 2E(r)y + y^2]$$
$$+ [(S/n)^{-1}\Sigma^{-1}x^2 - 2\mu x\Sigma^{-1}(S/n)^{-1} + \mu^2\Sigma^{-1}(S/n)^{-1}]\{\Sigma^{-1} + (S/n)^{-1}\}^{-1} \tag{3.148f}$$

$$= [\Sigma^{-1} + (S/n)^{-1}][E(r)^2 - 2E(r)y + y^2]$$
$$+ [\Sigma^{-1} + (S/n)^{-1}]^{-1}(x - \mu)[\Sigma^{-1}(S/n)^{-1}] \tag{3.148g}$$

The second term has no dependency on $E(r)$, thus it may be included in the proportionality factor and one is left with

$$\xi(x|\mu) \propto \exp[-\{[\Sigma^{-1} + (S/n)^{-1}]^{-1}[E(R) - y]^2\}] \tag{3.149}$$

Thus, the posterior mean is y and the variance is

$$[\Sigma^{-1} + (S/n)^{-1}]^{-1} \tag{3.150}$$

3.6.2.1 An Alternative Formulation of the Black–Litterman Formula

This is a derivation of the alternative formulation of the Black–Litterman master formula for the posterior expected return. Starting from (3.104), one will derive formula (3.105):

$$P(A|B) \sim N\{[(\tau\Sigma)^{-1}\Pi + P^T\Omega^{-1}Q][(\tau\Sigma)^{-1} + P^T\Omega^{-1}P]^{-1}, \ [(\tau\Sigma)^{-1} + P^T\Omega^{-1}P]^{-1}\} \tag{3.151}$$

$$\underline{\Pi} = \Pi + \tau\Sigma P^T[(P\tau\Sigma P^T) + \Omega]^{-1}[Q - P\Pi] \tag{3.152}$$

$$E(r) = \{(\tau\Sigma)^{-1} + P^T\Omega^{-1}P\}^{-1}\{(\tau\Sigma)^{-1}\Pi + P^T\Omega^{-1}Q\} \tag{3.152}$$

Separating the parts of the second term:

$$E(r) = [\{(\tau\Sigma)^{-1} + P^T\Omega^{-1}P\}^{-1}(\tau\Sigma)^{-1}\Pi] + [\{(\tau\Sigma)^{-1} + P^T\Omega^{-1}P\}^{-1}(P^T\Omega^{-1}Q)] \qquad (3.153)$$

Replacing the precision term in the first term with the alternative form:

$$E(r) = [\{\tau\Sigma - \tau\Sigma P^T(P\tau\Sigma P^T + \Omega)^{-1}P\tau\Sigma\}(\tau\Sigma)^{-1}\Pi] + [\{(\tau\Sigma)^{-1} + P^T\Omega^{-1}P\}^{-1}(P^T\Omega^{-1}Q)] \qquad (3.154a)$$

$$E(r) = [\Pi - \{\tau\Sigma P^T(P\tau\Sigma P^T + \Omega)^{-1}P\Pi\}] + [\{(\tau\Sigma)^{-1} + P^T\Omega^{-1}P\}^{-1}(P^T\Omega^{-1}Q)] \qquad (3.154b)$$

$$E(r) = [\Pi - \{\tau\Sigma P^T(P\tau\Sigma P^T + \Omega)^{-1}P\Pi\}] + [\{(\tau\Sigma)(\tau\Sigma)^{-1}\{(\tau\Sigma)^{-1} + P^T\Omega^{-1}P\}^{-1}(P^T\Omega^{-1}Q)] \qquad (3.154c)$$

$$E(r) = [\Pi - \{\tau\Sigma P^T(P\tau\Sigma P^T + \Omega)^{-1}P\Pi\}] + [(\tau\Sigma)(I_n + P^T\Omega^{-1}P\tau\Sigma)^{-1}(P^T\Omega^{-1}Q)] \qquad (3.154d)$$

$$E(r) = [\Pi - \{\tau\Sigma P^T(P\tau\Sigma P^T + \Omega)^{-1}P\Pi\}] + [\tau\Sigma(I_n + P^T\Omega^{-1}P\,\tau\Sigma)^{-1}(P^T\Omega^{-1})Q] \qquad (3.154e)$$

$$E(r) = [\Pi - \{\tau\Sigma P^T(P\tau\Sigma P^T + \Omega)^{-1}P\Pi\}] + [\tau\Sigma(I_n + P^T\Omega^{-1}P\tau\Sigma)^{-1}\{\Omega(P^T)^{-1}\}^{-1}Q] \qquad (3.154f)$$

$$E(r) = [\Pi - \{\tau\Sigma P^T(P\tau\Sigma P^T + \Omega)^{-1}P\Pi\}] + [\tau\Sigma\{\Omega(P^T)^{-1} + P\tau\Sigma\}^{-1}Q] \qquad (3.154g)$$

$$E(r) = [\Pi - \{\tau\Sigma P^T(P\tau\Sigma P^T + \Omega)^{-1}P\Pi\}] + [\tau\Sigma P^T(P^T)^{-1}\{+P\tau\Sigma\}^{-1}Q] \qquad (3.154h)$$

$$E(r) = [\Pi - \{\tau\Sigma P^T(P\tau\Sigma P^T + \Omega)^{-1}P\Pi\}] + \{\tau\Sigma P^T(\Omega + P\tau\Sigma P^T)\}^{-1}Q\} \qquad (3.154i)$$

which is the alternative form of the Black–Litterman formula for expected return:

$$E(r) = \Pi - \{\tau\Sigma P^T(P\tau\Sigma P^T + \Omega)^{-1}\}(\Omega - P\Pi) \qquad (3.155)$$

3.6.2.2 A Fundamental Relationship: $r_A \sim N\{\prod, (1 + \tau)\sum\}$
This is a derivation of Equation (3.88):

$$r_A \sim N\{\Pi, (1 + \tau)\Sigma\} \qquad (3.156)$$

Starting with the definition of the views.

$$Q = Q + \varepsilon \qquad (3.157)$$

for which

Q is the $k \times 1$ vector of the unknown mean returns to the views,
Q is the $k \times 1$ vector of the estimated mean returns to the views,
ε is the $n \times 1$ matrix of residuals from the regression where $E(\varepsilon) = 0$, $V(\varepsilon) = E$
  $(\varepsilon\varepsilon^T) = \Omega$, and
Ω is nonsingular.

One may rewrite (3.75) into a distribution of Q as follows:

$$r \sim N(\mu, \Sigma) \tag{3.75}$$

$$Q \sim N(Q, \Omega) \tag{3.208}$$

One may also write the definition of the unknown mean returns of the views based on the unknown mean returns of the assets and the portfolio pick matrix P:

$$P \; \underline{\Pi} = Q \tag{3.158}$$

where

P is the $k \times n$ vector of weights for the view portfolios, and
$\prod$ is the $n \times 1$ vector of the unknown returns of the assets.

Substituting Equation (3.157) into (3.158), one obtains the following:

$$\hat{P} \; \underline{\Pi} = Q + \varepsilon \tag{3.159}$$

Assuming that P is invertible, which requires it to be of full rank, then one may multiply both sides by P^{-1}. This is the projection of the view estimated means into asset space representing the Black–Litterman conditional distribution. If P is not invertible, then one would need a slightly different formulation here, adding another term to the right-hand side.

$$\underline{\Pi} = P^{-1}Q + P^{-1}\varepsilon \tag{3.160}$$

One would like to represent (3.160) as a distribution. In order to do this, one needs to compute the covariance of the random term. The variance of the unknown asset means about the estimated view means projected into asset space is calculated as follows:

$$
\begin{aligned}
\text{Variance} &= E\,[P^{-1}\varepsilon\varepsilon^T[P^{-1}]^T] \\
\text{Variance} &= P^{-1}E[\varepsilon\varepsilon^T][P^{-1}]T \\
\text{Variance} &= P^{-1}\Omega[P^{-1}]^T \\
&\quad \text{using } [P^T]^{-1} = [P-1]^T \\
\text{Variance} &= P^{-1}\Omega[P^T]^{-1} \\
&\quad \text{using } [AB]^{-1} = B^{-1}A^{-1} \\
\text{Variance} &= [P^T\Omega^{-1}P]^{-1}
\end{aligned}
\tag{3.161}
$$

So one arrives at the projection of the views into asset space as (3.162) follows:

$$\underline{\Pi} \sim N(P^{-1}Q, \quad [P^T\Omega^{-1}P]^{-1}) \tag{3.162}$$

The covariance term here is the covariance of the unknown mean returns about the estimated returns from the views, it is not the covariance of expected returns.

3.6.2.3 On Implementing the Black–Litterman Model

Given the following inputs:

w Equilibrium weights for each asset class; derived from capitalization-weighted CAPM Market portfolio

$\sum$ Matrix of covariances between the asset classes; can be computed from historical data

rf   Risk-free rate for base currency

δ The risk aversion coefficient of the market portfolio; this may be assumed, or can be computed if one knows the return and standard deviation of the market portfolio

τ A measure of uncertainty of the equilibrium variance; usually set to a small number on the order of 0.025–0.050.

First, one uses reverse optimization to compute the vector of equilibrium returns, $\prod$, using Equation (3.163):

$$\Pi = \delta \Sigma w \tag{3.163}$$

Then one formulates the investors views, and specify P, Ω, and Q. Given k views and n assets, then P is a $k \times n$ matrix where each row sums to 0 (relative view) or 1 (absolute view). Q is a $k \times 1$ vector of the excess returns for each view. Ω is a diagonal $k \times k$ matrix of the variance of the views, or the confidence in the views. To start with, most authors call for the values of ω_i to be set equal to $p^T \tau \sum_i p$ (where p is the row from P for the specific view).

Next, assuming one is uncertain in all the views, one applies the Black–Litterman "master formula" to compute the posterior estimate of the returns using (3.164).

$$\underline{\Pi} = \Pi + P^T \{(P\tau\Sigma P^T) + \Omega\}^{-1}(Q - P\Pi) \tag{3.164}$$

One calculates the posterior variance using (3.165):

$$M = \tau\Sigma - \tau\Sigma P^T \{(P\Sigma P^T + \Omega\}^{-1} P\tau\Sigma \tag{3.165}$$

Closely followed by the computation of the sample variance from (3.166):

$$\Sigma_p = \Sigma + M \tag{3.166}$$

And now one may calculate the portfolio weights for the optimal portfolio on the unconstrained efficient frontier from (3.167):

$$\underline{w} = \underline{\Pi}(\delta\Sigma_p)^{-1} \tag{3.167}$$

Worked Examples

Worked Example 1: The Black–Litterman (B-L) Asset Allocation Model

Overview   Combining information from *two* sources to create an estimate of expected returns:

Source 1: What does the current market tell us about the expected excess returns?

Implied excess equilibrium returns

Source 2: What views does the investment manager have about particular stocks, sectors, asset classes, or country?

The BL model combines these different sources to produce estimates of excess returns.

Combining Predicted and Implied Returns:

$$E(r - r_j) = [(\tau S)^{-1} + P^T \Omega P]^{-1}[(\tau S)^{-1}\Pi + P^T \Omega^{-1} Q]$$

Basic Notation

τ: A scalar (assume $\tau = 1$).
S: Variance–covariance matrix for all assets under consideration
Ω: Uncertainty surrounding your views
Π: Implied excess returns
Q: Views on expected excess returns for some or all assets
P: Link matrix identifying which assets you have views about

Understanding the Formula   Splitting into two sections for better understanding, consider the second section first:

$(\tau S)^{-1}$	$\Pi +$	$P^T \Omega^{-1}$	Q
↓	↓	↓	↓
Measures of Confidence	Implied Excess Returns	Confidence	Our Views

By combining implied excess returns with one's own views on excess returns: A weighted average
What are the weights?

$$E(r - rj) = [(\tau S) - 1 + PT\Omega P] - 1[(\tau S) - 1\Pi + PT\Omega - 1Q]$$

The investor about his views relative to the implied excess returns.

• How confident?

Ω will be large => Ω^{-1} will be small
Understanding the formula:

$$E(r - r_j) = [(\tau S)^{-1} + P^T \Omega P]^{-1}[(\tau S)^{-1}\Pi + P^T \Omega^{-1} Q]$$

The first part of this formula:

$$[(\tau S)^{-1} + P^T \Omega P]^{-1}$$

- τ implied excess returns and our views add up to 1.
- Namely, the formula is just a weighted average!

Consider a worked example:
Consider 1 stock: AMZN (Amazon.com)

- The implied excess returns are 0.74% per month and the variance is 2.018%.
- We predict excess returns of 2% per month. The uncertainty surrounding this view is reflected by a variance of 0.50%.

$$=> \Pi = 0.74\%$$
$$S \quad = 2.015\%$$
$$Q \quad = 2\%$$
$$\Omega \quad = 0.50\%$$

- Assume $\tau = 1$, $P = 1$

Substituting into the BL formula:

$$
\begin{aligned}
E(r - r_j) &= [(\tau S)^{-1} + P^T \Omega P]^{-1}[(\tau S)^{-1}\Pi + P^T \Omega^{-1} Q] \\
&= [0.02015^{-1} + 0.005 - 1]^{-1} \times [0.02015^{-1} \times 0.0 \times [0.3674 + 4]/074 \\
&\quad + 0.005^{-1} \times 0.02] \\
&= [49.63 + 200] - 1 \times [49.63 \times 0.0074 + 200 \times 0.02] \\
&= [249.64] - 1 \times [0.3673 + 4] \\
&= 4.3673/249.63 \\
&= 0.017495 \\
&= 1.749\% \text{ per month}
\end{aligned}
$$

Does this estimate make sense?
Since we are confident about our views, now, try a more complicated example:
Incorporating investor views:

Stock[a]	Historical excess returns	CAPM excess returns	Implied excess returns
INTC*	−0.0035	0.0135	0.0092
AEP	0.0026	0.0064	0.0032
AMZN	0.0326	0.0111	0.0074
MRK	0.0002	0.0057	0.0059
XOM	0.0111	0.0052	0.0064

[a]Intel, American Electric Power Company, Inc., Amazon, Merck, Exxon Mobil

Incorporating investor views:

1) Analysts tell you that AEP has found a way to store electricity. Based on this breakthrough, they expect AEP to outperform XOM by 1% per month.
2) Given the current economic conditions, we think that INTC will outperform AMZN by 1.75% per month.
   - Relative views are common in reality.
   - Absolute views, such as AEP having returns of 2% per month, are much less common.

Views versus Implied Excess Returns:
View 1: AEP outperforms XOM by 1% per month

- The difference in implied excess returns is −0.32% per month.
- One would expect that incorporating these views will lead to an increase in one's holdings of AEP and a decrease in XOM.

0.0092
View 2: INTC will outperform AMZN by 1.75% per month.

- The difference in implied excess returns is 0.18% per month.
- We would expect that incorporating our view will increase our holdings of INTC and reduce our holdings of AMZN.

The views expressed above are relative views of assets.
Incorporating our views:
To link our views to implied excess returns, we need a link matrix, P.
Matrix P is constructed in the following way:
Each row represents a view, each column represents a company:

$$
\begin{array}{ccccccc}
 & \text{INTC} & \text{AEP} & \text{AMZN} & \text{MRK} & \text{XOM} \\
P = \text{View 1} & [\;0 & 1 & 0 & 0 & -1 &] \\
\text{View 2} & [\;1 & 0 & -1 & 0 & 0 &] \\
\text{View 3} & [\;0.5 & 0 & 0.5 & -0.5 & -0.5 &]
\end{array}
$$

The View Vector and Uncertainty
What do Q and Ω look like?

$$
Q + \varepsilon = \begin{bmatrix} Q_1 \\ : \\ Q_k \end{bmatrix} + \begin{bmatrix} \varepsilon_1 \\ : \\ \varepsilon_k \end{bmatrix} \quad \rightarrow Q + \varepsilon = \begin{bmatrix} 0.01 \\ 0.0175 \end{bmatrix} + \begin{bmatrix} \varepsilon_1 \\ \varepsilon_2 \end{bmatrix}
$$

View 1
↓
[0.01]
[0.0175]
↑
View 2

where

$$\begin{bmatrix} \varepsilon_1 \\ : \\ \varepsilon_k \end{bmatrix} \sim N \begin{pmatrix} [0][\omega_{11} \dots \omega_{1k}] \\ [:], [: \dots :] \\ [0][\omega_{k1} \dots \omega_{kk}] \\ \uparrow \\ \Omega \end{pmatrix}$$

Calculating Ω There is no best way to calculate Ω. It will depend on how confident you are of your predictions.

Black and Litterman recommend:

$$\text{Base} => \Omega = \tau P S P^T$$

where

P is the link matrix, and
S is the variance–covariance matrix.

We have assumed that $\tau = 1$, and so we can ignore it!
These results are illustrated in Figures 3.32 and 3.33.

Worked Example 2

A worked example of the Black–Litterman approach in financial engineering, using the statistical program R, is given in Section 6.3.3.

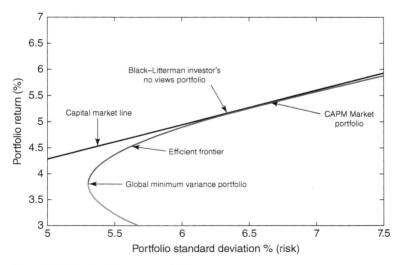

Figure 3.32 Black–Litterman model: 1.

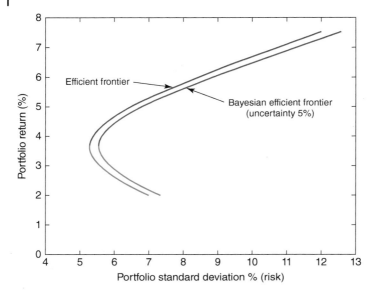

Figure 3.33 Black–Litterman model: 2.

Worked Example 3

In Pfaff (2013): *Financial Risk Modeling and Portfolio Optimization Using* R:

1) pp. 53–83, Chapter 6: "Suitable Distributions for Returns"
2) pp. 136–138, Chapter 9: "Modeling Dependence"are illustrations of the implementation of the foregoing BL approach using the BLCOP package in R.

3.7 The Black–Scholes Option Pricing Model

1) Nathan Coelen (2002): Black–Scholes Option Pricing Model

(Available at http://ramanujan.math.trinity.edu/tumath/research/studpapers/s11.pdf)

Introduction   Finance is one of the rapidly changing and fastest growing areas in the corporate business world. Owing to this rapid change, modern financial instruments are becoming very complex. New mathematical models are essential to implement and price these new financial instruments. The world of corporate finance, once managed by business students, is now controlled by applied mathematicians and computer scientists.

In the early 1970s, Myron Scholes, Robert Merton, and Fisher Black made an important breakthrough in the pricing of complex financial instruments by developing what has become known as the Black–Scholes model. In 1997, the importance of their model was recognized worldwide when Myron Scholes and Robert Merton received the Nobel Prize for Economic Sciences. (Fisher Black passed away in 1995, or he would have also received the award [Hull, 2000]). The Black–Scholes model displayed the importance that mathematics plays in the field of finance. It also led to the growth and success of the new field of mathematical finance or financial engineering.

Here, the Black–Scholes partial differential equation will be derived, and then It will be used to solve the equation for a European call option. First, one will discuss basic financial terms, such as stock and option, and review the arbitrage pricing theory. One will then derive a model for the movement of a stock, which will include a random component: Brownian motion. Then, one will discuss some basic concepts of stochastic calculus that will be applied to a stock model. From this model, one will derive the Black–Scholes partial differential equation, and it will be used, with boundary conditions for a European call option to solve the equation.

Definitions Financial assets are claims on certain issuers, such as the federal government or a corporation, such as Apple, or Microsoft. Financial assets also include real assets such as real estate, but one will be primarily concerned with common stock. Common stock represents an ownership in a corporation. Stocks provide a claim to the income and assets of the corporation. Anyone who buys a financial asset in hopes that it will increase in value has taken a long position. A person who sells a stock before he owns it hoping that it decreases in value is said to be short an asset. People who take short positions borrow the asset from large financial institutions, sell the asset, and buy the asset back at a later time.

A *derivative* is a financial instrument whose value depends on the value of other basic assets, such as common stock. In recent years, derivatives have become increasingly complex and important in the world of finance. Many individuals and corporations use derivatives to hedge against risk. Here, the derivative asset that will be most interested in is a European call *option*. A *call* option gives the owner the right to buy the underlying asset on a certain date for a certain price. The specified price is known as the *exercise* or *strike price* and will be denoted by E. The specified date is known as the *expiration date* or day until maturity. *European options can be exercised only on the expiration date itself.* Another common option is a *put* option, which gives the owner the right to sell the underlying asset on a certain date for a certain price.

For example, consider a November European call option contract on Microsoft with strike price $70. When the contract expires in November, if the price of Microsoft stock is $72 the owner will exercise the option and realize a profit of $2. He will buy the stock for $70 from the seller of the option and immediately sell the stock for $72. On the other hand, if a share of Microsoft is worth $69, or less, then the owner of the option will not exercise the option and it will expire worthless. In this case, the buyer would *lose* the purchase price of the option.

Arbitrage The *arbitrage pricing theory* is a fundamental theories in the world of finance. *Tise theory states that two otherwise identical assets cannot sell at different prices.* This means that there are no opportunities to make an instantaneous risk-free profit. Here one is assuming that the risk-free rate to be that of a bank account or a government bond, such as a Treasury bill.

To illustrate the application of arbitrage, consider a simple example of a stock that is traded in the U.S. and in London. In the U.S. the price of the stock is $200 and the asset sells for £150 in London, while the exchange rate is $1.50 per pound, £. A person could make an instantaneous profit by simultaneously buying 1,000 shares of stock in New York and selling them in London. An instantaneous profit of

$$1,000^* \{ (\$1.50^*150) - \$200 \} = \$25,000$$

is realized without risk!

Hedging In financial investments, it appears that traders are attracted to three classes of derivative securities: *speculators, arbitrageurs,* and *hedgers:*

1) Speculators take long or short positions in derivatives to increase their exposure to the market, betting that the underlying asset will go up or go down.
2) Arbitrageurs find mispriced securities and instantaneously lock in a profit by adopting certain trading strategies.
3) Hedgers who take positions in derivative securities opposite those taken in the underlying asset in order to help manage risk.

For example, consider an investor who owns 100 shares of Apple which is currently priced $100. The person is concerned that the stock might decline sharply in the next two months. The person could buy put options on Apple to sell 100 shares at a price of $80. The person would pay the price of the options, but this would ensure that he could sell the stock for $80 at expiration if the stock declines sharply. One very important hedging strategy is delta hedging. The delta, Δ, of the option is defined as the change of the option price $V = V(S, t)$ with respect to the change in the price S of the underlying asset, at time t. In other words, delta is the first derivative of the option price with respect to the stock price:

$$\Delta = \partial V / \partial S \tag{3.209}$$

For example: the delta of a call option is 0.85, the price of the Apple stock is $100 per share, and the price of a call option is $8. Thus, if an investor has sold 1 call option. The call option gives the buyer the right to buy 100 shares, since each option contract is for 100 shares.

The seller's position could be hedged by buying 0.85 x 100 = 85 shares. The gain (or loss) on the option position would then tend to be offset by the loss (or gain) on the stock position. If the stock price goes up by $1 (producing a gain of $85 on the shares purchased) the option price would tend to go up by 0.85^* $1 = $0.85 (producing a loss of $0.85 × 100 = $85 on the call option written) [Hull, 2000].

Stock Price Model   Stock prices move randomly because of the efficient market hypothesis. There are different forms of this hypothesis, but all refer to two factors:

1)   The history of the stock is fully reflected in the current price.
2)   The markets respond immediately to *new* information about the stock.

With the previous two assumptions, changes in a stock price follow a Markov process. *A Markov process is a stochastic process where only the present value of the variable is relevant for predicting the future.* Thus, the stock model states that predictions for the future price of the stock should be unaffected by the price one day, one week, one month, or one year ago.

As stated above, a Markov process is a stochastic process (viz., one based on probabilities). In the real world, stock prices are restricted to discrete values, and changes in the stock prices can only be realized during specified trading hours. Nevertheless, the continuous-variable, continuous-time model proves to more useful than a discrete model.

Another important observation is the *absolute* change in the price of a stock is by itself, not a useful quality. For example, an increase of one dollar in a stock is much more significant on a stock worth $10 than a stock worth $1000. The *relative* change of the price of a stock is the information that is more valuable. The relative change is defined as the change in the price divided by the original price. Now consider the price of a stock S at time t. Consider a small time interval dt during which the price of the underlying asset S changes by an amount dS. The most common model separates the return on the asset, dS/S into two parts. The first part is completely deterministic, and it is usually the risk free interest rate on a Treasury Bill issued by the government.

This part yields a contribution of μdt to dS/S. Here μ is a measure of the average growth rate of the stock, known as the drift, and μ is the assumed risk-free interest rate on a bond, and it may also be represented as a function of S and t. The second part of the model accounts for the random changes in the stock price owing to external effects, such as unanticipated catastrophic news. It may be modeled by a random sample drawn from a normal distribution with mean zero and contributes σdB to dS/S. In this formula σ is the *volatility* of the stock, which measures the standard deviation of the returns. Like the term μ, σ can be represented as a function of S and t. The B in dB denotes *Brownian motion*, which will be described in the next section. It is important to note that μ and σ may be estimated for individual stocks using statistical analysis of historical prices. It is important that μ and σ are functions of S and t. Combining this information together, one obtains the stochastic differential equation

$$dS/S = \mu dt + \sigma dB \qquad (3.210)$$

Notice that if the volatility σ is zero the model implies

$$dS/S = \mu dt \qquad (3.211)$$

When μ is constant this equation can be solved so that

$$\mathbf{S} = S_0 e^{\mu t} \qquad (3.212)$$

where S is the price of the stock at t and S_0 is the price of the stock at $t = 0$. This equation shows that when the variance of an asset is zero, the asset grows at a continuously compounded rate of μ per unit of time.

Brownian Motion   The term Brownian motion, which was originally used as a model for stock price movements in 1900 by L. Bachelier [Klebaner, 1998], is a stochastic process $B(t)$ characterized by the following three properties:

1) *Normal Increments:* $B(t) - B(s)$ has a normal distribution with mean 0 and variance $(t - s)$. Thus, if $s = 0$ so that $B(t) - B(0)$ has normal distribution with mean 0 and variance $(t - 0) = t$.
2) *Independence of Increments:* $B(t) - B(s)$ is independent of the past.
3) *Continuity of Paths:* $B(t)$, $t > 0$ are continuous functions of t.

These three properties alone define Brownian motion, but they also show why Brownian motion is used to model stock prices. Property (2) shows stock price changes will be independent of past price movements. This was an important *assumption* made in this stock price model.

An occurrence of Brownian motion from time 0 to T is called a *path* of the process on the interval $[0, T]$. There are five important properties of Brownian motion paths. The path $B(t)$, $0 < t < T$

1) is a continuous function of t,
2) is *not* monotonic on any interval, no matter how small the interval is,
3) is *not* differentiable at any point,
4) has infinite variation on any interval no matter how small it is,
5) has quadratic variation on $[0,t]$ equal to t, for any t.

Together, Properties (1) and (3) state that although Brownian motion paths are continuous, the $\Delta B(t)$ over interval Δt is much larger than Δt as $\Delta t \to 0$. Properties (4) and (5) show the distinction between functions of Brownian motion and normal, smooth functions. The variation of a function over the interval $[a, b]$ is defined as

$$V_g([a, b]) = \sup \sum_{i=1}^{n} \left| g(t_i) - g(t_{i-1}) \right| \tag{3.213}$$

where the supremum is taken over partitions:

$$a = t_0 < t_1 < t_2 < \cdots < t_n = b \tag{3.214}$$

It may be seen that smooth functions are of finite variation while Brownian motion is of infinite variation. Quadratic variation plays a very important role with Brownian motion and stochastic calculus. Quadratic variation is defined for a function g over the interval $[0, t]$ as

$$[g, g](t) = \lim \sum_{i=1}^{n} (g(t_i) - g(t_{i-1}))^2 \tag{3.215}$$

where the limit is taken over partitions:

$$0 = t_0 < t_1 < \cdots < t_n = t \tag{3.216}$$

Quadratic variation plays no role in standard calculus due to the fact that continuous functions of finite variation have quadratic variation of 0.

Stochastic or Probabilistic Calculus   In this section we will introduce concepts of stochastic integrals with respect to Brownian motion. These stochastic integrals are also called *Ito Integrals*. In order to proceed with the derivation of the Black–Scholes formula one need to define the stochastic integral $\int_0^T X(t)\, dB(t)$. If X(t) is a constant, c, then

$$\int_0^T c\, dB(t) = c\{B(T) - B(0)\} \tag{3.217}$$

The integral over $(0, T]$ should be the sum of integrals over subintervals

$$[0, a_1), (a_1, a_2), (a_2, a_3), \ldots, (a_{n-1}, T].$$

Thus, if $X(t)$ takes values c_i on each subinterval then the integral of X with respect

- to B is easily defined.

First, consider the integrals of simple processes $e(t)$ which depend on t and not on $B(t)$. A simple *deterministic process* $e(t)$ is a process for which there exist times

$$0 = t_0 < t_1 < t_2 < \cdots < t_n = T$$

and constants c_0, c_1, c_2, c_{n-1}, such that

$$e(t) = \{c_0, \quad \text{if } t = 0$$
$$= \{c_i, \quad \text{if } t_i < t \le t_{i+1}, \quad i = 0, \ldots, n-1$$

Hence, the Ito integral $\int_0^T X(t)dB(t)$ is defined as the sum

$$\int_0^T e(t)dB(t) = \sum_{i=0}^{n-1} c_i\{B(t_{i+1}) - B(t_i)\} \tag{3.218}$$

The Ito integral of simple processes is a random variable with the following four properties:

Property 1. Linearity: If $X(t)$ and $Y(t)$ are simple processes and α and β are constants, then

$$\int_0^T [\alpha X(t) + \beta Y(t)]dB(t) = \alpha \int_0^T X(t)dB(t) + \beta \int_{0n}^T Y(t)dB(t) \tag{3.219}$$

Property 2. The integral of the indicator function of an interval $I_{[a,b]}(t) = 1$ when $t \in [a, b]$, and zero otherwise, is just $B(b) - B(a)$, $0 < a < b < T$,

$$\int_0^T I[a, b](t)dB(t) = B(b) - B(a) \tag{3.220}$$

Property 3. Zero mean property:

$$E\left[\int_0^T X(t)dB(t)\right] = 0 \tag{3.221}$$

Property 4. Isometry property:

$$E\left[\int_0^T X(t)\, dB(t)\right]^2 = \int_0^T E[X^2(t)]\, dt \tag{3.222}$$

The definition of the Ito integral may be extended to processes $X(t)$ that may be approximated by sequences e^n of simple processes in the sense that

$$E\left[\int_0^T |e^n(t) - X(t)|^2\, dt\right] \to 0 \tag{3.223}$$

as $n \to \infty$. In that case, one defines

$$\int_0^T X(t)dB = \lim_{n\to\infty} \int_0^T e^n t dB \tag{3.224}$$

In this definition, the limit does *not* depend on the approximating sequence. Also, the integral that arises this way satisfies Properties 1–4.

Now that one has defined the Ito integral of simple processes, one may define the Ito integrals of other processes. It may be shown that if a general predictable process satisfies certain conditions, the general process is a limit in probability of simple predictable processes discussed earlier. The Ito integral of general predictable processes is defined as a limit of integrals of simple processes. If $X(t)$ is a predictable process such that

$$\int_0^T X^2(t)dt < \infty \tag{3.225}$$

then the Ito integral

$$\int_0^T X(t)\, dt(t)$$

is defined and satisfies the above four properties.

For example, to undertake the integral

$$\int_0^T B(t)dB(t)$$

one may let $0 = t_0 < t_1 < t_2 < \cdots < t_n = T$ be a partition of $[0, T]$, then

$$e^n(t) = \sum_{i=0}^{n-1} B(t_i)I_{[ti,ti+1]}(t) \tag{3.226}$$

Then, for any n, $e^n(t)$ is a simple process. One may take a sequence of the partitions such that $\max_i(t_{i+1} - t_i) \to 0$ as $n \to \infty$. The Ito integral of this simple function is given by

$$\int_0^T e^n(t)dB(t) = \sum_{i=0}^{n-1} B(t_i)[B(t_{i+1}) - B(t_i)] \tag{3.227}$$

One may show that this sequence of integrals converges in probability and identify the limit. Adding and subtracting $B^2(t_{i+1})$, one obtains

$$B(t_i)\{B(t_{i+1}) - B(t_i)\} = 1/2[B^2(t_{i+1}) - B^2(t_i) - \{B(t_{i+1}) - B(t_i)\}^2] \tag{3.228}$$

and

$$\int_0^T e^n(t)dB(t) = 1/2 \sum_{i=0}^{n-1} \{B^2(t_{i+1}) - B^2(t_i)\} - 1/2 \sum_{i=0}^{n-1} \{B(t_{i+1}) - B(t_i)\}^2$$

$$\tag{3.229}$$

$$= 1/2B^2(T) - 1/2B^2(0) - 1/2 \sum_{i=0}^{n-1} \{B(t_{i+1}) - B(t_i)\}^2 \tag{3.230}$$

since the first sum is a telescopic sum. Notice that from property 5 of Brownian motion path, the second sum converges to the limit T. Therefore, $\int_0^T e^n(t)dB(t)$ converges, and the limit is

$$\int_0^T B(t)dB(t) = \lim_{n \to \infty} \int_0^T e^n(t)dB(t) = 1/2B^2(T) - \tfrac{1}{2}T \tag{3.231}$$

This example illustrates the difference between deterministic calculus and stochastic calculus. The quadratic variation of continuous functions, $x(t)$, of finite variation in standard calculus is 0. Therefore, if one were calculating the integral of $\int_0^T x(t)dx(t)$ the same way as above the term $\sum_{i=0}^{n-1} \{x(t_{i+1}) - x(t_i)\}^2$ would to the limit 0. Thus, the integral $\int_0^T x(t)dx(t)$ equals $(1/2)x^2(T)$.

Now consider the main tools of stochastic calculus: Ito's formula, which is the stochastic counterpart of the chain rule in Calculus. Recall that Brownian motion has quadratic variation on $[0, t]$ equal to t, for any t. This also may be expressed as the following:

$$\int_0^t \{dB(s)\}^2 = \int_0^t ds = t \text{ or, in differential notation } \{dB(t)\}^2 = dt \tag{3.232}$$

Using this property and applying Taylor's formula, Ito's formula states that if $f(x)$ is a twice-differentiable function, then for any t

$$f[B(t)] = f(0) + \int_0^t f'\{B(s)\}dB(s) + (1/2) \int_0^t f''\{B(s)\}ds \tag{3.233}$$

and Ito's formula in differential notation becomes

$$d\{f(B(t))\} = f'B(t)dB(t) + (1/2)f''\{B(t)\}dt \tag{3.234}$$

The Ito Process   Next, one defines an Ito process: Let $Y(t)$ be an *Ito integral*, defined as

$$Y(t) = \int_0^t X(s)dB(s) \tag{3.235}$$

Then, an Ito process is an Ito integral plus an adapted continuous process of finite variation. A process Y is called an *Ito Process* if for any $0 \leq t \leq T$ it can be represented as

$$Y(t) = Y(0) + \int_0^t \mu(s)ds + \int_0^t \sigma(s)dB(s) \tag{3.236}$$

More generally, if Y is an Ito process represented above, then it has a stochastic differential on $[0, T]$ given by

$$dY(t) = \mu(t)dt + \sigma(t)dB(t) \tag{3.237}$$

for $0 \le t \le T$. The function μ is often called the *drift coefficient* and the function σ is called the *diffusion coefficient*—terms inherited from studies in mass and energy transfers in areas within classical physics and engineering.

One last important case to consider is for functions of the form $f\{X(t), t\}$. If $f(x, t)$ is a twice continuously differentiable function in x, as well as continuously differentiable in t, and $X(t)$ represents an Ito process, then

$$df\{x(t), t\} = (\partial f/\partial x)\{X(t), t\}dX(t) + (\partial f/\partial t)\{X(t), t\}dt$$
$$+ (1/2)\sigma^2\{X(t), t\}(\partial^2 f/\partial x^2)\{X(t), t|\}dt \tag{3.238}$$

This case of Ito's formula may be used to compute the Black–Scholes partial differential equation in the next section.

Derivation of the Black–Scholes Equation   Here, the price of a derivative security, $V(S, t)$, will be derived. The model for a stock derived previously satisfies an Ito process defined in (3.237).

$$dY(t) = \mu(t)dt + \sigma(t)dB(t) \tag{3.237}$$

Therefore, one may let the function $V(S, t)$ be twice differentiable in S and differentiable in t. Applying (3.238) one has

$$dV(S, t) = (\partial V/\partial S)dS + (\partial V/\partial t)dt + (1/2)\sigma^2 S^2(\partial^2 V/\partial S^2)dt \tag{3.239}$$

Substituting into (3.239) for dS with (3.210), one has

$$dS/S = \mu dt + \sigma dB \tag{3.210}$$

$$dV(S, t) = (\partial V/\partial S)(\mu Sdt + \sigma SdB) + (\partial V/\partial t)dt + (1/2)\sigma^2 S^2(\partial^2 V/\partial S^2)dt \tag{3.240}$$

This simplifies to

$$dV(S, t) = \sigma SdB(\partial V/\partial S) + \{\mu S(\partial V/\partial S) + (\partial V/\partial t) + (1/2)\sigma^2 S^2(\partial^2 V/\partial S^2)\}dt \tag{3.241}$$

Now set up a portfolio long one option, V, and short an amount $(\partial V/\partial S)$ stock. Note from above that this portfolio is hedged. The value of this portfolio, π, is

$$\pi = V - (\partial V/\partial S)S \tag{3.242}$$

The change $d\pi$ in the value of this portfolio over a small time interval dt is given by

$$d\pi = dV - (\partial V/\partial S)dS \tag{3.243}$$

Now substituting Equations (3.241) and (3.210):

$$dS/S = \mu dt + \sigma dB \tag{3.210}$$

into Equation (3.243) for dV and dS one obtains

$$d\pi = \sigma S d\dot{B}(\partial V/\partial S)dB + \{\mu S(\partial V/\partial S) + (\partial V/\partial t) + \tfrac{1}{2}\sigma^2 S^2(\partial^2 V/\partial S^2)\}dt \\ - (\partial V/\partial S)(\mu S dt + \sigma S dB) \tag{3.244}$$

This simplifies to

$$d\pi = \{(\partial V/\partial t) + (\tfrac{1}{2})\sigma^2 S^2(\partial^2 V/\partial S^2)\}dt \tag{3.245}$$

It is important to note that this portfolio is completely risk-free because it does not contain the random Brownian motion term. Since this portfolio contains no risk it must earn the same as other short-term risk-free securities. If it earned more than this, arbitrageurs could make a profit by shorting the risk-free securities and using the proceeds to buy this portfolio. If the portfolio earned less arbitrageurs could make a riskless profit by shorting the portfolio and buying the risk-free securities. It follows for a riskless portfolio that

$$d\pi = r\pi dt \tag{3.246}$$

where r is the risk-free interest rate. Substituting for $d\pi$ and π from (3.245) and (3.242) yields

$$\{(\partial V/\partial t) + (1/2)\sigma^2 S^2(\partial^2 V/\partial S^2)\}dt = r\{V - S(\partial V/\partial S\}dt \tag{3.247}$$

Further simplification yields the Black–Scholes differential equation:

$$\partial V/\partial t + (1/2)\sigma^2 S^2(\partial^2 V/\partial S^2) + rS(\partial V/\partial S) - rV = 0 \tag{3.248}$$

Solution for a European Call   To solve the Black–Scholes equation, consider final and boundary conditions, or else the partial differential equation does not have a unique solution. As an illustration, a European call, $C(S, t)$, will be solved given the exercise price E and the expiry date T.

The final condition at time $t = T$ may be derived from the definition of a call option. If at expiration $S > E$ the call option will be worth $S - E$ because the buyer of the option can buy the stock for E and immediately sell it for S. If at expiration $S < E$ the option will not be exercised and it will expire worthless. At

$t = T$, the value of the option is known for certain to be the payoff:

$$C(S, T) = \max(S - E, 0) \tag{3.249}$$

This is the final condition for the differential equation.

In order to find boundary conditions, one considers the value of C when $S = 0$ and as $S \to \infty$. If $S = 0$, then it is easy to see from (3.210) that $dS = 0$, and therefore S will never change. If at expiry $S = 0$, from (3.249) the payoff must be 0. Consequently, when $S = 0$ one has

$$C(0, T) = 0 \tag{3.250}$$

Now when $S \to \infty$, it becomes more and more likely the option will be exercised and the payoff will be $(S - E)$. The exercise price becomes less and less important as $S \to \infty$, so the value of the option is equivalent to

$$C(S, T) \approx S \text{ as } S \to \infty \tag{3.251}$$

In order to solve the Black–Scholes equation, one needs to transform the equation into an equation one can work with. The first step is to eliminate of the S and S^2 terms in (3.248) (16). To do this, consider the change of variables:

$$S = \underline{E} e^x \tag{3.252}$$

$$t = T - (\tau / 1 / 2\sigma^2) \tag{3.253}$$

$$V = Ev(x, \tau) \tag{3.254}$$

Using the chain rule from Calculus for transforming partial derivatives for functions of two variables, we have

$$\partial V / \partial S = (\partial V / \partial x)(\partial x / \partial S) + (\partial V / \partial \tau)(\partial \tau / \partial S) \tag{3.255}$$

$$\partial V / \partial t = (\partial V / \partial x)(\partial x / \partial t) + (\partial V / \partial t)(\partial \tau / \partial t). \tag{3.256}$$

Looking at (3.252)–(3.254), it may be shown that

$$\partial t / \partial t = -1 / 2\sigma^2, \quad \partial x / \partial t = 0, \quad \partial x / \partial S = 1/S, \quad \partial t / \partial S = 0 \tag{3.257}$$

Plugging these into (3.255) and (3.256) yields

$$\partial V / \partial S = (E/S) \partial V / \partial x \tag{3.258}$$

$$\partial V / \partial t = -(1/2)(\sigma^2 E) \tag{3.259}$$

$$\partial^2 V / \partial S^2 = (E/S^2)(\partial^2 v / \partial x^2) - (E/S^2)(\partial V / \partial x) \tag{3.260}$$

Substituting (3.255)–(3.260) into the Black–Scholes partial differential equation gives the differential equation:

$$\partial V / \partial t = \partial^2 v / \partial x^2 + (k - 1)(\partial V / \partial x) - kv \tag{3.261}$$

where

$$k = [r / (1/2)\sigma^2] \tag{3.262}$$

The initial condition $C(S, T) = \max(S - E, 0)$ is transformed into

$$v(x, 0) = \max(e^x - 1, 0) \tag{3.263}$$

Now we apply another change of variable and let

$$v = e^{(\alpha x + \beta \tau)} u(x, \tau) \tag{3.264}$$

Then by simple differentiation, one has

$$\partial V / \partial \tau = \beta e^{(\alpha x + \beta \tau)} u + e^{(\alpha x + \beta \tau)} (\partial u / \partial \tau) \tag{3.265}$$

$$\partial V / \partial x = \alpha e^{\alpha x + \beta \tau} u + e^{\alpha x + \beta \tau} \partial u / \partial x \tag{3.266}$$

$$\partial^2 v / \partial x^2 = \alpha(\alpha e^{\alpha x + \beta \tau} u + e^{\alpha x + \beta \tau} \partial u / \partial x) + \alpha e^{\alpha x + \beta \tau} \partial u / \partial x + e^{\alpha x + \beta \tau} \partial^2 u / \partial x^2 \tag{3.267}$$

Substituting these partials into (3.261) yields

$$\beta u + \partial u / \partial \tau = \alpha^2 u + 2\alpha(\partial u / \partial x) + \partial^2 u / \partial x^2 + (k - 1)(\alpha u + \partial u / \partial x) - ku \tag{3.268}$$

We may eliminate the u terms and the $\partial u / \partial x$ terms by carefully choosing values of α and β such that

$$\beta = \alpha^2 + (k - 1)\alpha - k \tag{3.269}$$

and

$$2\alpha + k - 1 = 0 \tag{3.270}$$

One may then rearrange these equations, so they can be written

$$\alpha = -1/2(k - 1) \quad \text{and} \quad \beta = -1/4(k + 1)^2 \tag{3.271}$$

We now have the transformation from v to u to be

$$v = \exp[-1/2(k - 1)x - 1/4(k + 1)^2 \tau] u(x, \tau) \tag{3.272}$$

resulting in the simple diffusion equation

$$du / d\tau = d^2 u / dx^2, \quad \text{for} \quad -\infty < x < \infty, \tau > 0. \tag{3.273}$$

The initial condition has now been changed as well to

$$u_0(x) = u(x, 0) = \max(e^{\frac{1}{2}(k+1)x} - e^{\frac{1}{2}(k-1)x}, 0) \tag{3.274}$$

The solution to the simple diffusion equation obtained above is well known to be

$$u(x, \tau) = [1 / \sqrt{(2\pi)}] \int_{-\infty}^{\infty} u_0(s) \exp\{-(x - s)^2 / 4\tau\} ds \tag{3.275}$$

where $u_0(x, 0)$ is given by (3.274). In order to solve this integral, it is convenient to make a change of variable as follows:

$$y = (s - x)/\sqrt{(2\tau)} \tag{3.276}$$

so that

$$u(x, \tau) = [1/\sqrt{(2\pi)}] \int_{-\infty}^{\infty} u_0\{y\sqrt{(2\tau)} + x)\}\exp(-y^2/2)dy \tag{3.277}$$

Substituting our initial condition into this equation results in

$$\begin{aligned} u(x, \tau) = [1/\sqrt{(2\pi)}] \int_{-x/\sqrt{(2\tau)}}^{\infty} \exp[1/2(k + 1)\{y\sqrt{(2\tau)} + x\}\exp(-y^2/2)dy \\ -[1/\sqrt{(2\pi)}] \int_{-x/\sqrt{(2\tau)}}^{\infty} \exp[1/2(k - 1)\{y\sqrt{(2\tau)} + x\}\exp(-y^2/2)dy \end{aligned} \tag{3.278}$$

In order to solve this, one will solve each integral separately. The first integral can be solved by completing the square in the exponent. The exponent of the first integral is $-1/2y^2 + 1/2(k + 1)\{x + y\sqrt{(2\pi)}\}$.

Factoring out the $-1/2$ gives us $-1/2\{y^2 - (k + 1)y\sqrt{(2\tau)} - (k + 1)x\}$.

Separating out the term that is not a function of y, and adding and subtracting terms to set up a perfect square yields

$$1/2(k + 1)x - 1/2(y^2 - (k + 1)y\sqrt{(2\tau)} + [\{(k + 1)\sqrt{(2\tau)}\}/2]^2 - [(k + 1)\sqrt{(2\tau)}\}/2]^2) \tag{3.279}$$

which may be written as

$$(1/2)(k + 1)x - (1/2)[y^2 - \{[(k + 1)\sqrt{(2\tau)}]/2\}]^2 + (1/2)[\{(k + 1)\sqrt{(2\tau)}\}/2]^2 \tag{3.280}$$

and simplified to

$$(1/2)(k + 1)x - (1/2)[y - \{[k + 1]\sqrt{(2\tau)}\}/2]^2 + \{(k + 1)^2\tau\}/4 \tag{3.281}$$

Thus, the first integral reduces to

$$I_1 = \exp\{1/2(k + 1)x\}/\sqrt{(2\pi)} \int_{-x/\sqrt{(2\tau)}}^{\infty} \exp\{1/4(k + 1)^2\tau\} \exp[-1/2\{y - 1/2(k + 1)\sqrt{(2\tau)}\}^2]dy$$

Now substituting

$$z = y - 1/2[k + 1]\sqrt{(2\tau)}$$

results in

$$I_1 = [\exp\{1/2(k+1)x\} + 1/4(k+1)^2\tau]/\sqrt{(2\pi)}\int_{-x/\sqrt{(2\tau)}-(1/2)(k+1)\sqrt{(2\tau)}}^{\infty} \exp(-1/2)z^2 dz \qquad (3.282)$$

$$= \{[\exp\{(1/2)(k+1)x\} + (1/4)(k+1)^2 r]/(\sqrt{2\pi})\}N(d1) \qquad (3.283)$$

where

$$d_1 = \{x/\sqrt{(2\tau)}\} + \{(1/2)(k+1)\sqrt{(2\pi)}\} \qquad (3.284)$$

and

$$N(x) = [1/\sqrt{(2\pi)}]\int_{-\infty}^{x} \exp\{-1/2y^2\}dy \qquad (3.285)$$

is the cumulative distribution function for the normal distribution.

The calculation of the second integral I_2 is identical to that of I_1, except that $(k-1)$ replaces $(k+1)$ throughout. Finally, one works the way backward with

$$v(x,\tau) = \exp\{-1/2(k-1)x - 1/4(k+1)^2\tau\}u(x,\tau) \qquad (3.286)$$

and then substituting the inverse transformations

$$x = \log(S/E) \qquad (3.287)$$
$$\tau = 1/2\sigma^2(T-t) \qquad (3.288)$$
$$C = Ev(x,\tau) \qquad (3.289)$$

one finally obtains the desired result:

$$C(S,t) = SN(d_1) - Ee^{-r}(T-t)N(d_2) \qquad (3.290)$$

where

$$d_1 = \{\log(S/E) + (r + 1/2\sigma^2)(T-t)\}/\{\sigma\sqrt{(T-t)}\} \qquad (3.291)$$

and

$$d_2 = \{\log(S/E) + (r - 1/2\sigma^2)(T-t)\}/\{\sigma\sqrt{(T-t)}\} \qquad (3.292)$$

Conclusion Note that in the derivation of the Black–Scholes differential equation, it was never restricted to a specific type of derivative security when trying to find the price for until the boundary conditions are specified for a European call. This means that a person can use the Black–Scholes differential equation to solve for the price of *any* type of option—by changing the boundary conditions only.

The Black–Scholes model revolutionized the world of finance. For the first time the model has given traders, hedgers, and investors a standard way to value options. The model has also caused a huge growth in the importance of financial engineering in the world of finance. Today, mathematicians are building models to maximize portfolio returns while minimizing risk. They are also building sophisticated computer programs to search for inefficiencies in the market. The world of finance is being built on mathematics and the Black–Scholes model was the beginning of this mathematical revolution.

3.8 Some Worked Examples

Worked Examples I Measuring VaR (Value-at-Risk)—American Options

Risk Management: Since the pioneering work of Markowitz in the 1950s, the theory and practice of risk management have developed to be a subfield of the theory of finance, attracting specialists from many applied mathematical, physical science and engineering fields—bringing with them enormous contributions made by quantitative disciplines, including mathematics, statistics, and computer science. Important principles of risk management have also come from the fields of accounting (including auditing, valuation, and management control), management (including audit, management control, and valuation), the theory of organization (organizational behavior), and law (pervading aspects of risk management). Thus, in risk management, while the foundation may be qualitative, the approach is heavily quantitative—much like the practice of engineering!

The R Package Dowd
March 11, 2016

Type	Package
Title	Functions Ported from 'MMR2' Toolbox Offered in Kevin Dowd's Book *Measuring Market Risk*
Version	0.12
Date	2015-08-20
Author	Dinesh Acharya <dines.acharya@gmail.com>
Maintainer	Dinesh Acharya <dines.acharya@gmail.com>

Description Kevin Dowd's book *Measuring Market Risk* is a widely read book in the area of risk measurement by students and practitioners alike. As he claims, MATLAB indeed might have been the most suitable language when he originally wrote the functions, but, with growing popularity of R, it is not entirely valid. As 'Dowd's' code was not intended to be error-free and were mainly for reference, some functions in this package have inherited those errors. An attempt will be made in future releases to identify and correct them.

'Dowd's' original code can be downloaded from www.kevindowd.org/measuring-marketrisk/.

It should be noted that 'Dowd' offers both 'MMR2' and 'MMR1' toolboxes.

Only 'MMR2' was ported to R. 'MMR2' is more recent version of 'MMR1' toolbox and they both have mostly similar function. The toolbox mainly contains different parametric and nonparametric methods for measurement of market risk as well as back-testing risk measurement methods.

Depends	R (>= 3.0.0), bootstrap, MASS, forecast
Suggests	PerformanceAnalytics, testthat
License	GPL
Needs Compilation	no
Repository	CRAN
Date/Publication	2016-03-11 00:45:03

Worked Example (a):

[6] AmericanPutESBinomial Estimates **Expected Shortfall** (**ES**) of American Vanilla *put* Using Binomial Tree.

Description

Estimates ES of American Put Option using binomial tree to price the option and historical method to compute the Value-at-Risk (VaR).

Usage

```
AmericanPutESBinomial(amountInvested, stockPrice, strike, r,
volatility, maturity, numberSteps, cl, hp)
```

Arguments

amountInvested	Total amount paid for the Put Option.
stockPrice	Stock price of underlying stock.
strike	Strike price of the option.
r	Risk-free rate.
volatility	Volatility of the underlying stock.
maturity	Time to maturity of the option, in days.
numberSteps	The number of time-steps considered for the binomial model.
cl	Confidence level for which VaR is computed.
hp	Holding period of the option in days.

Value
ES of the American Put Option

Author(s)
Dinesh Acharya

Examples

```
# Market Risk of American Put with given parameters.
AmericanPutESBinomial(0.20, 27.2, 25, .16, .05, 60, 20, .95, 30)
```

In the R domain:

```
R version 3.2.2 (2015-08-14) -- "Fire Safety"
Copyright (C) 2015 The R Foundation for Statistical Computing
Platform: i386-w64-mingw32/i386 (32-bit)
R is free software and comes with ABSOLUTELY NO WARRANTY.
You are welcome to redistribute it under certain conditions.
Type 'license()' or 'licence()' for distribution details.
  Natural language support but running in an English locale
R is a collaborative project with many contributors.
Type 'contributors()' for more information and
'citation()' on how to cite R or R packages in publications.
Type 'demo()' for some demos, 'help()' for on-line help, or
'help.start()' for an HTML browser interface to help.
Type 'q()' to quit R.
>
> install.packages("Dowd")
Installing package into 'C:/Users/Bert/Documents/R/win-
library/3.2'
(as 'lib' is unspecified)
--- Please select a CRAN mirror for use in this session ---
# A CRAN mirror is selected
also installing the dependencies 'stringi', 'magrittr',
'stringr', 'RColorBrewer', 'dichromat', 'munsell', 'labeling',
'quadprog', 'digest', 'gtable', 'plyr', 'reshape2', 'scales',
'zoo', 'timeDate', 'tseries', 'fracdiff', 'Rcpp', 'colorspace',
'ggplot2', 'RcppArmadillo', 'bootstrap', 'forecast'
  There is a binary version available but the source version is later:
      binary source needs_compilation
colorspace 1.2-6 1.2-7       TRUE
  Binaries will be installed
```

```
trying URL 'https://cran.cnr.berkeley.edu/bin/windows/
contrib/3.2/stringi_1.1.2.zip'
Content type 'application/zip' length 14229497 bytes (13.6 MB)
downloaded 13.6 MB
trying URL 'https://cran.cnr.berkeley.edu/bin/windows/
contrib/3.2/magrittr_1.5.zip'
Content type 'application/zip' length 149966 bytes (146 KB)
downloaded 146 KB
trying URL 'https://cran.cnr.berkeley.edu/bin/windows/
contrib/3.2/stringr_1.1.0.zip'
Content type 'application/zip' length 119831 bytes (117 KB)
downloaded 117 KB
trying URL 'https://cran.cnr.berkeley.edu/bin/windows/
contrib/3.2/RColorBrewer_1.1-2.zip'
Content type 'application/zip' length 26734 bytes (26 KB)
downloaded 26 KB
trying URL 'https://cran.cnr.berkeley.edu/bin/windows/
contrib/3.2/dichromat_2.0-0.zip'
Content type 'application/zip' length 147767 bytes (144 KB)
downloaded 144 KB
trying URL 'https://cran.cnr.berkeley.edu/bin/windows/
contrib/3.2/munsell_0.4.3.zip'
Content type 'application/zip' length 134334 bytes (131 KB)
downloaded 131 KB
trying URL 'https://cran.cnr.berkeley.edu/bin/windows/
contrib/3.2/labeling_0.3.zip'
Content type 'application/zip' length 40854 bytes (39 KB)
downloaded 39 KB
trying URL 'https://cran.cnr.berkeley.edu/bin/windows/
contrib/3.2/quadprog_1.5-5.zip'
Content type 'application/zip' length 51794 bytes (50 KB)
downloaded 50 KB
trying URL 'https://cran.cnr.berkeley.edu/bin/windows/
contrib/3.2/digest_0.6.10.zip'
Content type 'application/zip' length 172481 bytes (168 KB)
downloaded 168 KB
trying URL 'https://cran.cnr.berkeley.edu/bin/windows/
contrib/3.2/gtable_0.2.0.zip'
Content type 'application/zip' length 57917 bytes (56 KB)
downloaded 56 KB
trying URL 'https://cran.cnr.berkeley.edu/bin/windows/
contrib/3.2/plyr_1.8.4.zip'
Content type 'application/zip' length 1119520 bytes (1.1 MB)
downloaded 1.1 MB
```

```
trying URL 'https://cran.cnr.berkeley.edu/bin/windows/
contrib/3.2/reshape2_1.4.1.zip'
Content type 'application/zip' length 505413 bytes (493 KB)
downloaded 493 KB
trying URL 'https://cran.cnr.berkeley.edu/bin/windows/
contrib/3.2/scales_0.4.0.zip'
Content type 'application/zip' length 604312 bytes (590 KB)
downloaded 590 KB
trying URL 'https://cran.cnr.berkeley.edu/bin/windows/
contrib/3.2/zoo_1.7-13.zip'
Content type 'application/zip' length 900652 bytes (879 KB)
downloaded 879 KB
trying URL 'https://cran.cnr.berkeley.edu/bin/windows/
contrib/3.2/timeDate_3012.100.zip'
Content type 'application/zip' length 791098 bytes (772 KB)
downloaded 772 KB
trying URL 'https://cran.cnr.berkeley.edu/bin/windows/
contrib/3.2/tseries_0.10-35.zip'
Content type 'application/zip' length 321523 bytes (313 KB)
downloaded 313 KB
trying URL 'https://cran.cnr.berkeley.edu/bin/windows/
contrib/3.2/fracdiff_1.4-2.zip'
Content type 'application/zip' length 106748 bytes (104 KB)
downloaded 104 KB
trying URL 'https://cran.cnr.berkeley.edu/bin/windows/
contrib/3.2/Rcpp_0.12.7.zip'
Content type 'application/zip' length 3230471 bytes (3.1 MB)
downloaded 3.1 MB
trying URL 'https://cran.cnr.berkeley.edu/bin/windows/
contrib/3.2/colorspace_1.2-6.zip'
Content type 'application/zip' length 391490 bytes (382 KB)
downloaded 382 KB
trying URL 'https://cran.cnr.berkeley.edu/bin/windows/
contrib/3.2/ggplot2_2.1.0.zip'
Content type 'application/zip' length 2001613 bytes (1.9 MB)
downloaded 1.9 MB
trying URL 'https://cran.cnr.berkeley.edu/bin/windows/
contrib/3.2/RcppArmadillo_0.7.400.2.0.zip'
Content type 'application/zip' length 1745518 bytes (1.7 MB)
downloaded 1.7 MB
trying URL 'https://cran.cnr.berkeley.edu/bin/windows/
contrib/3.2/bootstrap_2015.2.zip'
```

```
Content type 'application/zip' length 104944 bytes (102 KB)
downloaded 102 KB
trying URL 'https://cran.cnr.berkeley.edu/bin/windows/
contrib/3.2/forecast_7.2.zip'
Content type 'application/zip' length 1388535 bytes (1.3 MB)
downloaded 1.3 MB
trying URL 'https://cran.cnr.berkeley.edu/bin/windows/
contrib/3.2/Dowd_0.12.zip'
Content type 'application/zip' length 396919 bytes (387 KB)
downloaded 387 KB
package 'stringi' successfully unpacked and MD5 sums checked
package 'magrittr' successfully unpacked and MD5 sums checked
package 'stringr' successfully unpacked and MD5 sums checked
package 'RColorBrewer' successfully unpacked and MD5 sums checked
package 'dichromat' successfully unpacked and MD5 sums checked
package 'munsell' successfully unpacked and MD5 sums checked
package 'labeling' successfully unpacked and MD5 sums checked
package 'quadprog' successfully unpacked and MD5 sums checked
package 'digest' successfully unpacked and MD5 sums checked
package 'gtable' successfully unpacked and MD5 sums checked
package 'plyr' successfully unpacked and MD5 sums checked
package 'reshape2' successfully unpacked and MD5 sums checked
package 'scales' successfully unpacked and MD5 sums checked
package 'zoo' successfully unpacked and MD5 sums checked
package 'timeDate' successfully unpacked and MD5 sums checked
package 'tseries' successfully unpacked and MD5 sums checked
package 'fracdiff' successfully unpacked and MD5 sums checked
package 'Rcpp' successfully unpacked and MD5 sums checked
package 'colorspace' successfully unpacked and MD5 sums checked
package 'ggplot2' successfully unpacked and MD5 sums checked
package 'RcppArmadillo' successfully unpacked and MD5 sums checked
package 'bootstrap' successfully unpacked and MD5 sums checked
package 'forecast' successfully unpacked and MD5 sums checked
package 'Dowd' successfully unpacked and MD5 sums checked
The downloaded binary packages are in
C:\Users\Bert\AppData\Local\Temp\Rtmp2n3pwv\downloaded_
packages
> library(Dowd)
Loading required package: bootstrap
Loading required package: MASS
Loading required package: forecast
Loading required package: zoo
Attaching package: 'zoo'
```

The following objects are masked from 'package:base':
  as.Date, as.Date.numeric
Loading required package: timeDate
This is forecast 7.2
Warning messages:
1: package 'Dowd' was built under R version 3.2.5
2: package 'bootstrap' was built under R version 3.2.5
3: package 'forecast' was built under R version 3.2.5
4: package 'zoo' was built under R version 3.2.5
5: package 'timeDate' was built under R version 3.2.5
> ls("package:Dowd")
  [1] "AdjustedNormalESHotspots"
  [2] "AdjustedNormalVaRHotspots"
  [3] "AdjustedVarianceCovarianceES"
  [4] "AdjustedVarianceCovarianceVaR"
  [5] "ADTestStat"
  [6] "AmericanPutESBinomial"
  [7] "AmericanPutESSim"
  [8] "AmericanPutPriceBinomial"
  [9] "AmericanPutVaRBinomial"
 [10] "BinomialBacktest"
 [11] "BlackScholesCallESSim"
 [12] "BlackScholesCallPrice"
 [13] "BlackScholesPutESSim"
 [14] "BlackScholesPutPrice"
 [15] "BlancoIhleBacktest"
 [16] "BootstrapES"
 [17] "BootstrapESConfInterval"
 [18] "BootstrapESFigure"
 [19] "BootstrapVaR"
 [20] "BootstrapVaRConfInterval"
 [21] "BootstrapVaRFigure"
 [22] "BoxCoxES"
 [23] "BoxCoxVaR"
 [24] "CdfOfSumUsingGaussianCopula"
 [25] "CdfOfSumUsingGumbelCopula"
 [26] "CdfOfSumUsingProductCopula"
 [27] "ChristoffersenBacktestForIndependence"
 [28] "ChristoffersenBacktestForUnconditionalCoverage"
 [29] "CornishFisherES"
 [30] "CornishFisherVaR"
 [31] "DBPensionVaR"
 [32] "DCPensionVaR"

```
[33] "DefaultRiskyBondVaR"
[34] "FilterStrategyLogNormalVaR"
[35] "FrechetES"
[36] "FrechetESPlot2DCl"
[37] "FrechetVaR"
[38] "FrechetVaRPlot2DCl"
[39] "GaussianCopulaVaR"
[40] "GParetoES"
[41] "GParetoMEFPlot"
[42] "GParetoMultipleMEFPlot"
[43] "GParetoVaR"
[44] "GumbelCopulaVaR"
[45] "GumbelES"
[46] "GumbelESPlot2DCl"
[47] "GumbelVaR"
[48] "GumbelVaRPlot2DCl"
[49] "HillEstimator"
[50] "HillPlot"
[51] "HillQuantileEstimator"
[52] "HSES"
[53] "HSESDFPerc"
[54] "HSESFigure"
[55] "HSESPlot2DCl"
[56] "HSVaR"
[57] "HSVaRDFPerc"
[58] "HSVaRESPlot2DCl"
[59] "HSVaRFigure"
[60] "HSVaRPlot2DCl"
[61] "InsuranceVaR"
[62] "InsuranceVaRES"
[63] "JarqueBeraBacktest"
[64] "KernelESBoxKernel"
[65] "KernelESEpanechinikovKernel"
[66] "KernelESNormalKernel"
[67] "KernelESTriangleKernel"
[68] "KernelVaRBoxKernel"
[69] "KernelVaREpanechinikovKernel"
[70] "KernelVaRNormalKernel"
[71] "KernelVaRTriangleKernel"
[72] "KSTestStat"
[73] "KuiperTestStat"
[74] "LogNormalES"
[75] "LogNormalESDFPerc"
```

```
 [76] "LogNormalESFigure"
 [77] "LogNormalESPlot2DCL"
 [78] "LogNormalESPlot2DHP"
 [79] "LogNormalESPlot3D"
 [80] "LogNormalVaR"
 [81] "LogNormalVaRDFPerc"
 [82] "LogNormalVaRETLPlot2DCL"
 [83] "LogNormalVaRFigure"
 [84] "LogNormalVaRPlot2DCL"
 [85] "LogNormalVaRPlot2DHP"
 [86] "LogNormalVaRPlot3D"
 [87] "LogtES"
 [88] "LogtESDFPerc"
 [89] "LogtESPlot2DCL"
 [90] "LogtESPlot2DHP"
 [91] "LogtESPlot3D"
 [92] "LogtVaR"
 [93] "LogtVaRDFPerc"
 [94] "LogtVaRPlot2DCL"
 [95] "LogtVaRPlot2DHP"
 [96] "LogtVaRPlot3D"
 [97] "LongBlackScholesCallVaR"
 [98] "LongBlackScholesPutVaR"
 [99] "LopezBacktest"
[100] "MEFPlot"
[101] "NormalES"
[102] "NormalESConfidenceInterval"
[103] "NormalESDFPerc"
[104] "NormalESFigure"
[105] "NormalESHotspots"
[106] "NormalESPlot2DCL"
[107] "NormalESPlot2DHP"
[108] "NormalESPlot3D"
[109] "NormalQQPlot"
[110] "NormalQuantileStandardError"
[111] "NormalSpectralRiskMeasure"
[112] "NormalVaR"
[113] "NormalVaRConfidenceInterval"
[114] "NormalVaRDFPerc"
[115] "NormalVaRFigure"
[116] "NormalVaRHotspots"
[117] "NormalVaRPlot2DCL"
[118] "NormalVaRPlot2DHP"
```

```
[119] "NormalVaRPlot3D"
[120] "PCAES"
[121] "PCAESPlot"
[122] "PCAPrelim"
[123] "PCAVaR"
[124] "PCAVaRPlot"
[125] "PickandsEstimator"
[126] "PickandsPlot"
[127] "ProductCopulaVaR"
[128] "ShortBlackScholesCallVaR"
[129] "ShortBlackScholesPutVaR"
[130] "StopLossLogNormalVaR"
[131] "tES"
[132] "tESDFPerc"
[133] "tESFigure"
[134] "tESPlot2DCL"
[135] "tESPlot2DHP"
[136] "tESPlot3D"
[137] "TQQPlot"
[138] "tQuantileStandardError"
[139] "tVaR"
[140] "tVaRDFPerc"
[141] "tVaRESPlot2DCL"
[142] "tVaRFigure"
[143] "tVaRPlot2DCL"
[144] "tVaRPlot2DHP"
[145] "tVaRPlot3D"
[146] "VarianceCovarianceES"
[147] "VarianceCovarianceVaR"
>
> # Market Risk of American Put with given parameters.
> AmericanPutESBinomial(0.20, 27.2, 25, .16, .05, 60, 20, .95,
+ 30)
> # Outputting:
[1] 0.2
> # The ES (Expected Shortfall) of this American Put Option is 0.2.
>
```

Worked Examples II

```
[8] "AmericanPutPriceBinomial"
```

Description   Estimates the price of an American Put, using the binomial approach.

Usage

```
AmericanPutPriceBinomial(stockPrice, strike, r, sigma,
maturity, numberSteps)
```

Arguments

stockPrice	Stock price of underlying stock
strike	Strike price of the option
r	Risk-free rate
sigma	Volatility of the underlying stock and is an annualized term
maturity	The term to maturity of the option in days
numberSteps	The number of time steps in the binomial tree

Value
Binomial American put price

Author(s)
Dinesh Acharya

Examples

```
# Estimates the price of an American Put
AmericanPutPriceBinomial(27.2, 25, .03, .2, 60, 30)
```

In the R domain:

```
>
> # Estimates the price of an American Put
> AmericanPutPriceBinomial(27.2, 25, .03, .2, 60, 30)
> # Outputting:
[1] 0.1413597
> # The ES (Expected Shortfall) of this American Put Option
is 0.1413597.
>
>
```

Worked Examples III

[9] "AmericanPutVaRBinomial" Estimates *Expected Shortfall (ES)* of American Vanilla **Put** Using Binomial Tree.

Description
Estimates VaR of American Put Option using binomial tree to price the option and historical method to compute the VaR.

Usage

```
AmericanPutVaRBinomial(amountInvested, stockPrice, strike, r,
volatility, maturity, numberSteps, cl, hp)
```

Arguments

amountInvested	Total amount paid for the Put Option.
stockPrice	Stock price of underlying stock.
strike	Strike price of the option.
r	Risk-free rate.
volatility	Volatility of the underlying stock.
maturity	Time to maturity of the option in days.
numberSteps	The number of time steps considered for the binomial model.
cl	Confidence level for which VaR is computed.
hp	Holding period of the option in days.

Value
VaR of the American Put Option

Author(s)
Dinesh Acharya

In the R domain:

```
>
> # Market Risk of American call with given parameters.
> BlackScholesCallESSim(0.20, 27.2, 25, .16, .2, .05, 60, 30,
+ .95, 30)
> # Outputting:
[1] 0.003958873
> # The Market Risk of this American Call Option is 0.003958873.
>
```

Review Questions and Exercises

1   Contrast the following three classical theories of portfolio allocation using

    i) The Markowitz model,
    ii) The Black–Litterman model, and
    iii) Capital asset pricing model (CAPM)

in terms of the following aspects of these models:

    a The assumptions of each model
    b The advantages and disadvantages of these assumptions

2   With respect to the Black–Litterman model for assets allocation and portfolio optimization,

    a state and discuss the basic assumptions and the strengths and weaknesses of this model;
    b state and discuss available improvements suggested for this model.

3   *The Black–Scholes equation*, and boundary conditions, for a European option call is

$$\partial C / \partial t + \tfrac{1}{2}\sigma^2 S^2 \partial^2 C / \partial S^2 + rS \partial C / \partial S - rC = 0 \qquad \text{(RE-3.1)}$$

where

C	=	the value of the call $= C(S, t)$,
S	=	the fixed Strike price,
t	=	time,
r	=	the risk-free interest rate,

and

σ	=	the volatility of the underlying asset.

Equation (RE-3.1) is a rather complicated second-order partial differential equation (pde).

It may be somewhat simplified, as follows:

If S is fixed and E is variable, show the European option price. $C = C(E, t)$ also satisfies the following partial differential equation:

$$\partial C / \partial t + \tfrac{1}{2}\sigma^2 E^2 \partial^2 C / \partial E^2 - rE \partial C / \partial E = 0 \qquad \text{(RE-3.2)}$$

which may be considered as a simplified form of the Black–Scholes equation.

As an Example:

In the analysis of a pde, often useful solutions may be found by judicial changes of variables. A typical case is the classical heat/mass diffusion equation, solvable by the method of Separation of Variables:

$$\partial u/\partial t = c^2 \partial^2 u/\partial x^2 \tag{RE-3.3}$$

with boundary conditions:

$$u(0,\ t) = 0 = u(L,t), \quad \forall t$$

$$u(x,0) = f(x), \quad @t = 0$$

and an initial condition: $u(x,\ 0) = f(x)$, a given function of x.

An elementary approach is to assume the separable of the independent variables: x and t, so that one may write

$$u(x,t) = X(x)\,T(t) \tag{RE-3.4}$$

and obtain, after some elementary steps, the final solution:

$$u(x,t) = \sum_{n=0}^{\infty} B_n \sin(n\pi/l)\, x \exp\left(-n^2\pi^2 c^2 t/l^2\right)$$

where

$$B_n = (2/l) \int_0^l f(x) \sin\left(n\pi/l\right) x dx, \quad \forall \text{ positive integers } n$$

4   The CRAN package BLCOP—implementing the Black–Litterman approach to Assets Allocation and Portfolio Optimization:

BLCOP: **Black–Litterman and Copula Opinion Pooling Frameworks**

An implementation of the Black–Litterman model and Atilio Meucci's copula opinion pooling framework.da

Version:	0.3.1
Depends:	methods, MASS, quadprog
Imports:	RUnit ($\geq$ 0.4.22), timeSeries, fBasics, fMultivar, fPortfolio ($\geq$3011.81)
Suggests:	sn, corpcor, mnormt
Published:	2015-02-04
Author:	Francisco Gochez, Richard Chandler-Mant, Suchen Jin, Jinjing Xie
Maintainer:	Ava Yang <ayang at mango-solutions.com>

License:	GPL-3
NeedsCompilation:	no
Materials:	NEWS
CRAN **checks**:	BLCOP results

Downloads:
Run the following example in the R **domain:**
optimalPortfolios
Calculates Optimal Portfolios under Prior and Posterior Distributions

Description

These are wrapper functions that calculate optimal portfolios under the prior and posterior return distributions. optimalPortfolios works with a user-supplied optimization function, although simple Markowitz minimum-risk optimization is done with solve.QP from quadprog if none is supplied.

optimalPortfolios.fPort is a generic utility function that calculates optimal portfolios using routines from the fPortfolio package.

Usage

```
optimalPortfolios(result, optimizer = .optimalWeights.
simpleMV,
..., doPlot = TRUE, beside = TRUE)
optimalPortfolios.fPort(result, spec = NULL,
constraints = "LongOnly",
optimizer = "minriskPortfolio",
inputData = NULL,
numSimulations = BLCOPOptions("numSimulations"))
```

Arguments

result	An object of class BL Result.
optimiser	For optimalPortfolios, an optimization function. It should take as arguments a vector of means and a variance–covariance matrix, and should return a vector of optimal weights. For optimalPortfolios, the name of a fPortfolio function that performs portfolio optimization.

spec	Object of class fPORTFOLIOSPEC. If NULL, will use a basic mean-variance spec for Black–Litterman results, and a basic CVaR spec for COP results.
inputData	Time series data (any form that can be coerced into a timeSeries object).
constraints	String of constraints that may be passed into fPortfolio optimization routines.
numSimulations	For COP results only—the number of posterior simulations to use in the optimization (large numbers here will likely cause the routine to fail).
...	Additional arguments to the optimization function.
doPlot	A logical flag. Should barplots of the optimal portfolio weights be produced?
beside	A logical flag. If a barplot is generated, should the bars appear side-by side? If FALSE, differences of weights will be plotted instead.

Details

By default, optimizer is a simple function that performs Markowitz optimization via solve.QP. In addition to a mean and variance, it takes an optional constraints parameter that if supplied should hold a named list with all of the parameters that solve.QP takes.

Value

optimalPortfolios	Will return a list with the following items.
priorPFolioWeights	The optimal weights under the prior distribution.
postPFolioWeights	The optimal weights under the posterior distribution.
optimalPortfolios. fPort	Will return a similar list with two elements of class fPORTFOLIO.

Note

It is expected that optimalPortfolios will be deprecated in future releases in favor of optimalPortfolios.fPort.

Author(s)

Francisco Gochez <fgochez@mango-solutions.com>

Examples

```
entries <- c(0.001005, 0.001328, -0.000579,-0.000675,
0.000121, 0.000128,
-0.000445, -0.000437, 0.001328, 0.007277, -0.001307,
-0.000610,
-0.002237, -0.000989, 0.001442,-0.001535, -0.000579,
-0.001307,
0.059852, 0.027588, 0.063497, 0.023036, 0.032967, 0.048039,
-0.000675, -0.000610, 0.027588, 0.029609, 0.026572, 0.021465,
0.020697, 0.029854, 0.000121,-0.002237, 0.063497, 0.026572,
0.102488, 0.042744, 0.039943, 0.065994, 0.000128,-0.000989,
0.023036, 0.021465, 0.042744, 0.032056, 0.019881, 0.032235,
-0.000445, 0.001442, 0.032967, 0.020697, 0.039943, 0.019881,
0.028355, 0.035064, -0.000437, -0.001535, 0.048039, 0.029854,
0.065994, 0.032235, 0.035064, 0.079958)
varcov <- matrix(entries, ncol = 8, nrow = 8)
mu <- c(0.08, 0.67,6.41, 4.08, 7.43, 3.70, 4.80, 6.60) / 100
pick <- matrix(0, ncol = 8, nrow = 3, dimnames = list(NULL,
letters[1:8]))
pick[1,7] <- 1
pick[2,1] <- -1; pick[2,2] <- 1
pick[3, 3:6] <- c(0.9, -0.9, .1, -.1)
confidences <- 1 / c(0.00709, 0.000141, 0.000866)
views <- BLViews(pick, c(0.0525, 0.0025, 0.02), confidences,
letters[1:8])
posterior <- posteriorEst(views, tau = 0.025, mu, varcov)
optimalPortfolios(posterior, doPlot = TRUE)
optimalPortfolios.fPort(posterior, optimizer =
"tangencyPortfolio")
# An example based on one found in "Beyond Black-Litterman:
# Views on Non-normal Markets"
dispersion <-
c(.376,.253,.360,.333,.360,.600,.397,.396,.578,.775) / 1000
sigma <- BLCOP:::.symmetricMatrix(dispersion, dim = 4)
caps <- rep(1/4, 4)
mu <- 2.5 * sigma
dim(mu) <- NULL
marketDistribution <- mvdistribution("mt", mean = mu, S =
sigma, df = 5)
pick <- matrix(0, ncol = 4, nrow = 1, dimnames = list(NULL,
c("SP", "FTSE", "CAC", "DAX")))
pick[1,4] <- 1
```

```
vdist <- list(distribution("unif", min = -0.02, max = 0))
views <- COPViews(pick, vdist, 0.2, c("SP", "FTSE", "CAC",
"DAX"))
posterior <- COPPosterior(marketDistribution, views)
optimalPortfolios.fPort(myPosterior, spec = NULL, optimizer =
"minriskPortfolio",
inputData = NULL, numSimulations = 100)
## End
```

5 The CRAN package PerformanceAnalytics—providing a number of useful programs for investment risk analysis and portfolio performance.

Package 'PerformanceAnalytics'	February 19, 2015
Type	Package
Title	Econometric tools for performance and risk analysis
Version	1.4.3541
Date	2014-09-15 04:39:58–0500 (Mon, Sep 15, 2014)
Description	Collection of econometric functions for performance and risk analysis. This package aims to aid practitioners and researchers in utilizing the latest research in analysis of nonnormal return streams. In general, it is most tested on return (rather than price) data on a regular scale, but most functions will work with irregular return data as well, and increasing numbers of functions will work with P&L or price data where possible
Imports	zoo
Depends	R (>= 3.0.0), xts (>= 0.9)
Suggests	Hmisc, MASS, quantmod, gamlss, gamlss.dist, robustbase, quantreg, gplots
License	GPL-2 | GPL-3
URL	http://r-forge.r-project.org/projects/returnanalytics/
Copyright	(c) 2004–2014
Authors	Brian G. Peterson [cre, aut, cph],
	Peter Carl [aut, cph],

Kris Boudt [ctb, cph],

Ross Bennett [ctb],

Joshua Ulrich [ctb],

Eric Zivot [ctb],

Matthieu Lestel [ctb],

Kyle Balkissoon [ctb],

Diethelm Wuertz [ctb]

Maintainer	Brian G. Peterson <brian@braverock.com>
Needs Compilation	yes
Repository	CRAN
Date/Publication	2014-09-16 09:47:58

****

chart.TimeSeries Creates a time series chart with some extensions.

Description

Draws a line chart and labels the x-axis with the appropriate dates. This is really a "primitive", since it extends the base plot and standardizes the elements of a chart. Adds attributes for shading areas of the timeline or aligning vertical lines along the timeline. This function is intended to be used inside other charting functions.

Usage

```
chart.TimeSeries(R, auto.grid = TRUE, xaxis = TRUE, yaxis = TRUE,
yaxis.right = FALSE, type = "l", lty = 1, lwd = 2, las = par
("las"),
main = NULL, ylab = NULL, xlab = "", date.format.in = "%Y-%m-%d",
date.format = NULL, xlim = NULL, ylim = NULL,
element.color = "darkgray", event.lines = NULL, event.labels =
NULL,
period.areas = NULL, event.color = "darkgray",
period.color = "aliceblue", colorset = (1:12), pch = (1:12),
legend.loc = NULL, ylog = FALSE, cex.axis = 0.8, cex.legend = 0.8,
cex.lab = 1, cex.labels = 0.8, cex.main = 1, major.ticks = "auto",
minor.ticks = TRUE, grid.color = "lightgray", grid.lty =
"dotted",
xaxis.labels = NULL, ...)
charts.TimeSeries(R, space = 0, main = "Returns", ...)
```

Arguments
R an xts, vector, matrix, data frame, timeSeries or zoo object of
asset returns
auto.grid if true, draws a grid aligned with the points on the x
and y axes
xaxis if true, draws the x axis
yaxis if true, draws the y axis
yaxis.right if true, draws the y axis on the right-hand side of the
plot
type set the chart type, same as in plot
lty set the line type, same as in plot
lwd set the line width, same as in plot
las set the axis label rotation, same as in plot
main set the chart title, same as in plot
ylab set the y-axis label, same as in plot
xlab set the x-axis label, same as in plot
date.format.in allows specification of other date formats in the
data object, defaults to "%Y-
%m-%d"
date.format re-format the dates for the xaxis; the default is "%m/
%y"
chart.TimeSeries 73
·xlim set the x-axis limit, same as in plot
ylim set the y-axis limit, same as in plot
element.color provides the color for drawing chart elements, such
as the box lines, axis lines,
etc. Default is "darkgray"
event.lines If not null, vertical lines will be drawn to indicate
that an event happened during
that time period. event.lines should be a list of dates (e.g.,
c("09/03","05/06"))
formatted the same as date.format. This function matches the
re-formatted row
names (dates) with the events.list, so to get a match the
formatting needs to be
correct.
event.labels if not null and event.lines is not null, this will
apply a list of text labels (e.g.,
c("This Event", "That Event") to the vertical lines drawn. See the
example
below.
period.areas these are shaded areas described by start and end
dates in a vector of xts date

rangees, e.g., c("1926-10::1927-11","1929-08::1933-03") See the examples below.

event.color draws the event described in event.labels in the color specified

period.color draws the shaded region described by period.areas in the color specified

colorset color palette to use, set by default to rational choices

pch symbols to use, see also plot

legend.loc places a legend into one of nine locations on the chart: bottomright, bottom,

bottomleft, left, topleft, top, topright, right, or center.

ylog TRUE/FALSE set the y-axis to logarithmic scale, similar to plot, default FALSE

cex.axis The magnification to be used for axis annotation relative to the current setting

of 'cex', same as in plot.

cex.legend The magnification to be used for sizing the legend relative to the current setting

of 'cex'.

cex.lab The magnification to be used for x- and y-axis labels relative to the current

setting of 'cex'.

cex.labels The magnification to be used for event line labels relative to the current setting

of 'cex'.

cex.main The magnification to be used for the chart title relative to the current setting of

'cex'.

major.ticks Should major tickmarks be drawn and labeled, default 'auto'

minor.ticks Should minor tickmarks be drawn, default TRUE

grid.color sets the color for the reference grid

grid.lty defines the line type for the grid

xaxis.labels Allows for non-date labeling of date axes, default is NULL

space default 0

... any other passthru parameters

Author(s)

Peter Carl

74 chart.TimeSeries

See Also

plot, par, axTicksByTime

```
Examples
# These are start and end dates, formatted as xts ranges.
## http://www.nber.org-cycles.html
cycles.dates<-c("1857-06/1858-12",
"1860
Examples
# These are start and end dates, formatted as xts ranges.
## http://www.nber.org-cycles.html
cycles.dates<-c("1857-06/1858-12",
"1860-10/1861-06",
"1865-04/1867-12",
"1869-06/1870-12",
"1873-10/1879-03",
"1882-03/1885-05",
"1887-03/1888-04",
"1890-07/1891-05",
"1893-01/1894-06",
"1895-12/1897-06",
"1899-06/1900-12",
"1902-09/1904-08",
"1907-05/1908-06",
"1910-01/1912-01",
"1913-01/1914-12",
"1918-08/1919-03",
"1920-01/1921-07",
"1923-05/1924-07",
"1926-10/1927-11",
"1929-08/1933-03",
"1937-05/1938-06",
"1945-02/1945-10",
"1948-11/1949-10",
"1953-07/1954-05",
"1957-08/1958-04",
"1960-04/1961-02",
"1969-12/1970-11",
"1973-11/1975-03",
"1980-01/1980-07",
"1981-07/1982-11",
"1990-07/1991-03",
"2001-03/2001-11",
"2007-12/2009-06"
)
# Event lists - FOR BEST RESULTS, KEEP THESE DATES IN ORDER
```

```
risk.dates = c (
"Oct 87",
"Feb 94",
"Jul 97",
"Aug 98",
"Oct 98",
"Jul 00",
"Sep 01")
risk.labels = c (
"Black Monday",
chart.VaRSensitivity 75
"Bond Crash",
"Asian Crisis",
"Russian Crisis",
"LTCM",
"Tech Bubble",
"Sept 11")
data(edhec)
R=edhec[, "Funds of Funds",drop=FALSE]
Return.cumulative = cumprod(1+R) - 1
chart.TimeSeries(Return.cumulative)
chart.TimeSeries(Return.cumulative, colorset = "darkblue",
legend.loc = "bottomright",
period.areas = cycles.dates,
period.color = "lightblue",
event.lines = risk.dates,
event.labels = risk.labels,
event.color = "red", lwd = 2)
```

Solutions to Exercise 3: The Black–Scholes Equation

3. For the Black–Scholes equation (RE-3.1),
    Let

$$C = Ef(S/E) \tag{RE-3.5}$$

for some function f, then

$$C/E = f(S/E) \tag{RE-3.6}$$

Apply the change of variables stated in (RE-3.5) to reduce the Black–Scholes equation (RE-3.1) to its simpler form:

$$\partial C/\partial t + 1/2\sigma^2 E^2 \partial^2 C/\partial E^2 - rE\partial C/\partial E = 0 \tag{RE-3.7}$$

which is (RE-3.2).

Proof:

Since $C = Ef(S/E)$, from (RE-3.5),

$$\partial C/\partial E = \partial/\partial E\{Ef(S/E)\}$$
$$= f(S/E)\partial\{E\}/\partial E + E\partial\{f(S/E)\}/\partial E$$
$$= \{f(S/E)\} \cdot 1 + E.S\{(-1/E^2)\partial F/\partial E\}$$
$$= f(S/E) - (S/E)F'(S/E)$$

therefore, upon substituting for $f(S/E)$ from (RE-3.6):

$$\partial C/\partial E = C/E - (S/E)F'(S/E) \tag{RE-3.8}$$

Again, since $C = Ef(S/E)$, from (RE-3.4),

$$\partial C/\partial S = (\partial/\partial S)[C]$$
$$= (\partial/\partial S)[Ef(S/E)]$$
$$= E(\partial/\partial S)[f(S/E)]$$
$$= E(1/E)(\partial/\partial S)[f(S/E)]$$
$$= (1)f'(S/E)]$$
$$= f\prime(S/E)]$$

Therefore,

$$\partial C/\partial S = f'(S/E)] \tag{RE-3.9}$$

or

$$\partial C/\partial S = f'(S/E)] = f'$$

Hence,

$$S\partial C/\partial S = Sf'(S/E)] = Sf'$$

Again, from (RE-3.5),

$$C = Ef(S/E)$$

Therefore,

$$\partial C/\partial S = (\partial/\partial S)[Ef(S/E)]$$
$$= (1/E)\partial/\partial(S/E)[Ef(S/E)] \tag{RE-3.10}$$
$$= \partial/\partial(S/E)[f(S/E)]$$
$$= f\prime(S/E)$$

Therefore,

$$E\partial/\partial E(C) = \partial/\partial E[Ef(S/E)]$$

upon substituting for C from (RE-3.5).
Also, from (RE-3.8),

$$\partial C/\partial E = C/E - (S/E)F'(S/E) \tag{RE-3.11}$$

Hence,

$$\partial C/\partial E = \partial/\partial E[Ef(S/E)]$$
$$= f(S/E) \tag{RE-3.12}$$

Now,

$$E\partial C/\partial E = E[C/E - (S/E)F'(S/E)], \text{ from (RE-3.11)}$$
$$= C - SF'(S/E)$$
$$= C - S\partial C/\partial S, \text{ from (RE-3.10)}$$

Hence,

$$S\partial C/\partial S = C = E\partial C/\partial E \tag{RE-3.13}$$

which is an important *intermediate* result.
    Similarly, it may be shown that

$$S^2\partial^2 C/\partial S^2 = E^2\partial^2 C/\partial E^2 \tag{RE-3.14}$$

which is a *final* result that may be readily obtained as follows:
    Since $S\partial C/\partial S = C - E\partial C/\partial E$, from (RE-3.13)

$$\partial C/\partial S = C/S - (E/S)\partial C/\partial E$$

and

$$\partial^2 C/\partial S^2 = \partial/\partial S\{C/S - (E/S)\partial C/\partial E\}$$
$$= \partial/\partial S\{C/S\} - \partial/\partial S\{(E/S)(\partial C/\partial E)\}$$
$$= -(C/S2) + (E/S^2)(\partial C/\partial E)$$

namely,

$$S^2\partial^2 C/\partial S^2 = -C + E(\partial C/\partial E) \tag{RE-3.15}$$

Moreover, since

$$\partial C/\partial E = F(S/E), \text{ from (RE-3.12)}$$
$$E\partial C/\partial E = EF(S/E) = C, \text{ from (RE-3.5)}$$

Therefore,

$$\partial C / \partial E = C/E \tag{RE-3.16}$$

and

$$(\partial^2 C / \partial E^2) = \partial / \partial E [\partial C / \partial E]$$

by definition

$$= \partial / \partial E [C/E], \text{ from (RE-3.16)}$$
$$= -(C/E^2) + (1/E)(\partial C / \partial E)$$

by the differentiation of a quotient rule

or

$$E^2 (\partial^2 C / \partial E^2) = -C + E(\partial C / \partial E) \tag{RE-3.17}$$

Combining (RE-3.15) and (RE-3.17), one finally obtains (RE-3.14), as required.

And now, upon substituting for the terms $S^2 \partial^2 C / \partial S^2$ and $S \partial C / \partial S$, from (RE-3.14) and (RE-3.13), respectively, into the Black–Scholes equation (RE-3.1), the result is

$$\partial C / \partial t + 1/2 \sigma^2 S^2 \partial^2 C / \partial S^2 + rS \partial C / \partial S - rC = 0 \tag{RE-3.1}$$

namely,

$$\partial C / \partial t + 1/2 \sigma^2 (S^2 \partial^2 C / \partial S^2) + r(S \partial C / \partial S) - rC = 0$$

namely,

$$\partial C / \partial t + 1/2 \sigma^2 (E^2 \partial 2C / \partial E^2) + r(C - E \partial C / \partial E) - rC = 0$$

namely,

$$\partial C / \partial t + 1/2 \sigma^2 (E^2 \partial^2 C / \partial E^2) - rE \partial C / \partial E = 0 \tag{RE-3.2}$$

which is (RE-3.7), as required.

4

Data Analysis Using R Programming

In an Internet online advertisement, a job vacancy advertisement for a statistician reads as follows:

Job Summary

Statistician I

Salary: Open

Employer: XYZ Research and Statistics

Location: City X, State Y

Type: Full time – entry level

Category: Financial analyst/Statistics,
Data analysis/processing, Statistical organization
& administration

Required Education: Masters Degree preferred

XYZ Research and Statistics is a national leader in designing, managing, and analyzing financial data. XYZ partners with other investigators to offer respected statistical expertise supported by sophisticated web-based data management systems. XYZ services assure timely and secure implementation of trials and reliable data analyses.

Job Description

Position Summary: An exciting opportunity is available for a statistician to join a small but growing group focused on financial investment analysis and related translational research. XYZ, which is located in downtown City XX, is responsible for the design, management, and analysis of a variety of investment and financial, as well as the analysis of associated market data. The successful candidate will collaborate with fellow statistics staff and financial investigators to design, evaluate, and interpret investment studies.

Applied Probabilistic Calculus for Financial Engineering: An Introduction Using R, First Edition.
Bertram K. C. Chan.
© 2017 John Wiley & Sons, Inc. Published 2017 by John Wiley & Sons, Inc.
Companion website: www.wiley.com/go/chan/appliedprobabilisticcalculus

Primary Duties and Responsibilities: Analyzes investment situations and associated ancillary studies in collaboration with fellow statisticians and other financial engineers. Prepares tables, figures, and written summaries of study results; interprets results in collaboration with other financial; and assists in preparation of manuscripts. Provides statistical consultation with collaborating staff. Performs other job-related duties as assigned.

Requirements

Required Qualifications: Masters Degree in Statistics, Applied Mathematics, or related field. Sound knowledge of applied statistics. Proficiency in statistical computing in R.

Preferred Responsibilities/Qualifications: Statistical consulting experience. S-Plus or R programming language experience. Experience with analysis of high-dimensional data. Ability to communicate well orally and in writing. Excellent interpersonal/teamwork skills for effective collaboration. Spanish language skills a plus.

*In your cover letter, describe how your skills and experience match the qualifications for the position.

To learn more about XYZ, visit www.XYZ.org.

Clearly, one should be cognizant of the overt requirement of an acceptable level of professional proficiency in data analysis using R programming!

Even if one is not in such a job market, as a statistician working in the fields of Finance, Asset Allocations, Portfolio Optimization, and so on, a skill set that would include R programming would be helpful and interesting.

4.1 Data and Data Processing

Data are facts or figures from which conclusions can be drawn. When the data have been recorded, classified, organized, related, or interpreted within a framework so that meaning emerges, they become *information.* There are several steps involved in turning data into information, and these steps are known as *data processing.* This section describes data processing and how computers perform these steps efficiently and effectively.

It will be indicated that many of these processing activities may be undertaken using R programming, or performed in an R environment with the aid of available R packages – where R functions and data sets are stored.

4.1.1 Introduction

4.1.1.1 Coding

4.1.1.1.1 *Automated coding systems*

The simplified flowchart shows how raw *data* are transformed into information:

Data → Collection → Processing → Information

Data processing takes place once all of the relevant data have been collected. They are gathered from various sources and entered into a computer where they can be processed to produce *information* – the output.

Data processing includes the following steps:

- Data coding
- Data capture
- Editing
- Imputation
- Quality control
- Producing results

Data Coding   First, before raw data can be entered into a computer, they must be coded. To do this, survey responses must be labeled, usually with simple, numerical codes. This may be done by the interviewer in the field or by an office employee. The data coding step is important because it makes data entry and data processing easier.

Surveys have two types of questions – closed questions and open questions. The responses to these questions affect the type of coding performed.

A *closed question* means that only a *fixed* number of predetermined survey responses are permitted. These responses will have already been coded.

The following question, in a survey on sporting activities, is an example of a closed question:

To what degree is sport important in providing you with the following benefits?

1) Very important
2) Somewhat important
3) Not important

An *open question* implies that *any* response is allowed, making subsequent coding more difficult. In order to code an open question, the processor must sample a number of responses, and then design a code structure that includes all possible answers.

The following code structure is an example of an open question:

What sports do you participate in?
Specify (28 characters)_____

In the Census and almost all other surveys, the codes for each question field are premarked on the questionnaire. To process the questionnaire, the codes are entered directly into the database and are prepared for data capturing. The following is an example of premarked coding:

What language does this person speak most often at home?

1) English
2) French
3) Other – Specify_____

Automated Coding Systems There are programs in use that will automate repetitive and routine tasks. Some of the advantages of an automated coding system are that the process increasingly becomes

- faster,
- more consistent, and
- more economical.

The next step in data processing is inputting the coded data into a computer database. This method is known as *data capture*.

Data Capture This is the process by which data are transferred from a paper copy, such as questionnaires and survey responses, to an electronic file. The responses are then put into a computer. Before this procedure takes place, the questionnaires must be groomed (prepared) for data capture. In this processing step, the questionnaire is reviewed to ensure that the entire minimum required data have been reported, and that they are decipherable. This grooming is usually performed during extensive automated edits.

There are several methods used for capturing data:

- *Tally charts* are used to record data, such as the number of occurrences of a particular event, and to develop frequency distribution tables.
- *Batch keying* is one of the oldest methods of data capture. It uses a computer keyboard to type in the data. This process is very practical for high-volume entry where fast production is a requirement. No editing procedures are necessary but there must be a high degree of confidence in the editing program.
- *Interactive capture* is often referred to as intelligent keying. Usually, captured data are edited before they are imputed. However, this method combines data capture and data editing in one function.
- *Optical character readers* or bar code scanners are able to recognize alpha or numeric characters. These readers scan lines and translate them into the program. These bar code scanners are quite common and often seen in department stores. They can take the shape of a gun or a wand.

- *Magnetic recordings* allow for both reading and writing capabilities. This method may be used in areas where data security is important. The largest application for this type of data capture is the PIN number found on automatic bank cards. A computer keyboard is one of the best known input (or data entry) devices in current use. In the past, people performed data entry using punch cards or paper tape.

Some modern examples of data input devices are

- optical mark reader
- bar code reader
- scanner used in desktop publishing
- light pen
- trackball
- mouse

Once data have been entered into a computer database, the next step is ensuring that all of the responses are accurate. This method is known as data editing.

Data Editing   Data should be edited before being presented as information. This action ensures that the information provided is accurate, complete, and consistent. There are two levels of data editing – *micro-* and *macroediting*.

Microediting corrects the data at the record level. This process detects errors in data through checks of the individual data records. The intent at this point is to determine the consistency of the data and correct the individual data records.

Macroediting also detects errors in data, but does this through the analysis of aggregate data (totals). The data are compared with data from other surveys, administrative files, or earlier versions of the same data. This process determines the compatibility of data.

Imputations   Editing is of little value to the overall improvement of the actual survey results, if no corrective action is taken when items fail to follow the rules set out during the editing process. When all of the data have been edited using the applied rules and a file is found to have missing data, then imputation is usually done as a separate step.

Nonresponse and invalid data definitely impact the quality of the survey results.

Imputation resolves the problems of missing, invalid, or incomplete responses identified during editing, as well as any editing errors that might have occurred.

At this stage, all of the data are screened for errors because respondents are not the only ones capable of making mistakes; errors can also occur during coding and editing.

Some other types of imputation methods include the following:

- *Hot deck* uses other records as "donors" in order to answer the question (or set of questions) that needs imputation.
- *Substitution* relies on the availability of comparable data. Imputed data can be extracted from the respondent's record from a previous cycle of the survey, or the imputed data can be taken from the respondent's alternative source file (e.g., administrative files or other survey files for the same respondent).
- *Estimator* uses information from other questions or from other answers (from the current cycle or a previous cycle), and through mathematical operations derives a plausible value for the missing or incorrect field.
- *Cold deck* makes use of a fixed set of values, which covers all of the data items. These values can be constructed with the use of historical data, subject matter expertise, and so on.
- The donor can also be found through a method called *nearest neighbor imputation*. In this case, some sort of criteria must be developed to determine which responding unit is "most like" the unit with the missing value in accordance with the predetermined characteristics. The closest unit to the missing value is then used as the donor.

Imputation methods can be performed automatically, manually, or in combination.

Data Quality

- Quality assurance
- Quality control
- Quality management in statistical agencies

Quality is an essential element at all levels of processing. To ensure the quality of a product or service in survey development activities, both quality assurance and quality control methods are used.

Quality Assurance   Quality assurance refers to all planned activities necessary in providing confidence that a product or service will satisfy its purpose and the users' needs. In the context of survey conducting activities, this can take place at any of the major stages of survey development: planning, design, implementation, processing, evaluation, and dissemination.

This approach anticipates problems prior to their unexpected occurrences, and uses all available information to generate improvements. It is not restricted to any specific quality, the planning stage, and is all-encompassing in its activities standards. It is applicable mostly at the planning stage, and is all-encompassing in its activities.

Quality Control   Quality control is a regulatory procedure through which one may measure quality, with preset standards, and then act on any differences.

Examples of this include controlling the quality of the coding operation, the quality of the survey interviewing, and the quality of the data capture.

Quality control responds to observed problems, using ongoing measurements to make decisions on the processes or products. It requires a prespecified quality for comparability. It is applicable mostly at the processing stage following a set procedure that is a subset of quality assurance.

Quality Management in Statistical Agencies   The quality of the data must be defined and assured in the context of being "fit for use," which will depend on the intended function of the data and the fundamental characteristics of quality. It also depends on the users' expectations of what is considered to be useful information.

There is no standard definition among statistical agencies for the term "official Statistics." There is a generally accepted, but evolving, range of quality issues underlying the concept of "fitness for use". These elements of quality need to be considered and balanced in the design and implementation of an agency's statistical program.

The following is a list of the elements of quality:

- Relevance
- Accuracy
- Timeliness
- Accessibility
- Interpretability
- Coherence

These elements of quality tend to overlap. Just as there is no single measure of accuracy, there is no effective statistical model for bringing together all these characteristics of quality into a single indicator. Also, except in simple or one-dimensional cases, there is no general statistical model for determining whether one particular set of quality characteristics provides higher overall quality than another.

Producing Results   After editing, data may be processed further to produce a desired output. The computer software used to process the data will depend on the form of output required. Software applications for word processing, desktop publishing, graphics (including graphing and drawing), programming, databases, and spreadsheets are commonly used. The following are some examples of ways that software can produce data:

- *Spreadsheets* are programs that automatically add columns and rows of figures, calculate means, and perform statistical analyses.
- *Databases* are electronic filing cabinets. They systematically store data for easy access, and produce summaries, aggregates, or reports.
- *Specialized programs* can be developed to edit, clean, impute, and process the final tabular output.

Review Questions for Section 4.1

1   In the job description for an entry level statistician today, from the viewpoint of a prospective applicant for that position, what basic statistical computing languages are important in order to meet the requirement? Why?

2   For a typical MBA (Master of Business Administration) program in Business and Finance, should the core curriculum include the development of proficient skill in the use of R programming in Statistics? Why?

3   **a** Contrast the concepts of data and information.
    b How does the process of data processing convert data to information?

4   In the steps which convert data into information, how are statistics and computing applied to the various data processing steps.

5   **a** Describe and delineate quality assurance and quality control in computer data processing.
    b In what way does statistics feature in these phases of data processing?

4.2   Beginning R

R is an open-source, freely available, integrated software environment for data manipulation, computation, analysis, and graphical display.

The R environment consists of the following:

- A data handling and storage facility
- Operators for computations on arrays and matrices
- A collection of tools for data analysis
- Graphical capabilities for analysis and display
- An efficient and continuing developing programming algebra-like programming language that consists of loops, conditionals, user-defined functions, and input and output capabilities.

The term "environment" is used to show that it is indeed a planned and coherent system.

R and Statistics

R was initially written by Robert Gentleman and Ross Ihaka of the Statistics Department of the University of Auckland, New Zealand, in 1997. Since then there has been the R-development core group of about 20 people with write access to the R source code.

The original introduction to the R environment, evolved from the S/S-Plus languages, was *not* primarily directed toward statistics. However, since its development in the 1990s, it appeared to have been "hijacked" by many working in the areas of classical and modern statistical techniques, including many applications in financial engineering, econometrics, biostatistics with respect to epidemiology, public health and preventive medicine. These applications have led to the *raison d'état* for writing this book.

As of this writing, the latest version of R is R-3.3.2, officially released on October 31, 2016. The primary source of R packages is the Comprehensive R Archive Network, CRAN, at http://cran.r-project.org/. Another source of R packages may be found in numerous publications, for example, the *Journal of Statistical Software*, now at its 45th volume, is available at http://www.jstatsoft.org/v45.

Let us get started (the R-3.3.2 version environment is being used here). Recall in Section 4.1, the R environment was obtained as follows:

```
Here is R:
Let us download the open-source high-level program R from the
Internet and take a first look at the R computing environment.
Remark:   Access the Internet at the website of CRAN (The
Comprehensive
R Archive Network:  http://cran.r-project.org/
To install R:   R-3.3.2-win32.exe http://www.r-project.org/
=> download R
=> Select:  USA
http://cran.cnr.Berkeley.edu <http://cran.cnr.berkeley.edu/ >
University of California, Berkeley, CA
=>  http://cran.cnr.berkeley.edu/
=> Windows (95 and later)
=> base
=> R-3.3.2-win32.exe
AFTER the down-loading:
=> Double-click on:   R-3.3.2-win32.exe
(on the DeskTop) to un-zip & install R
=> An icon (Script R 3.3.2) will appear on ones Computer
"desktop" as follows: Figure 4.1
On the computer "desktop" is the R icon:
```

In this book, the following special color scheme legend will be used for all statements during the computational activities in the R environment, to clarify the various inputs to and outputs from the computational process:

1) Texts in this book (WarnockPro-Regular font)
2) Line input in R code (CourierStd)
3) Line output *in* R code (CourierStd)
4) *Line comment statements in* R *code (WarnockPro-Italic font)*

Note: The # sign is the comment character: All text in the line following this sign is treated as a comment by the R program, that is, no computational action

R 2.9.1.lnk

Figure 4.1 The R icon on the computer desktop (The R 3.3.2 looks *exactly* the same as that for R 2.9.1).

will be taken regarding such a statement. That is, the computational activities will proceed as though the comment statements are ignored. These comment statements help the programmer and user by providing some clarification of the purposes involved in the remainder of the R environment. The computations will proceed even if these comment statements are eliminated.

is known as the number sign, it is also known as the pound sign/key, the hash key, and, less commonly, as the octothorp, octothorpe, octathorp, octotherp, octathorpe, and octatherp.

To use R under Windows: **Double-click on the R 3.3.2 icon . . .**

Upon selecting and clicking on R, the R window opens with the following declaration:

R version 3.3.2 (2016-10-31)
Copyright 2016 The R Foundation for Statistical Computing
ISBN 3-900051-07-0
R is free software and comes with ABSOLUTELY NO WARRANTY.
You are welcome to redistribute it under certain conditions.
Type 'license()' or 'licence()' for distribution details.
R is a collaborative project with many contributors.
Type 'contributors()' for more information and
'citation()' on how to cite R or R packages in publications.
Type 'demo()' for some demos, 'help()' for on-line help, or

'help.start()' for an HTML browser interface to help.
Type 'q()' to quit R.

```
> # This is the R computing environment.
> # Computations may begin now!
>
> # First, use R as a calculator, and try a simple arithmetic
> # operation, say:  1 + 1
> 1+1
> [1] 2   # This is the output!
>         # WOW!  It's really working!
> # The [1] in front of the output result is part of R's way of printing
> # numbers and vectors. Although it is not so useful here, it does
> # become so when the output result is a longer vector
```

From this point on, this book is most beneficially read with the R environment at hand. It will be a most effective learning experience if one practices each R command as one goes along the textual materials.

4.2.1 A First Session Using R

This section introduces some important and practical features of the R environment (Figure 4.2). Login and start an R session in the Windows system of the computer

```
>
> # This is the R environment.
> help.start()  # Outputting the page shown in Figure 4.1
>                  # Statistical Data Analysis Manuals
starting httpd help server ... done
If nothing happens, you should open
'http://127.0.0.1:28103/doc/html/index.html' yourself
At this point, explore the HTML interface to on-line help
right from the desktop, using the mouse pointer to note
the various features of this facility available within
the R environment.  Then, returning to the R environment:

> help.start()
Carefully read through each of the sections under
"Manuals" - to obtain an introduction to the basic
language of the R environment.  Then look through the
items under "Reference" to reach beyond the elementary
level, including access to the available "R Packages" - all
R functions and datasets are stored in packages.
```

Statistical Data Analysis

Manuals

An introduction to R   *The R language definition*

Writing R extensions   *R installation and administration*

R data import/export   *R internals*

Reference

Packages   *Search engine and keywords*

Miscellaneous Material

About R   *Authors*   *Resources*

License   *Frequently asked questions*   *Thanks*

News   *User manuals*   *Technical papers*

Material specific to the windows port

Changes   *Windows FAQ*

Figure 4.2 Output of the R Command.

For example, if one selects the Packages Reference, the following R Package Index window will open up, showing Figure 4.3, listing a collection of R program packages under the R library: C:\Program Files\R\R-2.14.1\library

One may now access each of these R program packages, and use them for further applications as needed.

Returning to the R environment:

```
>
> x <- rnorm(100)
> # Generating a pseudo-random 100-vector x
> y <- rnorm(x)
> # Generating another pseudo-random 100-vector y
> plot (x, y)
> # Plotting x vs. y in the plane, resulting in a graphic
> # window: Figure 4.4.
```

base	The R base package
boot	Bootstrap functions (originally by Angelo Canty for S)
class	Functions for classification
cluster	Cluster analysis extended Rousseeuw et al.
codetools	Code analysis tools for R
compiler	The R compiler package
datasets	The R datasets package
foreign	Read data stored by Minitab, S, Sas, SPSS, Stata, Systat, dBase, ...
graphics	The R graphics package
grDevices	The R graphics devices and support for colours and fonts
grid	The grid graphics package
KernSmooth	Functions for kernel smoothing for Wand & Jones (1995)
lattice	Lattice graphics
MASS	Support functions and datasets for Venables and Ripley's MASS
Matrix	Sparse and dense matrix classes and methods
methods	Formal methods and classes
mgcv	GAMs with GCV/AIC/REML smoothness estimation and GAMMs by PQL
nlme	Linear and nonlinear mixed effects models
nnet	Feed-forward neural networks and multinomial log-linear models
parallel	Support for parallel computation in R
rpart	Recursive partitioning
spatial	Functions for kriging and point pattern analysis
splines	Regression spline functions and classes
stats	The R stats package
stats4	Statistical functions using S4 classes
survival	Survival analysis, including penalised likelihood.
tcltk	Tcl/Tk interface
tools	Tools for package development
utils	The R utils package

Figure 4.3 **Package Index.**

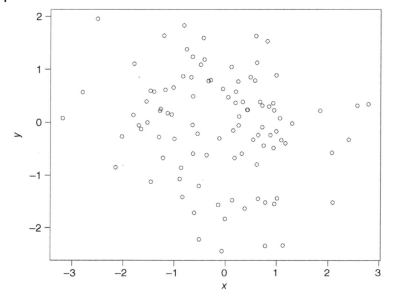

Figure 4.4 Graphical Output for `plot (x, y)`.

Remark: For reference, Appendix 1 contains the CRAN documentation of the R function `plot ()`, available for graphic outputting, which may be found by the R code segment:

```
> ?plot
```

CRAN has documentations for many R functions and packages.
Again, returning to the R workspace, and enter

```
>
>
> ls()   # (This is a lower-case "L" followed by "s", viz., the 'list'
>        #  command.)
>        #  (NOT 1 = "ONE" followed by "s")
>        #  This command will list all the R objects now in the
>        #  R workspace:
>        #  Outputting:
[1] "E" "n" "s" "x" "y" "z"
```

Again, returning to the R workspace, and enter

```
>
> rm (x, y) # Removing all x and all y from the R workspace
> x         # Calling for x
```

```
Error: object 'x' not found
>          # Of course, the xs have just been removed!
> y        # Calling for y
Error: object 'y' not found # Because the ys have also been
                            # removed!
>
> x <- 1:10  # Let x = (1, 2, 3, 4, 5, 6, 7, 8, 9, 10)
> x          # Outputting x (just checking!)
 [1]  1 2 3 4 5 6 7 8 9 10
> w <- 1 + sqrt(x)/2  # w is a weighting vector of
>                     # standard deviations
> dummy <- data.frame (x = x, y = x + rnorm(x)*w)
> # Making a data frame of 2 columns, x, and y, for inspection
> dummy  # Outputting the data frame dummy
   x      y
1  1  1.311612
2  2  4.392003
3  3  3.669256
4  4  3.345255
5  5  7.371759
6  6 -0.190287
7  7 10.835873
8  8  4.936543
9  9  7.901261
10 10 10.712029
>
> fm <- lm(y~x, data=dummy)
> # Doing a simple Linear Regression
> summary(fm) # Fitting a simple linear regression of y on x,
>             # then inspect the analysis, and outputting:
Call:
lm(formula = y ~ x, data = dummy)
Residuals:
Min        1Q      Median     3Q       Max
-6.0140   -0.8133  -0.0385   1.7291   4.2218
Coefficients:
             Estimate  Std. Error  t value  Pr(>|t|)
(Intercept)  1.0814    2.0604      0.525    0.6139
x            0.7904    0.3321      2.380    0.0445 *
---
Signif. codes: 0 '***' 0.001 '**' 0.01 '*' 0.05 '.' 0.1 ' ' 1
Residual standard error: 3.016 on 8 degrees of freedom
Multiple R-squared: 0.4146, Adjusted R-squared: 0.3414
F-statistic: 5.665 on 1 and 8 DF, p-value: 0.04453
```

```
> fm1 <- lm(y~x, data=dummy, weight=1/w^2)
> summary(fm1) # Knowing the standard deviation, then doing a
>                 # weighted regression and outputting:
Call:
lm(formula = y ~ x, data = dummy, weights = 1/w^2)
Residuals:
    Min      1Q     Median     3Q        Max
-2.69867 -0.46190 -0.00072 0.90031 1.83202
Coefficients:
        Estimate Std. Error t value Pr(>|t|)
(Intercept)  1.2130   1.6294   0.744  0.4779
x            0.7668   0.3043   2.520  0.0358 *
---
Signif. codes:  0 '***' 0.001 '**' 0.01 '*' 0.05 '.' 0.1 ' ' 1
Residual standard error: 1.356 on 8 degrees of freedom
Multiple R-squared: 0.4424,   Adjusted R-squared: 0.3728
F-statistic: 6.348 on 1 and 8 DF, p-value: 0.03583
> attach(dummy) # Making the columns in the data
>                 # frame as variables
The following object(s) are masked _by_ '.GlobalEnv': x
> lrf <- lowess(x, y) # a non-parametric local
>                       # regression function lrf
> plot (x, y) # Making a standard point plot, outputting: Figure 4.5.
.
```

Remark: For reference, Appendix 1 contains the CRAN documentation of the R function `plot()`, available for graphic outputting, which may be found by the R code segment:

```
> ?plot
> # CRAN has documentations for many R functions and packages.
```

Again, returning to the R workspace, and enter

```
>
> ls() # (This is a lower-case "L" followed by "s", viz., the 'list'
>        # command.)
>        #  (NOT 1 = "ONE" followed by "s")
>        # This command will list all the R objects now in the
>        # R workspace:
>        # Outputting:
[1] "E" "n" "s" "x" "y" "z"
```

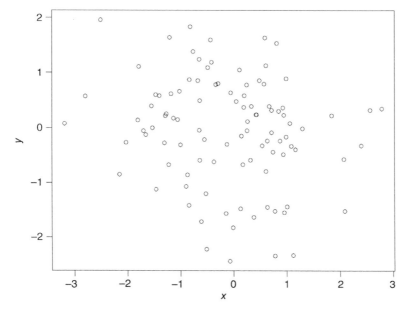

Figure 4.5 Graphical Output for `plot (x, y)`.

Again, returning to the R workspace, and enter

```
>
> rm (x, y) # Removing all x and all y from the R workspace
> x          # Calling for x
Error: object 'x' not found
>          # Of course, the xs have just been removed!
>
> y       # Calling for y
Error: object 'y' not found
>  # Because the y s have been removed too!
>
> x <- 1:10  # Let x = (1, 2, 3, 4, 5, 6, 7, 8, 9, 10)
> x            # Outputting x (just checking!)
[1] 1 2 3 4 5 6 7 8 9 10
> w <- 1 + sqrt (x)/2  # w is a weighting vector of
>                       # standard deviations
> dummy <- data.frame (x = x, y = x + rnorm(x) *w)
> # Making a data frame of 2 columns, x, and y, for inspection
> dummy  # Outputting the data frame dummy
```

```
     x      y
1    1   1.311612
2    2   4.392003
3    3   3.669256
4    4   3.345255
5    5   7.371759
6    6  -0.190287
7    7  10.835873
8    8   4.936543
9    9   7.901261
10  10  10.712029
> fm <- lm(y~x, data=dummy)
>   # Doing a simple Linear Regression
> summary(fm)
> # Fitting a simple linear regression of y on x,
> # then inspect the analysis, and outputting:
Call:
lm(formula = y ~ x, data = dummy)
Residuals:
   Min      1Q Median      3Q     Max
 -6.0140 -0.8133 -0.0385  1.7291  4.2218
Coefficients:
            Estimate Std. Error t value Pr(>|t|)
(Intercept)  1.0814     2.0604   0.525   0.6139
x            0.7904     0.3321   2.380   0.0445 *
---
Signif. codes:  0 '***' 0.001 '**' 0.01 '*' 0.05 '.' 0.1 ' ' 1
Residual standard error: 3.016 on 8 degrees of freedom
Multiple R-squared: 0.4146,    Adjusted R-squared: 0.3414
F-statistic: 5.665 on 1 and 8 DF,  p-value: 0.04453
> fm1 <- lm(y~x, data=dummy, weight=1/w^2)
> summary(fm1) # Knowing the standard deviation,
>              # then doing a weighted
>              # regression and outputting:
>              # regression and outputting:
Call:
lm(formula = y ~ x, data = dummy, weights = 1/w^2)
Residuals:
   Min       1Q  Median       3Q     Max
 -2.69867 -0.46190 -0.00072  0.90031  1.83202
Coefficients:
            Estimate Std. Error t value Pr(>|t|)
(Intercept)  1.2130     1.6294   0.744   0.4779
x            0.7668     0.3043   2.520   0.0358 *
```

```
---
Signif. codes:  0 '***' 0.001 '**' 0.01 '*' 0.05 '.' 0.1 ' ' 1
Residual standard error: 1.356 on 8 degrees of freedom
Multiple R-squared: 0.4424,    Adjusted R-squared: 0.3728
F-statistic: 6.348 on 1 and 8 DF,  p-value: 0.03583
> attach(dummy)  # Making the columns in the data frame as
>                 # variables
> lrf <- lowess(x, y)
> lrf
> plot (x, y) # Making a standard point plot,
>                 # outputting: Figure 4.6
```

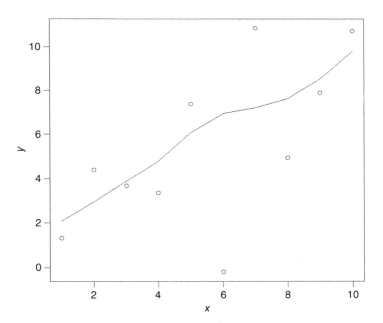

Figure 4.6 Adding in the local regression line.

```
> abline(0, 1, lty=3) # adding in the true regression line:
>                      # (Intercept = 0, Slope = 1),
>                      # outputting: Figure 4.7.

> abline(coef(fm)) # adding in the unweighted regression line:
>                   # outputting Figure 4.8
```

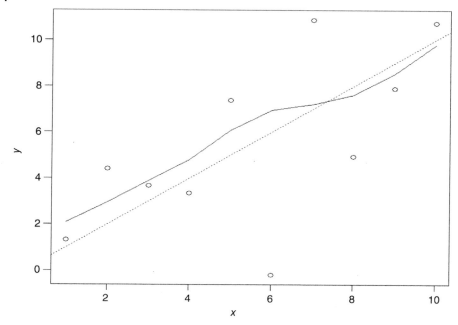

Figure 4.7 Adding in the true regression line (intercept = 0, slope =1).

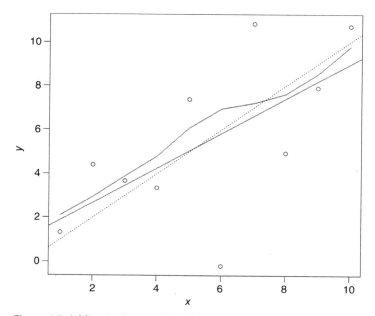

Figure 4.8 Adding in the unweighted regression line.

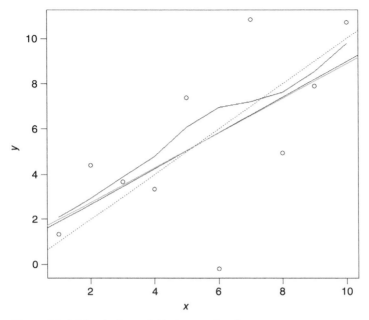

Figure 4.9 Adding in the weighted regression line.

```
> abline(coef(fm1), col="red")
> # adding in the weighted regression line:
> # outputting Figure 4.9.

> detach() # Removing data frame from the search path
> plot(fitted(fm), resid(fm), # Doing a standard diagnostic
+                            # plot
+ xlab="Fitted values",   # to check for heteroscedasticity**,
+ ylab="residuals",     # viz., checking for differing variance.
+ main="Residuals vs Fitted")
# Outputting Figure 4.10.
**Heteroskedasticity occurs when the variance of the error terms
differ across observations.

> qqnorm(resid(fm), main="Residuals Rankit Plot")
> # Doing a normal scores plot to check for skewness, kurtosis,
+ # and
> # outliers.
> # (Not very useful here.) Outputting
Figure 4.11.
```

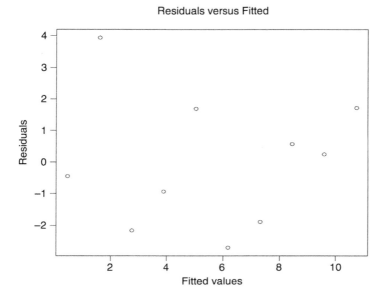

Figure 4.10 A standard diagnostic plot to check for heteroscedasticity.

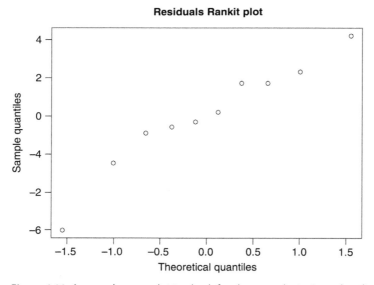

Figure 4.11 A normal scores plot to check for skewness, kurtosis, and outliers.

```
>
> rm(fm, fm1, lrf, x, dummy) # Removing these 5 objects
> fm
Error: object 'fm' not found   # Checked!
> fm1
Error: object 'fm1' not found      # Checked!
> lrf
Error: object 'lrf' not found        # Checked!
> x
Error: object 'x' not found        # Checked!
> dummy
Error: object 'dummy' not found    # Check!
# END OF THIS PRACTICE SESSION!
```

4.2.2 The R Environment – This is Important!

Getting through the First Session in Section 4.2.1 shows:

Technically, R is an expression language with a simple syntax, which is almost self-explanatory. It is case sensitive, so x and X are different symbols and refer to different variables. All alphanumeric symbols are allowed, plus '.' and '-', with the restriction that a name must start with '.' or a letter, and if it starts with '.' the second character must not be a digit. The command prompt > indicates when R is ready for input.

This is where one types commands to be processed by R, which will happen when one hits the ENTER key. Commands consist of either expressions or assignments. When an expression is given as a command, it is immediately evaluated, printed, and the value is discarded. An assignment evaluates an expression and passes the value to a variable – but the value is not automatically printed. To print the computed value, simple enter the variable again at the next command.Commands are separated either by a new line, or separated by a semicolon (';'). Several elementary commands may be grouped together into one compound expression by braces ('{' and '}').

Comments, starting with a hashmark/number sign ('#'), may be put almost anywhere: everything to the end of the line following this sign is a comment. Comments may not be used in an argument list of a function definition or inside strings. If a command is not complete at the end of a line, R will give a different prompt, a "+" sign, by default.

On the second and subsequent lines, continue to read input until the command is completed syntactically. The result of a command is printed to the output device: If the result is an array, such as a vector or a matrix, then the elements are formatted with line break (wherever necessary) with the indices of the leading entries labeled in square brackets: [index].

For example, an array of 15 elements may be outputted as:

```
> array (8, 15)
[1]   8 8 8 8 8 8 8 8 8 8
[11] 8 8 8 8 8
The labels '[1]' and '[11]' indicate the 1st  and
11th elements in the output.
These labels are not part of the data itself!
Similarly, the labels for a matrix are placed at the
start of each row and column in the output.
For example, for the 3 x 5 matrix M, it is outputted
as:
>
> M <- matrix (1:15, nrow=3)
> M
     [,1] [,2] [,3] [,4] [,5]
[1,]   1   4   7   10   13
[2,]   2   5   8   11   14
[3,]   3   6   9   12   15
>
```

Note that the storage is a column-major, namely, the elements of the first column are printed out first followed by those of the second column, and so on. To cause a matrix to be filled in a row-wise manner, rather than the default column-wise fashion, the additional switch byrow=T will cause the matrix to be filled row-wise rather than column-wise:

```
>
> M <- matrix (1:15, nrow=3, byrow=T)
> M
     [,1]  [,2]  [,3]  [,4]   [,5]
[1,]   1    2    3    4    5
[2,]   6    7    8    9    10
[3,]  11   12   13   14   15
>
```

The First Session also shows that there is a host of helpful resources imbedded in the R environment that one can readily access, using the online help provided by CRAN.

Review Questions for Section 4.2

1   Let us get started!
    Please follow the step-by-step instructions given in Section 4.2 to set up
    an R environment. The R window show looks like this:

```
>
Great!
```

Now enter the following arithmetic operations: press "Enter" after
each entry

```
(a)   2 + 3   <Enter>
(b)   13 - 7  <Enter>
(c)   17 * 23 <Enter)
(d)   100/25  <Enter>10/25
(e)   Did you obtain the following results:
      5, 6, 391, 4?
```

2   Here is a few more: The <Enter> prompt will be omitted from now on.
    a 2^4
    b sqrt(3)
    c $1i$ [$1i$ is used for the complex unit i, where $i = -1$]
    d $(2 + 3i) + (4 + 5i)$
    e $(2 + 3i) \times (4 + 5i)$

3   Here is a short session on using R to do complex arithmetic, just enter the
    following commands into the R environment, and report the results:

```
> th <- seq(-pi, pi, len=20)
> th  (a)  How many numbers are printed out?
> z <- exp(1i*th)
> z   (b)  How many complex numbers are printed out?
> par(pty="s")
(c)  Along the menu-bar at the top of the R environment:
*   Select and left-click on "Window":, then
*   Move downwards and select the 2nd option:
R Graphic Device 2 (ACTIVE)
*   Go to the "R Graphic Device 2 (ACTIVE) Window"
(d)  What is there?
> plot(z)
(e)  Describe what is in the Graphic Device 2 Window.
```

4.3   R as a Calculator

4.3.1   Mathematical Operations Using R

To learn to do statistical analysis and computations, one may start by considering the R programming language as a simple calculator.

Start from here: just enter an arithmetic expression, press the <Enter> key, and the answer from the machine is found in the next line

```
>
> 2 + 3
[1] 5
>
```

OK! What about other calculations? Such as: 13 - 7, 3 x 5, 12 /4, 72, $\sqrt{2}$, e^3, $e^{i\pi}$, ln 5 = loge5, $(4 + \sqrt{3})$ $(4 - \sqrt{3})$, $(4 + i\sqrt{3})$ $(4 - i\sqrt{3})$, . . .
and so on. Just try:

```
>
> 13 - 7
[1] 6
> 3*5
[1] 15
> 12/4
[1] 3
> 7^2
[1] 49
> sqrt(2)
[1] 1.414214
>
> exp(3)
[1] 20.08554
>
> exp(1i*pi)    [1i is used for the complex number
i = √-1.]
[1] -1-0i       [ This is just the famous Euler's Identity
equation:   $e^{i\pi} + 1 = 0$.]
> log(5)
[1] 1.609438
> (4 + sqrt(3))*(4 - sqrt(3))
[1] 13
[Checking: $(4+\sqrt{3})$ $(4-\sqrt{3})$ = 42 - $(\sqrt{3})2$ = 16 - 3 = 13 (Checked!)]
> (4 + 1i*sqrt(3))*(4 - 1i*sqrt(3))
[1] 19+0i  [Checking: $(4+i\sqrt{3})$ $(4-i\sqrt{3})$ = 42 - $(i\sqrt{3})2$
= 16 - (-3) = 19 (Checked!)
```

Remark: The [1] in front of the computed result is R's way of outputting numbers. It becomes useful when the result is a long vector. The number N in the brackets $[N]$ is the index of the first number on that line. For example, if one generated 23 random numbers from a normal distribution

```
>
> x <- rnorm(23)
> x
 [1] -0.5561324  0.2478934 -0.8243522  1.0697415  1.5681899
 [6] -0.3396776 -0.7356282  0.7781117  1.2822569 -0.5413498
[11]  0.3348587 -0.6711245 -0.7789205 -1.1138432 -1.9582234
[16] -0.3193033 -0.1942829  0.4973501 -1.5363843 -0.3729301
[21]  0.5741554 -0.4651683 -0.2317168
>
```

Remark: After the random numbers have been generated, there is no output until one calls for x, namely, x has become a vector with 23 elements, call that a 23-vector.

The [11] on the third line of the output indicates that 0.3348587 is the 11th element in the 23-vector x. The numbers of outputs per line depends on the length of each element as well as the width of the page.

4.3.2 Assignment of Values in R and Computations Using Vectors and Matrices

R is designed to be a dynamically typed language, namely, at any time one may change the data type of any variable. For example, one can first set x to be numeric as has been done so far, say $x = 7$; next one may set x to be a vector, say $x = c$ (1, 2, 3, 4); then again one may set x to a word object, such as "Hi!". Just watch the following R environment:

```
>
> x <- 7
> x
[1] 7
> x <- c(1, 2, 3, 4) # x is assigned to be a 4-vector.
> x
[1] 1 2 3 4
> x <- c("Hi!") # x is assigned to be a character string.
> x
[1] "Hi!"
> x <- c("Greetings & Salutations!")
> x
[1] "Greetings & Salutations!"
```

```
> x <- c("The rain in Spain falls mainly on the
+       plain.")
[1] "The rain in Spain falls mainly on the plain."
> x <- c("Calculus", "Financial", "Engineering", "R")
> x
[1] "Calculus", "Financial", "Engineering", "R"
>
```

4.3.3 Computations in Vectors and Simple Graphics

The use of arrays and matrices was introduced in Section 4.2.2.

In finite mathematics, a matrix is a two-dimensional array of elements, which are usually numbers. In R, the use of the matrix extends to elements of any type, such as a matrix of character strings. Arrays and matrices may be represented as vectors with dimensions. In statistics in which most variables carry multiple values, therefore, computations are usually performed between vectors of many elements. These operations among multivariates result in large matrices. To demonstrate the results, graphical representations are often useful. The following simple example illustrates these operations being readily accomplished in the R environment:

```
>
> weight <- c(73, 59, 97)
> height <- c(1.79, 1.64, 1.73)
> bmi <- weight/height^2
> bmi  # Read the BMI Notes below
[1] 22.78331  21.93635  32.41004
> # To summarize the results proceed to compute as follows:
> cbind(weight, height, bmi)
weight   height     bmi
[1,]   73    1.79    22.78331
[2,]   59    1.64    21.93635
[3,]   97    1.73    32.41004
>
> rbind(weight, height, bmi)
[,1]          [,2]           [,3]
weight 73.00000    59.00000    97.00000
height  1.79000     1.64000     1.73000
bmi    22.78331    21.93635    32.41004
>
```

Clearly, the functions cbind and rbind bind (namely, join, link, glue, concatenate) by column and row, respectively, the vectors to form new vectors or matrices.

4.3.4 Use of Factors in R Programming

In the analysis of, for example, health science data sets, categorical variables are often needed. These categorical variables indicate subdivisions of the origin data set into various classes, for example: age, gender, disease stages, degrees of diagnosis, and so on. Input of the original data set is generally delineated into several categories using a numeric code: $1 = $ age, $2 = $ gender, $3 = $ disease stage, and so on. Such variables are specified as factors in R, resulting in a data structure that enables one to assign specific names to the various categories. In certain analyses, it is necessary for R to distinguish among categorical codes and variables whose values have direct numerical meanings.

A factor has four levels, consisting of two items

1) a vector of integers between 1 and 4 and
2) a character vector of length four containing strings which describe the four levels.

Consider the following example:

- A certain type of cancer is being categorized into 4 levels: Levels 1, 2, 3, and 4.
- The corresponding pain levels corresponding with these diagnoses are: none, mild, moderate, and severe, respectively.
- In the data set, five case subjects have been diagnosed in terms of their respective levels.

The following R code segment delineates the data set:

```
> cancerpain <- c(1, 4, 3, 3, 2, 4)
> fcancerpain <- factor(cancerpain, level=1:4)
> levels(fcancerpain) <- c("none", "mild",
+                 "moderate", "severe")
```

The first statement creates a numerical vector cancerpain that encodes the pain levels of six case subjects. This is being considered as a categorical variable for which, using the factor function, a factor fcancerpain is created. This may be called with one argument in addition to cancerpain, namely, levels 1–4, which indicates that the input coding uses the values 1–4. In the final line, the pain level names are changed to the four specified character strings. The result is

```
> fcancerpain
[1] none    severe  moderate moderate mild    severe
Levels: none mild moderate severe
> as.numeric(fcancerpain)
[1] 1 4 3 3 2 4
> levels(fcancerpain)
[1] "none"   "mild"   "moderate"   "severe"
```

Remarks: The function as.numeric outputs the numerical coding as numbers 1–4, and the function levels outputs the names of the respective levels.

The original input coding in terms of the numbers 1–4 is no longer needed. There is an additional option using the function ordered that is similar to the function factor used here.

BMI: The body mass index (BMI), is a useful measure for human body fat based on an individual's weight and height – it does not actually measure the percentage of fat in the body. Invented in the early nineteenth century, BMI is defined as a person's body weight (in kilograms) divided by the square of the height (in meters). The formula universally used in health science produces a *unit of measure* of kg/m²:

$$BMI = Body\ mass(kg)/\{Height(m)\}^2$$

A BMI chart may be used displaying BMI as a function of weight (horizontal axis) and height (vertical axis) with contour lines for different values of BMI or colors for different BMI categories (Figure 4.12).

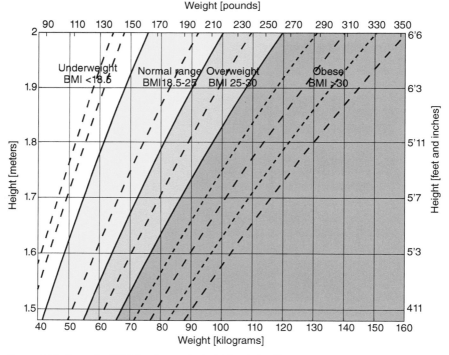

Figure 4.12 A graph of BMI (body mass index): The dashed lines represent subdivisions within a major class. The "underweight" classification is further divided into "severe," "moderate," and "mild" subclasses. World Health Organization data.

4.3.5 Simple Graphics

Generating graphical presentations is an important aspect of statistical data analysis. Within the R environment, one may construct plots that allow production of plots and control of the graphical features. Thus, with the previous example, the relationship between body weight and height may be considered by first plotting one versus the other by using the following R code segments:

```
>
> plot (weight, height)
> # Outputting: Figure 4.13.
```

Remarks:

1) Note the order of the parameters in the plot (x, y) command: the first parameter is x (the independent variable – on the horizontal axis), and the second parameter is y (the dependent variable – on the vertical axis).
2) Within the R environment, there are many plotting parameters that may be selected to modify the output. To get a full list of available options, return to the R environment and call for

```
> ?plot # This is a call for "Help!" within the R environment.
> # The output is the R documentation for:
> plot {graphics} # Generic X-Y plotting
```

This is the official documentation of the R function plot, within the R package graphics – note the special notations used for plot and {graphics}. To fully make use of the provisions of the R environment, one should carefully investigate all such documentations. (R has many available packages, each containing a number of useful functions.) This document shows all the plotting options available with the R environment. A copy of this documentation is shown in Appendix 1 for reference.

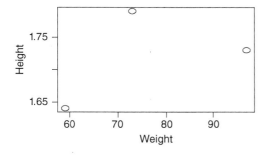

Figure 4.13 An *x–y* plot for > plot (weight, height).

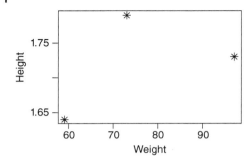

Figure 4.14 An *x–y* plot for > plot
(weight, height, pch = 8).

For example, to change the plotting symbol, one may use the keyword pch (for "plotting character") in the following R command:

```
> plot (weight, height, pch=8)
> # Outputting: Figure 4.14.
```

Note that the output is the same as that shown in Figure 4.13, except that the points are marked with little "8-point stars", corresponding to plotting character pch = 8.

In the documentation for pch, a total of 26 options are available, providing different plotting characteristics for points in R graphics. They are shown in Figure 4.15.

The parameter BMI was chosen in order that this value should be independent of a person's height, thus expressing as a single number or index indicative of whether a case subject is overweight, and by what relative amount. Of course, one may plot "height" as the abscissa (namely, the horizontal "*x*-axis") and "weight" as the ordinate (namely, the vertical "*y*-axis"), as follows:

```
> plot (height, weight, pch=8) # Outputting: Figure 4.16.
```

Since a normal BMI is between 18.5 and 25, averaging $(18.5 + 25)/2 = 21.75$. For this BMI value, the weight of a typical "normal" person would be $(21.75 \times \text{height}^2)$. Thus, one can superimpose a line of "expected" weights at BMI = 21.75 on Figure 4.16. This line may be accomplished in the R environment by the following code segments:

```
> ht <- c(1.79, 1.64, 1.73)
> lines (ht, 21.75*ht^2) # Outputting: Figure 4.17.
```

0   1   2   3   4   5   6   7   8   9   10 11 12 13 14 15 16 17 18 19 20 21 22
□   ○   △   +   ×   ◇   ▽   ⊠   ✳   ⊕   ⊕   ⊠   ⊞   ⊠   ⊠   ■   ●   ▲   ◆   ●   •   ⊙   ▣

Figure 4.15 Plotting symbols in R: pch = *n*, *n* = 0, 1, 2, . . . , 25.

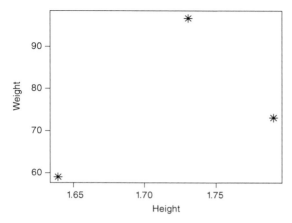

Figure 4.16 An *x–y* plot for > plot (height, weight, pch = 8).

In the last plot, a new variable for heights (ht) was defined instead of the original (height) because

i) The relation between height and weight is a quadratic one, and hence nonlinear. Although it may not be obvious on the plot, it is preferable to use points that are spread evenly along the *x*-axis than to rely on the distribution of the original data.

ii) As the values of height are not sorted, the line segments would not connect neighboring points but would run back and forth between distant points.

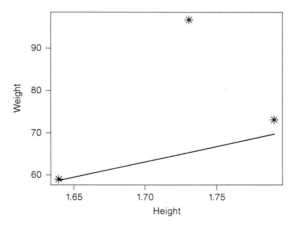

Figure 4.17 Superimposed reference curve using line (ht, 21.75×ht^2).

Remarks:

1) In the final example above, R was actually doing the arithmetic of vectors.
2) Notice that the two vectors weight and height are both 3-vectors, making it reasonable to perform the next step.
3) The cbind statement, used immediately after the computations have been completed, forms a new matrix by binding together matrices horizontally or column-wise. It results in a multivariate response variable. Similarly, the rbind statement does a similar operation vertically, or row-wise.
4) But, if for some reason (such as mistake in one of the entries) the two entries weight and height have different number of elements, then R will output an error message. For example

```
>
> weight <- c(73, 59, 97)               # a 3-vector
> height <- c(1.79, 1.64, 1.73, 1.48)   # a 4-vector !
> bmi <- weight/height^2                # Outputting:
Warning message:                        # An Error message!
In weight/height^2 :
longer object length is not a multiple of shorter object length
>
```

4.3.6 x as Vectors and Matrices in Statistics

It has just been shown that a variable, such as x or M may be assigned as

1) a number, such as $x = 7$,
2) a vector or an array, such as $x = c(1, 2, 3, 4)$, and
3) a matrix, such as $x =$

	[,1]	[,2]	[,3]	[,4]	[,5]
[1,]	1	4	7	10	13
[2,]	2	5	8	11	14
[3,]	3	6	9	12	15

4) a character string, such as x = "The rain in Spain falls mainly on the plain."
5) in fact, in R, a variable x may be assigned a complete data set that may consist of a multiple-dimensional set of elements each of which may in turn be anyone of the above kinds of variables. For example, besides being a numerical vector, such as in (2) above, x may be the following:
   a) A character vector, which is a vector of text strings whose elements are expressed in quotes, using double-, single-, or mixed-quotes:

   ```
   > c("one", "two", "three", "four", "five")
   ```

```
> # Double-quotes
[1] "one"   "two"   "three" "four" "five"
>
> c('one', 'two', 'three', 'four', 'five')
> # Single-quotes
[1] 'one'   'two'   'three'   'four'   'five'
>
> c("one", 'two', "three", 'four', "five")
> # Mixed-quotes
[1] "one"  'two'  "three"  'four'  "five"
```

However, if there is a mixed pair of quotes such as "xxxxx," it will not be accepted. For example:

```
> c("one", "two", "three", "four", "five')
```

b) A logical vector, which takes the value TRUE or FALSE (or NA). For inputs, one may use the abbreviations T or F.

These vectors are similarly specified using the c function:

```
> c(T, F, T, F, T)
[1] TRUE FALSE TRUE FALSE TRUE
```

In most cases, there is no need to specify repeated logical vectors. It is acceptable to use a single logical value to provide the needed options, as vectors of more than one value will respond in terms of relational expressions. Observe

```
> weight <- c(73, 59, 97)
> height <- c(1.79, 1.64, 1.73)
> bmi <- weight/height^2
> bmi # Outputting:
[1] 22.78331  21.93635  32.41004
> bmi > 25 # A single logical value will suffice!
[1] FALSE FALSE TRUE
>
```

4.3.7   Some Special Functions that Create Vectors

Three functions that create vectors are c, seq, and rep

1) c for "concatenate", or the joining of objects end-to-end (this was introduced earlier) – for example

```
> x <- c(1, 2, 3, 4) # x is assigned to be a 4-vector.
```

```
> x
[1] 1 2 3 4
```

2) seq for "sequence", for defining an equidistant sequence of numbers – for example

```
> seq(1, 20, 2) # To output a sequence from 1 to 20, in steps of 2
[1]  1 3 5 7 9 11 13 15 17 19
> seq(1, 20) # To output a sequence from 1 to 20, in steps of 1
>             # (which may be omitted)
[1]  1 2 3 4 5 6 7 8 9 10 11 12 13 14 15 16 17 18 19 20
> 1:20 # This is a simplified alternative to writing seq(1,
20).
[1]  1 2 3 4 5 6 7 8 9 10 11 12 13 14 15 16 17 18 19 20
> seq(1, 20, 2.5) # To output a sequence from 1 to 20, in steps
of
>                 # 2.5 .
[1]  1.0 3.5 6.0 8.5 11.0 13.5 16.0 18.5
```

3) rep for " replicate", for generating repeated values. This function takes two forms, depending on whether the second argument is a single number or a vector – for example

```
> rep(1:2, c(3,5)) # Replicating the first element (1) 3 times,
+                    # and
>                  # then replicating the second element (2) 5 +
# times
[1] 1 1 1 2 2 2 2 2 # This is the output.
> vector <- c(1, 2, 3, 4)
> vector # Outputting the vector
[1] 1 2 3 4
> rep(vector, 5) # Replicating vector 5 times:
[1] 1 2 3 4 1 2 3 4 1 2 3 4 1 2 3 4 1 2 3 4
```

4.3.8 Arrays and Matrices

In finite mathematics, a matrix M is a two-dimensional array of elements, generally numbers, such as

$$M = \begin{matrix} 1 & 4 & 7 & 10 & 13 \\ 2 & 5 & 8 & 11 & 14 \\ 3 & 6 & 9 & 12 & 15 \end{matrix}$$

and the array is usually placed inside parenthesis () or some brackets {}, [], and so on. In R, the use of a matrix is extended to elements of many types: numbers as well as character strings. For example, in R, the matrix M is expressed as

	[,1]	[,2]	[,3]	[,4]	[,5]
[1,]	1	4	7	10	13
[2,]	2	5	8	11	14
[3,]	3	6	9	12	15

4.3.9   Use of the Dimension Function dim() in R

In R, the above 3×5 matrix may be set up as vectors with dimension dim(x) using the following code segment:

```
> x <- 1:j
> x
 [1]  1 2 3 4 5 6 7 8 9 10 11 12 13 14 15
> dim(x) <- c(3, 5) # a dimension of 3 rows by 5 columns
..,
```

Remark: Here a total of 15 elements, 1–15, are set to be the elements of the matrix x. Then the dimension of x is set as c(3, 5), making x to become a 3×5 matrix. The assignment of the 15 elements follows a column-wise procedure, namely, the elements of the first column are allocated first followed by those of the second column, then the third column, and so on.

4.3.10   Use of the Matrix Function matrix() In R

Another way to generate a matrix is using the function matrix().
    The above 3×5 matrix may be created by the following 1-line code segment:

```
> matrix (1:15, nrow=3)
> matrix # Outputting:
     [,1] [,2] [,3] [,4] [,5]
[1,]    1    4    7   10   13
[2,]    2    5    8   11   14
[3,]    3    6    9   12   15
```

However, if the 15 elements should be allocated by row, then the following code segment should be used:

```
> matrix (1:15, nrow=3, byrow=T)
> matrix # Outputting:
     [,1] [,2] [,3] [,4] [,5]
```

```
[1,]    1    2    3    4    5
[2,]    6    7    8    9   10
[3,]   11   12   13   14   15
```

4.3.11 Some Useful Functions Operating on Matrices in R: Colnames, Rownames, and t (for transpose)

Using the previous example

i) the five columns of the 3×5 matrix x is first assigned the names C1, C2, C3, C4, and C5, respectively, then
ii) the transpose is obtained, and finally
iii) one takes the transpose of the transpose to obtain the original matrix x:

```
> matrix (1:15, nrow=3, byrow=T)
> matrix # Outputting:
[,1] [,2] [,3] [,4] [,5]
[2,]    6    7    8    9   10
[3,]   11   12   13   1 4   15
> colnames(x) <- c("C1", "C2", "C3", "C4", "C5")
> x # Outputting:
C1 C2  C3 C4 C5
[1,]    1    4    7   10 .gf
[2,]    2    5    8   11 14
[3,]    3    6    9   12 15
> t (x)
[,1]  [,2] [,3]
C1    1    2    3
C2    4    5    6
C3    7    8    9
C4   10   11   12
C5   13   14   15
> t(t(x)) # which is just x, as expected!
C1 C2  C3 C4 C5
[1,]    1    4    7   10 13
[2,]    2    5    8   11 14
[3,]    3    6    9   12 15
```

Yet another way is to use the function LETTERS, which is a built-in variable containing the capital letters A through Z. Other useful vectors include letters, month.name, and month.abb for lower-case letters, month names, and abbreviated names of months, respectively. Take a look:

```
> X <-LETTERS
> X # Outputting:
```

```
[1] "A" "B" "C" "D" "E" "F" "G" "H" "I" "J" "K" "L" "M" "N" "O"
[16] "P" "Q" "R" "S" "T" "U" "V" "W" "X" "Y" "Z"
> M # Outputting:
[1] "January"  "February" "March"    "April"    "May"
[6] "June"     "July"     "August"   "September" "October"
[11] "November" "December"
> m <- month.abb
> m # Outputting:
[1] "Jan" "Feb" "Mar" "Apr" "May" "Jun" "Jul" "Aug" "Sep" "Oct"
[11] "Nov" "Dec"
```

4.3.12   NA "Not Available" for Missing Values in Data sets

NA is a logical constant of length 1 that contains a missing value indicator. NA may be forced to any other vector type except raw. There are also constants NA integer, NA real, NA complex, and NA character of the other atomic vector types that support missing values: all of these are reserved words in the R language.

The generic function .na indicates which elements are missing.

The generic function .na<- sets elements to NA.

The reserved words in R's parser are

if, else, repeat, while, function, for, in next, break, NA complex, NA character, and 1, 2, and so on, which are used to refer to arguments passed down from an enclosing function.

Reserved words outside quotes are always parsed to be references to the objects linked to in the foregoing list, and are not allowed as syntactic names. They are allowed as nonsyntactic names.

4.3.13   Special Functions That Create Vectors

There are three useful R functions that are often used to create vectors:

1) c for "concatenate", which was introduced in Section 4.3.2 for joining items together end-to-end, for example:

```
> c(2, 3, 5, 7, 11, 13, 17, 19, 23, 29)
> # The first 10 prime numbers
[1]  2  3  5  7 11 13 17 19 23 29
```

2) seq for "sequence" is used for listing equidistant sequences of numbers, for example:

```
> seq(1, 20) # Sequence from 1 to 20
[1]  1  2  3  4  5  6  7  8  9 10 11 12 13 14 15 16 17 18 19 20
> seq(1, 20, 1) # Sequence from 1 to 20, in steps of 1
```

```
[1]  1 2 3 4 5 6 7 8 9 10 11 12 13 14 15 16 17 18 19 20
> 1:20 # Sequence from 1 to 20, in steps of 1
[1]  1 2 3 4 5 6 7 8 9 10 11 12 13 14 15 16 17 18 19 20
> seq(1, 20, 2) # Sequence from 1 to 20, in steps of 2
[1]  1 3 5 7 9 11 13 15 17 19
> seq(1, 20, 3) # Sequence from 1 to 20, in steps of 3
[1]  1 4 7 10 13 16 19
> seq(1, 20, 10) # Sequence from 1 to 20, in steps of 10
[1]  1 11
> seq(1, 20, 20) # Sequence from 1 to 20, in steps of 20
[1] 1
> seq(1, 20, 21) # Sequence from 1 to 20, in steps of 21
[1] 1
>
```

3) rep for "replicate" is used to generate repeated values, and may be expressed in two ways, for example:

```
> x <- c(3, 4, 5)
> rep(x, 4) # Replicate the vector x 4-times.
[1] 3 4 5 3 4 5 3 4 5 3 4 5
> rep(x, 1:3) # Replicate the elements of x: the first element
>               # once, the second element twice,
>               # and the third element three times.
[1] 3 4 4 5 5 5
> rep(1:3, c(3,4,5)) # For the sequence (1, 2, 3), replicate its
>                     # elements 3-, 4-, and 5-times,
respectively
[1] 1 1 1 2 2 2 2 3 3 3 3 3
```

Review Questions for Section 4.3

1   A Tower of Powers – by computations using R
There is an interesting challenge in arithmetic that goes like this

$$\sqrt{2}^{\sqrt{2}^{\cdots}}$$

What is the value of $\sqrt{2}^{\sqrt{2}^{\cdots}}$? Namely, an infinity of ascending tower of powers of the square root of 2.

Solution: Let x be the value of this "Tower of Powers", then it is easily seen that $\sqrt{2}^{x} = x$ itself ! Agree?
Watch the lowest $\sqrt{2}$.

And clearly it follows that $x = 2$, because $\sqrt{2^2} = 2$.

This shows that the value of this "Infinite Tower of Powers of $\sqrt{2}$" is just 2.

Now use the R environment to verify this interesting result:

(a) Compute $\sqrt{2}$
```
> sqrt(2)
```
(b) Compute $\sqrt{2}\sqrt{2}$
```
> sqrt(2)^sqrt(2)                        [a 2-Towers of √2-s]
(c) > sqrt(2)^sqrt(2)^sqrt(2)            [a 3-Towers of √2-s]
(d) > sqrt(2)^sqrt(2)^sqrt(2)^sqrt(2) [a 4-Towers of √2-s]
(e) > sqrt(2)^sqrt(2)^sqrt(2)^sqrt(2)^sqrt(2)
[a 5-Towers of √2-s]
```
(f) Now try the following computations of 10-, 20-, 30-, and finally 40-Towers of Powers of $\sqrt{2}$, and finally reach the result of 2 (accurate to 6 places of decimal!).
```
> sqrt(2)^sqrt(2)^sqrt(2)^sqrt(2)^sqrt(2)^sqrt(2)^sqrt(2)
^sqrt(2)^sqrt(2)^sqrt(2)
[1] 1.983668    [a 10-Towers of Powers of √2-s]
> sqrt(2)^sqrt(2)^sqrt(2)^sqrt(2)^sqrt(2)^sqrt(2)^sqrt(2)
^sqrt(2)^sqrt(2)^sqrt(2)^
+ sqrt(2)^sqrt(2)^sqrt(2)^sqrt(2)^sqrt(2)^sqrt(2)^sqrt(2)
^sqrt(2)^sqrt(2)^sqrt(2)
[1] 1.999586      [a 20-Towers of Powers of √2-s]
>sqrt(2)^sqrt(2)^sqrt(2)^sqrt(2)^sqrt(2)^sqrt(2)^sqrt(2)
^sqrt(2)^sqrt(2)^sqrt(2)^
+sqrt(2)^sqrt(2)^sqrt(2)^sqrt(2)^sqrt(2)^sqrt(2)^sqrt(2)
^sqrt(2)^sqrt(2)^sqrt(2)^
+ sqrt(2)^sqrt(2)^sqrt(2)^sqrt(2)^sqrt(2)^sqrt(2)^sqrt(2)
^sqrt(2)^sqrt(2)^sqrt(2)
[1] 1.999989      [a 30-Towers of Powers of √2-s]
>sqrt(2)^sqrt(2)^sqrt(2)^sqrt(2)^sqrt(2)^sqrt(2)^sqrt(2)
^sqrt(2)^sqrt(2)^sqrt(2)^
+sqrt(2)^sqrt(2)^sqrt(2)^sqrt(2)^sqrt(2)^sqrt(2)^sqrt(2)
^sqrt(2)^sqrt(2)^sqrt(2)^
+sqrt(2)^sqrt(2)^sqrt(2)^sqrt(2)^sqrt(2)^sqrt(2)^sqrt(2)
^sqrt(2)^sqrt(2)^sqrt(2)^
+ sqrt(2)^sqrt(2)^sqrt(2)^sqrt(2)^sqrt(2)^sqrt(2)^sqrt(2)
^sqrt(2)^sqrt(2)^sqrt(2)
[1] 2             [a 40-Towers of Powers of √2-s]
```

Thus, this R computation verifies the solution.

2 **a** What are the equivalents in R for the basic mathematical operations of $+, -, \times, /$ (division), $\sqrt{}$, squaring of a number

   b Describe the use of factors in R programming. Give an example.

3 If $x = (0, 1, 2, 3, 4, 5)$ and $y = (0, 1, 4, 9, 16, 25)$ in R, plot:

   a y versus x,
   b x versus y,
   c $\sqrt{y}$ versus x,
   d y versus $\sqrt{x}$,
   e $\sqrt{y}$ versus $\sqrt{x}$, and
   f $\sqrt{x}$ versus $\sqrt{y}$.

4 Explain, given an example, how the following functions may be used to combine matrices to form new ones:

   a cbind and
   b rbind.

5 **a** Describe is the R function factor().
   b Give an example of using factor() to create new arrays.

6 Using examples, illustrate two procedures for creating

   a a vector and
   b a matrix.

7 Describe, using examples, the following three functions for creating vectors:

   a c,
   b seq, and
   c rep.

8 **a** Use the function dim() to set up a matrix. Give an example.
   b Use the function matrix() to set up a matrix. Give an example.

9 Describe, using an example, the use of the following functions operating on a matrix in R: t(), colnames(), and rownames().

10 **a** What are reserved word in the R environment?
   b In R, how is the logical constant NA used? Give an example.

Exercises for Section 4.3

Enter the R environment, and do the following exercises using R programming:

1   Perform the following elementary arithmetic exercises:

    (a)   7 + 31;   (b)   87 - 23;   (c)   3.1417 X (7) 2;   (d)   22/7;
    (e)   e√2

2   Body mass index (BMI) is calculated from your weight in kilograms and your height in meters:

    BMI = kg / m2
    Using 1 kg ≈ 2.2 lb, and 1m ≈ 3.3 ft ≈ 39.4 in
    (a)   Calculate your BMI.
    (b)   Is it in the "normal" range 18.5 ≤ BMI ≤ 25?

3   In the MPH program, five graduate students taking the class in Introductory Epidemiology measured their weight (in kg) and height (in meters). The result is summarized in the following matrix:

	John	Chang	Michael	Bryan	Jose
WEIGHT	69.1	62.5	74.3	70.9	96.6
HEIGHT	1.81	1.46	1.69	1.82	1.74

    (a)   Construct a matrix showing their BMI as the last row.
    (b)   Plot: (i) WEIGHT (on the y-axis) vs HEIGHT (on the x-axis)
          (ii) HEIGHT vs WEIGHTi
          (iii) Assuming that the weight of a typical "normal":
                person is
    (21.75 x HEIGHT2), superimpose a line-of-
    "expected"-weight at BMI = 21.75 on the plot in (i).

4   **a** To convert between temperatures in degrees Fahrenheit (F) and Celsius (C), the following conversion formulas are used:

    $F = (9/5)C + 32$

    $C = (5/9) \times (F - 32)$

    At standard temperature and pressure, the freezing and boiling points of water are 0 and 100 °C, respectively. What are the freezing and boiling points of water in degrees Fahrenheit?

    b For C = 0, 5, 10, 15, 20, 25, ..., 80, 85, 90, 95, 100, compute a conversion table that shows the corresponding F temperatures.

    Note: To create the sequence of Celsius temperatures use the R function seq(0, 100, 5).

5 Use the data in Table A below. Assume a person is initially HIV-negative. If the probability of getting infected per act is p, then the probability of not getting infected per act is $(1-p)$.

The probability of not getting infected after two consecutive acts is $(1-p)^2$, and the probability of not getting infected after three consecutive acts is $(1-p)^3$. Therefore, the probability of not getting infected after n consecutive acts is $(1-p)^n$, and the probability of getting infected after n consecutive acts is $1-(1-p)^n$.

For the nonblood transfusion transmission probability (per act risk) in Table A, calculate the risk of being infected after one year (365 days) if one carries out the needle sharing injection drug use (IDU) once daily for one year.

Do these cumulative risks seem reasonable? Why? Why not?

Table A

Estimated per-act risk (transmission probability) for acquisition of HIV, by exposure route to an infected source. Source: CDC

Exposure route	Risk per 10,000 exposures
Blood Transfusion (BT)	9,000
Needle-sharing Injection-Drug Use (IDU)	67

Solution:

```
> p <- 67/10000
> p
[1] 0.0067
> q <- (1 - p)
> q
[1] 0.9933
> q365 <- q^365
> q365
[1] 0.08597238
> p365 <- 1 - q365
> p365
[1] 0.9140276
# => Probability of being infected in a year = 91.40%.
#    A high risk, indeed!
```

Straddle in the Black–Scholes Model

What is a Straddle? A straddle, also known as disambiguation, is a type of financial investment strategy involving *simultaneously* both the put and call of a given stock, to provide additional opportunities for profiting (with the concomitant risk for losing!).

In finance, a straddle refers to *two* transactions that share the *same* security, with positions that offset one another. One holds long risk, the other short. Thus, it involves the purchase or sale of particular option derivatives that allow the holder to profit based on how much the price of the underlying security moves, regardless of the direction of price movement.

A straddle involves buying a call and put with same strike price and expiration date:

i) If the stock price is close to the strike price at expiration of the options, the straddle leads to a loss.

ii) However, if there is a sufficiently large move in either direction, a significant profit will result.

A straddle is appropriate when an investor is expecting a large move in a stock price but does not known in which direction the move will be.

The purchase of particular option derivatives is known as *a long straddle*, while the sale of the option derivatives is known as a *short straddle*.

A *long straddle* buys *both* a call option and a put option on some stock, interest rate, index, or other underlying. The two options are bought at the *same* strike price and expire at the *same* time. The owner of a long straddle makes a profit if the underlying price moves a long way from the strike price, either above or below. Thus, an investor may take a long straddle position if he thinks the market is highly volatile, but does not know in which direction it is going to move. This position is a limited risk, since the most a purchaser may lose is the cost of both options. At the same time, there is unlimited profit potential.

For example, company ABC is set to release its quarterly financial results in 2 weeks. An investor believes that the release of these results will cause a large movement in the price of the ABC's stock, but does not know whether the price will go up or down. The investor may enter into a long straddle, where one gets a profit no matter which way the price of ABC stock moves, if the price changes enough either way:

- If the price goes up enough, the investor may use the option and ignores the put option.
- If the price goes down, the investor uses the put option and ignores the call option.
- If the price does not change enough, he loses money, up to the total amount paid for the two options.

  Thus, the risk is limited by the total premium paid for the options, as opposed to the short straddle where the risk is virtually unlimited.
- If the stock is sufficiently volatile and option duration is long, the investor could profit from *both* options. This would require the stock to move both below the put option's strike price and above the call option's strike price at different times before the option expiration date.

A *short straddle* is a nondirectional option trading strategy that involves simultaneously selling a put and a call of the same underlying security, strike price, and expiration date. The profit is limited to the premium received from the sale of put and call. The risk is virtually unlimited as large moves of the underlying security's price either up or down will cause losses proportional to the magnitude of the price move. A maximum profit upon expiration is achieved if the underlying security trades exactly at the strike price of the straddle. Thus, both puts and calls comprising the straddle expire worthless allowing straddle owner to keep full credit received as their profit. This strategy is also called "nondirectional" because the short straddle profits when the underlying security changes little in price before the expiration of the straddle. The short straddle may be classified as a credit spread because the sale of the short straddle results in a credit of the premiums of the put and call.

(A risk for holder of a short straddle position is unlimited due to the sale of the call and the put options that expose the investor to unlimited losses (on the call) or losses limited to the strike price (on the put), whereas maximum profit is limited to the premium gained by the initial sale of the options.)

An example of a Straddle Financial Investment, using R:

In CRAN, the package, FinancialMath: *Financial Mathematics for Actuaries* provides a numerical example for assessing a Straddle investment, using the Black–Scholes equation for estimating the call and put prices:

```
traddle.bls                    Straddle Spread - Black Scholes
```

Description

Gives a table and graphical representation of the payoff and profit of a long or short straddle for a range of future stock prices. Uses the Black–Scholes equation for estimating the call and put prices.

Usage

```
straddle.bls(S,K,r,t,sd,position,plot=TRUE/FALSE)
```

Arguments

S	spot price at time 0
K	strike price of the put and call
r	continuously compounded yearly risk free rate
t	time of expiration (in years)
sd	standard deviation of the stock (a measure of its volatility)
position	either buyer or seller of option ("long" or "short")
plot	specifying whether or not to plot the payoff and profit

Details

Stock price at time t = S_t
For $S_t \leq$ K1: payoff = K1 - S_t
For K1 < $S_t \leq$ K2: payoff = 0
For S_t > K2: payoff = S_t - K2
profit = payoff - ($price_{K1}$ + $price_{K2}$) * er*t
For $S_t \leq$ K: payoff = S_t - K
For St > K: payoff = K - St
Profit = Payoff + ($price_{call}$ + $price_{put}$) * er*t

Value

A list of two components.

Payoff A data frame of different payoffs and profits for given stock prices.

Premiums A matrix of the premiums for the call and put options, and the net cost.

See Also
option.put
option.call
strangle.bls

Examples

```
straddle.bls(S=100,K=110,r=.03,t=1,sd=.2,position="short")
straddle.bls(S=100,K=110,r=.03,t=1,sd=.2,position="long",
plot=TRUE)
```

In the R domain:
```
>
> install.packages("FinancialMath")
Installing package into 'C:/Users/Bert/Documents/R/win-
library/3.2'
(as 'lib' is unspecified)
--- Please select a CRAN mirror for use in this session ---
A CRAN domain is selected.
trying URL
'https://cran.ism.ac.jp/bin/windows/contrib/3.2/
FinancialMath_0.1.1.zip'
Content type 'application/zip' length 168463 bytes (164 KB)
downloaded 164 KB package 'FinancialMath' successfully unpacked
and MD5 sums checked
```

The downloaded binary packages are in
C:\Users\Bert\AppData\Local\Temp\RtmpUtnvRV
\downloaded_packages
> library(FinancialMath)
> ls("package:FinancialMath")
[1] "amort.period"   "amort.table"   "annuity.arith"
[4] "annuity.geo"   "annuity.level"   "bear.call"
[7] "bear.call.bls"   "bls.order1"   "bond"
[10] "bull.call"   "bull.call.bls"   "butterfly.spread"
[13] "butterfly.spread.bls" "cf.analysis"   "collar"
[16] "collar.bls"   "covered.call"   "covered.put"
[19] "forward"   "forward.prepaid"   "IRR"
[22] "NPV"   "option.call"   "option.put"
[25] "perpetuity.arith"   "perpetuity.geo"   "perpetuity.
level"
[28] "protective.put"   "rate.conv"   "straddle"
[31] "straddle.bls"   "strangle"   "strangle.bls"
[34] "swap.commodity"   "swap.rate"   "TVM"
[37] "yield.dollar"   "yield.time"
> straddle.bls
function (S, K, r, t, sd, position, plot = FALSE)
{
all = list(S, K, r, t, sd, plot, position)
if (any(lapply(all, is.null) == T))
stop("Cannot input any variables as NULL.")
if (any(lapply(all, length) != 1) == T)
stop("All inputs must be of length 1.")
num2 = list(S, K, r, t, sd)
na.num2 = num2[which(lapply(num2, is.na) == F)]
if (any(lapply(na.num2, is.numeric) == F))
stop("S, K, r, t, and sd must be numeric.")
nalist = list(S, K, r, t, sd, plot, position)
if (any(lapply(nalist, is.na) == T))
stop("Cannot input any variables, as NA.")
stopifnot(is.logical(plot))
NA.Neg = array(c(S, K, r, t, sd))
NA.Neg.Str = c("S", "K", "r", "t", "sd")
app = apply(NA.Neg, 1, is.na)
na.s = which(app == F & NA.Neg <= 0)
if (length(na.s) > 0) {

```
errs = paste(NA.Neg.Str[na.s], collapse = " & ")
stop(cat("Error: '", errs, "' must be positive real number(s).
\n"))
}
na.s2 = which(app == F & NA.Neg == Inf)
if (length(na.s2) > 0) {
errs = paste(NA.Neg.Str[na.s2], collapse = " & ")
stop(cat("Error: '", errs, "' cannot be infinite.\n"))
}
if (position != "long" & position != "short")
stop("Position must be either short or long.")
d1 = (log(S/K) + (r + sd^2/2) * t)/(sd * sqrt(t))
d2 = d1 - sd * sqrt(t)
callP = S * pnorm(d1) - K * exp(-r * t) * pnorm(d2)
putP = K * exp(-r * t) * pnorm(-d2) - S * pnorm(-d1)
stock = seq(0, K, length.out = 6)
stock = c(stock, round(seq(K, K * 2, length.out = 6)))
stock = unique(round(stock))
payoff = rep(0, length(stock))
profit = rep(0, length(stock))
if (position == "long") {
for (i in 1:length(stock)) {
if (stock[i] == K)
payoff[i] = 0
if (stock[i] < K)
payoff[i] = K - stock[i]
if (stock[i] > K)
payoff[i] = stock[i] - K
profit[i] = payoff[i] - (callP + putP) * exp(r *
t)
}
cost = callP + putP
}
if (position == "short") {
for (i in 1:length(stock)) {
if (stock[i] == K)
payoff[i] = 0
if (stock[i] < K)
payoff[i] = stock[i] - K
if (stock[i] > K)
payoff[i] = K - stock[i]
```

```
profit[i] = payoff[i] + (callP + putP) * exp(r *
t)
}
cost = -(callP + putP)
}
if (plot == T) {
if (position == "long") {
lpos = "bottomright"
m = "Purchased Straddle\nPayoff and Profit"
}
else {
lpos = "topright"
m = "Written Straddle\nPayoff and Profit"
}
plot(stock, profit, type = "l", xlab = "Stock Price",
main = m, ylab = "$", ylim = c(min(profit, payoff),
max(profit, payoff)), xaxt = "n", yaxt = "n",
col = "steelblue", lwd = 2)
lines(stock, payoff, lty = 2, lwd = 2, col = "firebrick")
abline(h = 0, lty = 2, col = "gray")
x = round(seq(min(payoff, profit), max(payoff, profit),
length.out = 8))
axis(2, at = x, labels = x, las = 2)
axis(1, at = stock)
legend(lpos, c("Profit", "Payoff"), lty = c(1, 2), col =
c("steelblue",
"firebrick"), lwd = c(2, 2))
}
out1 = data.frame(stock, payoff, profit)
names(out1) = c("Stock Price", "Payoff", "Profit")
out2 = matrix(c(callP, putP, cost), nrow = 3)
rownames(out2) = c("Call", "Put", "Net Cost")
colnames(out2) = c("Premiums")
out = list(Payoff = out1, Premiums = out2)
return(out)
}
<environment: namespace:FinancialMath>
> straddle.bls(S=100,K=110,r=.03,t=1,sd=.2,position="short")
$Payoff
Stock Price Payoff   Profit
1      0 -110 -92.136241
2     22  -88 -70.136241
```

```
3       44   -66 -48.136241
4       66   -44 -26.136241
5       88   -22 -4.136241
6      110     0 17.863759
7      132   -22 -4.136241
8      154   -44 -26.136241
9      176   -66 -48.136241
10     198   -88 -70.136241
11     220  -110 -92.136241
```

```
$Premiums
Premiums
Call     5.293398
Put     12.042407
Net Cost -17.335805
```

```
> straddle.bls(S=100,K=110,r=.03,t=1,sd=.2,position="long",
+           plot=TRUE)
```

```
$Payoff
Stock Price Payoff   Profit
1        0  110 92.136241
2       22   88 70.136241
3       44   66 48.136241
4       66   44 26.136241
5       88   22 4.136241
6      110    0 -17.863759
7      132   22 4.136241
8      154   44 26.136241
9      176   66 48.136241
10     198   88 70.136241
11     220  110 92.136241
```

```
$Premiums
Premiums
Call     5.293398
Put     12.042407
Net Cost 17.335805
```

```
> # Outputting: Figure 4.18a
```

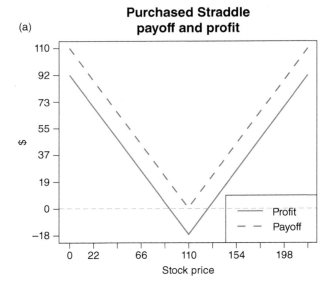

(a)

Purchased Straddle payoff and profit

Figure 4.18(a) Straddle investment in the Black–Scholes model.

4.4 Using R in Data Analysis in Financial Engineering

In financial investigations, after preparing the collected data sets (as discussed in Section 3.1) to undertake financial analysis, the first step is to enter the data sets into the R environment. Once the data sets are placed within the R environment, analysis will process the data to obtain results leading to creditable conclusions, and likely recommendations for definitive courses of actions to improve pertinent aspects of public and personal data. Several methods for data set entry will be examined.

4.4.1 Entering Data at the R Command Prompt

Data Frames and Data sets: For many financial investigators, the terms data frame and data set may be used interchangeably.

Data sets: In many applications, a complete data set contain several data frames, including the real data that have been collected.

Data Frames: Rules for data frames are similar to those for arrays and matrices, introduced earlier. However, data frames are more complicated than arrays. In an array, if just one cell is a character, then all the columns will be characters. On the other hand, a data frame can consist of the following:

- A column of "IDnumber," which is numeric
- A column of "Name," which is character.

In a data frame:

a) each variable can have long variable descriptions and
b) a factor can have "levels" or value levels.

These properties can be transferred from the original data set in other software formats (such as SPS, Stata, etc.). They can also be created in R.

4.4.1.1 Creating a Data Frame for R Computation Using the EXCEL Spreadsheet (on a Windows Platform)

As an example, using a typical set of real case control epidemiologic research data, consider the data set in Table B, from a clinical trial to evaluate the efficacy of maintenance chemotherapy for case subjects with acute myelogenous leukemia (AML), conducted at Stanford University, California, U.S.A., in 1977. After reaching a status of remission through treatment by chemotherapy, the patients who entered the study were assigned randomly to two groups:

1) Maintained: This group received maintenance chemotherapy
2) Nonmaintained: This group did not and is the control group.

The clinical trial was to ascertain whether maintenance chemotherapy prolonged the time until relapse (= "death").

```
Procedure: (1)  To create an acute myelogenous
leukemia (AML) data file,
called AML.csv, in Windows
(2)  To input it into R as a data file AML
(1) Creating a data frame for R computation
1. Data Input Using EXCEL:
(a)  Open the Excel spreadsheet
(b)  Type in data such that the variable names are
in the row 1 of the Excel spreadsheet.
(c)  Consider each row of data as an individual in
the study.
(d)  Start with column A.
2.  Save as a .csv file:
(a)  Click: "File" → "Save as" → and then, in the
file name box (the upper box at the bottom) type: AML
(b)  In the "Save in:" Box (at the top), choose
"Local Disc (C:)"
The file AML will then be saved in the top
level of the C:Drive, but another level may also
be chosen.
In the "Save as Type" Box (the lower box at the bottom), scroll
down, select, and click on CSV
```

(Comma delimited = Comma Separated Values)

To close out of Excel using the big "X" at the top right-hand corner: Click X.

Table B**

Data for the AML maintenance Clinical Study (a + indicates a censored value)

Group   Duration for Complete Remission
(weeks)   .
1=Maintained (11)   9,13,13+,18,23,28+,31,34,45+,48,161+ }
1=Uncensored
0=Nonmaintained (12)   5, 5, 8, 8, 12, 16+,23,27,30,33,43,45}
0=Censored (+) .

NB: The Nonmaintained group may be considered as MBD***

The AML Clinical Study Data: Tableman & Kim (2004).- Table B-1: 23 data points, taken from -

"Survival Analysis Using S: Analysis of Time-to-Event Data" by Mara Tableman and Jong Sung Kimz, published by Chapman & Hall/CRC, Boca Raton, 2004

The cancer epigenome is characterised by specific DNA methylation and chromatin modification patterns. The proteins that mediate these changes are encoded by the epigenetics genes, here defined as DNA methyltransferases (DNMT), methyl-CpG-binding domain (MBD) proteins, histone acetyltransferases (HAT), histone deacetylases (HDAC), histone methyltransferases (HMT), and histone demethylases.

3. In Windows, check the C:Drive for the AML.csv file., namely, C: AML

4. Read AML into R:

(a) Open R

(b) Use the read.csv() function:

```
> aml <- read.csv("C:\\AML.csv", header = T, sep=
+                  ",")
```

(c) Actually, it can also be done by

```
> aml <- read.csv("C:\\AML.csv")
> # Read in the AML.csv file from the C:Drive of the
> # Computer, and call it
> # aml
```

5. Output the AML.csv file for inspection

```
> aml # Outputting:
weeks group status
1   9   1   1
2   13  1   1
```

```
3     13   1   0
4     18   1   1
5     23   1   1
6     28   1   0
7     31   1   1
8     34   1   1
9     45   1   0
10    48   1   1
11   161   1   0
12     5   0   1
13     5   0   1
14     8   0   1
15     8   0   1
16    12   0   1
17    16   0   0
18    23   0   1
19    27   0   1
20    30   0   1
21    33   0   1
22    43   0   1
23    45   0   1
>
```

Later, in Section 6.3, this data set will be revisited and further processed for survival analysis.

4.4.1.2 Obtaining a Data Frame from a Text File

Data from various sources are often entered using many different software programs.

They may be transferred from one format to another through the ASCII file format.

For example, in Windows, a text file is the most common ASCII file, usually having a ".txt" extension. There are other files in ASCII format, including the ".R" command file.

Data from most software programs can be outputted or saved as an ASCII file. From Excel, a very common spreadsheet program, the data can be saved as ".csv" (comma separated values) format. This is an easy way to interface between Excel spreadsheet files and R. Open the Excel file and "save as" the csv format.

- Files with field separators:
  As an example, suppose the file "csv1.xls" is originally an Excel spreadsheet. After "save as" into csv format, the output file is called "csv1.csv", the contents of which is

```
"name","gender","age"
"A",      "F",      20
"B",      "M",      30
"C",      "F",      40
```

The characters are enclosed in quotes and the delimiters (variable separators) are commas. Sometimes the file may not contain quotes, as in the file "csv2.csv".

```
name, gender, age
A,        F,        20
B,        M,        30
C,        F,        40
```

For both files, the R command to read in the data set is the same.

```
> a <- read.csv("csv1.csv", as.is=TRUE)
> a
    name    gender    age
1   A       F         20
2   B       M         30
3   C       F         40
```

The argument 'as.is=TRU ' keeps all characters as they are, otherwise the characters would have been coerced into factors. The variable "name" should not be factored but "gender" should. The following command should, therefore, be entered

```
> a$gender <- factor(a$gender)
```

Note that the object "a" has class data frame and that the names of the variables within the data frame "a" must be referenced using the dollar sign notation $. Otherwise, R will state that the object "gender" cannot be found.

For files with white space (spaces and tabs) as the separator, such as in the file

```
"data1.txt", the command to use is read.table():
> a <- read.table("data1.txt", header=TRUE,
+                as.is=TRUE)
```

- Files without field separators:
  Consider the file "data2.txt" which in fixed field format without field separators.

```
      name  gender   age
1      A       F      20
2      B       M      30
3      C       F      40
```

To read in such a file, use the function read.fwf():
i) Skip the first line, which is the header.
ii) The width of each variable and the column names must be specified.

```
> a <- read.fwf ("data2.txt", skip=1, width=c(1,1,2),
+          col.names = c("name", "gender", "age"),
+          as.is=TRUE)
```

4.4.1.3   Data Entry and Analysis Using the Function data.entry()

The previous section deals with creating data frames by reading in data created from programs outside R, such as Excel. It is also possible to enter data directly into R by using the function data.entry(). However, if the data size is large (say more than 15 columns and/or more than 25 rows), the chance of human error is high with the spreadsheet or text mode data entry. A software program specially designed for data entry, such as Epidata, is more appropriate. (http://www.epidata.dk)

4.4.1.4   Data Entry Using Several Available R Functions

The data set, in Table C, lists deaths among subjects who received a dose of tolbutamide or a placebo in the University Group Diabetes Program (1970), stratifying by age:

Table C**
Deaths Among Subjects Who Received Tolbutamide or a Placebo in the University Group Diabetes Program (1970)

	Age < 55		Age ≥ 55		Combined	
	Tolbutamide	Placebo	Tolbutamide	Placebo	Tolbutamide	Placebo
Deaths	8	5	22	16	30	21
Survivors	98	115	76	69	174	184

**Available at http://www.medepi.net/data/ugdp.txt

The R functions that can be used to import the data frame have been previously introduced in Sections 4.3.3–4.3.13 A convenient way to enter data at the command prompt is to use the R functions:

```
c(), matrix(), array(), apply(), list(), data.frame(), and
odd.ratio(),
```

as shown by the following examples and using the data in Table C.

```
> #Entering data for a vector
> vector1 <- c(8, 98, 5, 115) # Using data from Table C.
> vector1
[1]   8 98  5 115
>
> vector2 <- c(22, 76, 16, 69); vector2
> # Data from Table C.
[1] 22 76 16 69
>
> # Entering data for a matrix
> matrix1 <- matrix(vector1, 2, 2)
> matrix1
      [,1] [,2]
[1,]    8    5
[2,]   98  115
> matrix2 <- matrix(vector2, 2, 2); matrix2
      [,1] [,2]
[1,]   22   16
[2,]   76   69
>
> # Entering data for an array
>  udata <- array(c(vector1, vector2), c(2, 2, 2))
>  udata
, , 1
      [,1] [,2]
[1,]    8    5
[2,]   98  115
, , 2
      [,1] [,2]
[1,]   22   16
[2,]   76   69
> apply(udata, c(1, 2), sum); udata.tot
      [,1]  [,2]
[1,]   30    21
[2,]  174   184
>
> # Entering a list
> x <- list(crude.data = udata.tot, stratified.data =
+        udata)
> x$crude.data
      [,1] [,2]
[1,]   30   21
```

```
[2,] 174  184
> x$stratified
, , 1
     [,1]  [,2]
[1,]   8    5
[2,]  98  115
, , 2
     [,1]   [,2]
[1,]  22    16
[2,]  76    69
>
> # Entering a simple data frame
> subjectname <- c("Peter", "Paul", "Mary")
> subjectnumber <- 1:length(subjectname)
> age <- c(26, 27, 28) # These are their true ages,
>                       # respectively, in 1964!
> gender <- c("Male", "Male", "Female")
> data1 <- data.frame(subjectnumber, subjectname,
+               age, gender)
> data1
   subjectnumber subjectname  age  gender
1             1       Peter    26    Male
2             2        Paul    27    Male
3             3        Mary    28  Female
>
> # Entering a simple function
> odds.ratio <- function(aa, bb, cc, dd){ aa*dd /
+                 (bb*cc)}
> odds.ratio(30, 174, 21, 184) # Data from Table C.
[1] 1.510673
```

4.4.1.5 Data Entry and Analysis Using the Function scan()

The R function scan() is taken from the CRAN package base.

This function reads data into a vector or list from the console or file. This function takes the following usage form:

```
scan(file = ", what = double(), nmax = -1, n = -1,
          sep = "",
          quote = if(identical(sep, "\n")) "" else
          "'\"", dec = ".",
          skip = 0, nlines = 0, na.strings = "NA",
          flush = FALSE, fill = FALSE, strip.white =
          FALSE,
```

```
quiet = FALSE, blank.lines.skip = TRUE,
multi.line = TRUE,
comment.char = "", allowEscapes =
FALSE,
fileEncoding = "", encoding = "unknown",
text)
```

Argument

what   The type of what gives the type of data to be read.
        The supported types are logical, integer, numeric, complex, character,
        raw, and *list*. If what is a list, it is assumed that the lines of the data
        file are records each containing length (what) items (fields) and the
        list components should have elements that are one of the first six
        types listed or NULL.

The what argument describes the tokens that scan() should expect in the
input file.
For a detailed description of this function, execute

```
> ?scan
```

The methodology of applying scan() is similar to c(), as described in
Section 4.4.1.4, except that it does not matter the numbers are being entered
on different lines, it will still be a vector.

- Use scan() when accessing data from a file that has an irregular or a complex
  structure.
- Use scan() to read individual tokens and use the argument what to describe
  the stream of tokens in the file.
- scan() converts tokens into data, and then assemble the data into records.
- Use scan() along with readLines(), especially when one attempts to read an
  unorthodox file format. Together, these two functions will likely result in a
  successful processing of the individual lines and tokens of the file!

The function readLines() reads lines from a file, and returns them to a list of
character strings:

```
> lines <- readLines("input.text")
```

One may limit the number of lines to be read, per pass, by using the n
parameter that gives the maximum number of lines to be read:

```
> lines <- readLines("input.text, n=5)
> # read 5 lines and stop
```

The function scan() reads one token at a time, and handles it accordingly as instructed.

An example:

Assume that the file to be scanned and read contains triplets of data (like the dates, and the corresponding daily highs and lows of financial markets):

15-Oct-1987	2439.78	2345.63
16-Oct-1987	2396.21	2207.73
19-Oct-1987	2164.16	1677.55
20-Oct-1987	2067.47	1616.23
21-Oct-1987	2087.07	1951.76

Use a list to operate scan() that it should expect a repeating, 3-token sequence:

```
> triplets <- scan("triples.txt, what=list(character(0),
+            numeric(0), numeric(0)))
```

Give names to the list elements, and scan() will assign those names to the data:

```
> triplets <- scan("triples.txt,
+            what=list(date=character(0),
+            high=numeric(0), low=numeric(0)))
Reads 5 records.
> triples # Outputs:
$date
[1] "15-Oct-1987" "15-Oct-1987" "19-oct-1987" "20-Oct-1987"
"21-oct-1987"
$high
[1] 2439-78 2396.21 2164.16 2067.47 2081.07
$low
[1] 2345.63 2207.73 1677.55 1616.21 1951.76
```

4.4.1.6 Data Entry and Analysis Using the Function Source()

The R function source() is also taken from the CRAN package base. This function reads data into a vector or list from the console or file. It takes the following usage form:

source() causes R to accept its input from the named file or URL or connection. Input is read and *parsed* from that file until the end of the file is reached, then the parsed expressions are evaluated sequentially in the chosen environment:

```
source (file, local = FALSE, echo = verbose, print.eval
        = echo,
        verbose = getOption ("verbose"),
        prompt.echo = getOption ("prompt"),
        max.deparse.length = 150, chdir = FALSE,
        encoding = getOption ("encoding"),
        continue.echo = getOption ("continue"),
        skip.echo = 0, keep.source =
        getOption ("keep.source"))
```

For commands that are stored in an external file, such as "commands.R" in the working directory "work," they can be executed in an R environment with the command

```
> source ("command.R")
```

The function source() instructs R to read the text and execute its contents. Thus, when one has a long, or frequently used, piece of R code, one may capture it inside a text file. This allows one to rerun the code without having to retype it, and use the function source() to read and execute the code.

For example, suppose the file howdy.R contains the familiar greeting:

```
Print ("Hi, My Friend!")
```

Then by sourcing the file, one may execute the content of the file, as in the follow R code segment:

```
> source ("howdy.R")
[1] "Hi, My Friend!"
```

Setting echo-TRUE will echo the same script lines before they are executed, with the R prompt shown before each line:

```
> source ("howdy.R", echo=TRUE)
> Print ("Hi, My Friend!")
[1] "Hi, My Friend!"
```

4.4.1.7 Data Entry and Analysis Using the Spreadsheet Interface in R

This method consists of the following R functions in the package Utils.

- *Spreadsheet Interface for Entering Data*
  This is a spreadsheet-like editor for entering or editing data, with the following R functions:

```
data.entry(..., Modes = NULL, Names = NULL)
dataentry(data, modes)
de(..., Modes = list(), Names = NULL)
```

The arguments of these R functions are:

A list of variables: currently these should be numerals or character vectors or a list containing such vectors.

Modes   The modes to be used for the variables.

Names   The names to be used for the variables.

data   A list of numeric and/or character vectors.

modes   A list of length up to that of data giving the modes of (some of) the variables. list() is allowed.

The function data.entry() edits an existing object, saving the changes to the original object name.

However, the function edit() edits an existing object but not saving the changes to the original object name so that one must assign it to an object name (even if it is the original name). To enter a vector, one needs to initialize a vector and then use the function data.entry(). For example:

```
Start by entering the R environment, and set
> x <- c(2, 4, 6, 8, 10)
> # X is initially defined as an array of 5 elements.
> x # Just checking - to make sure!
[1]  2 4 6 8 10 # x is indeed set to be an array of 5elements
>
> data.entry(x) # Entering the Data Editor:
> # The Data Editor window pops up, and looking at the first
> # column: it is now named "x", with the first 5 rows (all on first
> # column) filled, respectively, by the numbers 2, 4, 6, 8, 10
> # One can now edit this dataset by, say, changing all the
> # entries to 2, then closing the Data Editor window, and
> # returning to the R console window:
> x
[1] 2 2 2 2 2 # x is indeed changed!
> # Thus one can change the entries for x via the Data Editor,
> # and save the changes.
```

When using the functions data.entry(x) and edit() for data entry, there are a number of limitations:

i) Arrays and nontabular lists cannot be entered using a spreadsheet editor.
ii) When using the function edit() to create a new data frame, one must assign an object name in order to save the data frame.

iii) This approach is not a preferred method of entering data because one often prefers to have the original data to be in a text editor or available to be read in from a data file.

4.4.1.8   Financial Mathematics Using R: The CRAN Package FinancialMath

To illustrate the ease of use of R in financial mathematics, consider a simple example of the repayment process of a loan, such as a mortgage on a piece of real estate.

Two examples are used for calculating the loan repayment process:

Example (A): A loan of one million dollars, at an interest rate of 2.5%, to be repaid by equal monthly installments over 30 years or 360 months. One should also consider the rate and total amount of interests to be repaid over the life of the loan

```
> amort.table(Loan=1000000,n=360,
+             i=0.025,pf=360,plot=TRUE)
```

Example (B): Example A shows that the monthly payment is $2,812.31. Suppose the borrower can afford to pay more that this monthly amount, say, $5,000.00 monthly, how does this alternate repayment scheme affect the overall repayment process, particularly in terms of the total interest payable over the whole life of the loan?

Package:	'FinancialMath'
Date:	December 16, 2016
Type:	Package
Title:	Financial Mathematics for Actuaries
Version:	0.1.1
Author **Kameron Penn**	[aut, cre],
Jack Schmidt	[aut]
Maintainer	Kameron Penn <kameron.penn.financialmath@gmail.com>
Description	Contains financial math functions and introductory derivative functions included in the Society of Actuaries and Casualty Actuarial Society 'Financial Mathematics' exam and some topics on the "Models for Financial Economics' exam."
License	GPL-2
Encoding	UTF-8
LazyData	true
Needs Compilation	no
Repository	CRAN
Date/Publication	2016-12-16 22:51:34
amort.table **Amortization Table**	

Description

Produces an amortization table for paying off a loan while also solving for either the number of payments, loan amount, or the payment amount. In the amortization table, the payment amount, interest paid, principal paid, and balance of the loan are given for each period. If n ends up not being a whole number, outputs for the balloon payment, drop payment, and last regular payment are provided. The total interest paid, and total amount paid is also given. It can also plot the percentage of each payment toward interest versus period.

Usage

```
amort.table(Loan=NA, n=NA, pmt=NA, i=0.025,
        ic=1, pf=1, plot=TRUE)
```

Arguments

Loan   loan amount

n   the number of payments/periods

pmt   value of level payments

i   nominal interest rate convertible ic times per year

ic   interest conversion frequency per year

pf   the payment frequency- number of payments per year

plot   tells whether or not to plot the percentage of each payment
        toward interest vs.period

Details

Effective Rate of Interest: $\text{eff.i} = [1 + (i/ic)]^{ic} - 1$
$$j = (1 + \text{eff.i})^{(1/pf)} - 1$$
$$\text{Loan} = pmt * a_{n\rceil j}$$
Balance at the end of period t: $B_t = pmt * a_{n - t\rceil j}$
Interest paid at the end of period t: $i_t = B_{t-1} * j$
Principal paid at the end of period t: $p_t = pmt - i_t$
Total Paid $= pmt * n$
Total Interest Paid $= pmt * n - \text{Loan}$
If $n = n* + k$ where $n*$ is an integer and $0 < k < 1$:
Last regular payment (at period $n*$) $= pmt * s_{k/j}$
Drop payment (at period $n* + 1$) $= \text{Loan} * (1 + j)^{n*+1} - pmt * s_{n\rceil j}$
Balloon payment (at period $n*$) $= \text{Loan} * (1 + j)^{n*} - pmt * s_{n*\rceil j} + pmt$

Value
A list of two components.

Schedule	A data frame of the amortization schedule.
Other	A matrix of the input variables and other calculated variables.

Note
Assumes that payments are made at the end of each period.

One of n, Loan, or pmt must be NA (unknown).

If pmt is less than the amount of interest accumulated in the first period, then the function will stop because the loan will never be paid off due to the payments being too small.

If pmt is greater than the loan amount plus interest accumulated in the first period, then the function will stop because one payment will pay off the loan.

Author(s)
K. Penn and J. Schmidt
See Also
amort.period
annuity.level

Examples

A. **A $1M loan, repayable in equal monthly installments, over 30 years, at 2.5% interest p.a., using the following R code segment:**

```
> amort.table(Loan=1000000, n=360,
+                          i=0.025, pf=360, plot=TRUE)
```

B. **The same $1M loan, repayable in equal monthly installments of $5,000.00, with the loan being amortized over 30 years, at 2.5% interest p.a., using the following R code segment:**

```
>   amort.table(Loan=1000000, pmt=5000, n=360,
+                          i=0.025, plot=TRUE)
In the R domain:
> install.packages("FinancialMath")
Installing package into 'C:/Users/Bert/Documents/R/win-
library/3.2'
(as 'lib' is unspecified)
--- Please select a CRAN mirror for use in this session ---
A CRAN mirror is selected.
trying URL 'https://cran.ism.ac.jp/bin/windows/contrib/3.2/
FinancialMath_0.1.1.zip'
```

```
Content type 'application/zip' length 168463 bytes (164 KB)
downloaded 164 KB
package 'FinancialMath' successfully unpacked and MD5 sums
checked
The downloaded binary packages are in
        C:\Users\Bert\AppData\Local\Temp\Rtmp48HxfA
\downloaded_packages
> library(FinancialMath)
Warning message:
package 'FinancialMath' was built under R version 3.2.5
> ls("package:FinancialMath")
 [1] "amort.period"        "amort.table"       "annuity.arith"
 [4] "annuity.geo"         "annuity.level"     "bear.call"
 [7] "bear.call.bls"       "bls.order1"        "bond"
[10] "bull.call"           "bull.call.bls"     "butterfly.spread"
[13] "butterfly.spread.bls" "cf.analysis"      "collar"
[16] "collar.bls"          "covered.call"      "covered.put"
[19] "forward"             "forward.prepaid"   "IRR"
[22] "NPV"                 "option.call"       "option.put"
[25] "perpetuity.arith"    "perpetuity.geo"    "perpetuity.level"
[28] "protective.put"      "rate.conv"         "straddle"
[31] "straddle.bls"        "strangle"          "strangle.bls"
[34] "swap.commodity"      "swap.rate"         "TVM"
[37] "yield.dollar"        "yield.time"
>
(A)    Loan Amount = $1,000,000,00, Interest = 2.5% p.a.
        Repayment Plan:  To repay, by a monthly fixed amount,
               both the capital and interests, in 30 years
               time - in 360 (30 x12) monthly payments
> amort.table(Loan=1000000, n=360,
+             i=0.025, pf=360, plot=TRUE)
>
>   # Outputting:
$Schedule
Year    Payment        Interest Paid   Principal   Paid Balance
1    0.00 2812.31          68.59         2743.72     997256.28
2    0.01 2812.31          68.40         2743.91     994512.38
3    0.01 2812.31          68.22         2744.09     991768.28
4    0.01 2812.31          68.03         2744.28     989024.00
5    0.01 2812.31          67.84         2744.47     986279.53
6    0.02 2812.31          67.65         2744.66     983534.87
7    0.02 2812.31          67.46         2744.85     980790.02
8    0.02 2812.31          67.28         2745.04     978044.99
```

9	0.02	2812.31	67.09	2745.22	975299.76
10	0.03	2812.31	66.90	2745.41	972554.35
11	0.03	2812.31	66.71	2745.60	969808.75
12	0.03	2812.31	66.52	2745.79	967062.96
13	0.04	2812.31	66.33	2745.98	964316.99
14	0.04	2812.31	66.15	2746.17	961570.82
15	0.04	2812.31	65.96	2746.35	958824.47
16	0.04	2812.31	65.77	2746.54	956077.93
17	0.05	2812.31	65.58	2746.73	953331.19
18	0.05	2812.31	65.39	2746.92	950584.28
19	0.05	2812.31	65.20	2747.11	947837.17
20	0.06	2812.31	65.01	2747.30	945089.87
21	0.06	2812.31	64.83	2747.48	942342.39
22	0.06	2812.31	64.64	2747.67	939594.72
23	0.06	2812.31	64.45	2747.86	936846.86
24	0.07	2812.31	64.26	2748.05	934098.81
25	0.07	2812.31	64.07	2748.24	931350.57
26	0.07	2812.31	63.88	2748.43	928602.14
27	0.08	2812.31	63.70	2748.62	925853.53
28	0.08	2812.31	63.51	2748.80	923104.72
29	0.08	2812.31	63.32	2748.99	920355.73
30	0.08	2812.31	63.13	2749.18	917606.55
31	0.09	2812.31	62.94	2749.37	914857.18
32	0.09	2812.31	62.75	2749.56	912107.62
33	0.09	2812.31	62.56	2749.75	909357.88
34	0.09	2812.31	62.38	2749.94	906607.94
35	0.10	2812.31	62.19	2750.12	903857.82
36	0.10	2812.31	62.00	2750.31	901107.50
37	0.10	2812.31	61.81	2750.50	898357.00
38	0.11	2812.31	61.62	2750.69	895606.31
39	0.11	2812.31	61.43	2750.88	892855.44
40	0.11	2812.31	61.24	2751.07	890104.37
41	0.11	2812.31	61.05	2751.26	887353.11
42	0.12	2812.31	60.87	2751.44	884601.67
43	0.12	2812.31	60.68	2751.63	881850.03
44	0.12	2812.31	60.49	2751.82	879098.21
45	0.12	2812.31	60.30	2752.01	876346.20
46	0.13	2812.31	60.11	2752.20	873594.00
47	0.13	2812.31	59.92	2752.39	870841.61
48	0.13	2812.31	59.73	2752.58	868089.04
49	0.14	2812.31	59.54	2752.77	865336.27
50	0.14	2812.31	59.36	2752.95	862583.32
51	0.14	2812.31	59.17	2753.14	859830.17

52	0.14	2812.31	58.98	2753.33	857076.84
53	0.15	2812.31	58.79	2753.52	854323.32
54	0.15	2812.31	58.60	2753.71	851569.61
55	0.15	2812.31	58.41	2753.90	848815.71
56	0.16	2812.31	58.22	2754.09	846061.62
57	0.16	2812.31	58.03	2754.28	843307.35
58	0.16	2812.31	57.84	2754.47	840552.88
59	0.16	2812.31	57.66	2754.65	837798.23
60	0.17	2812.31	57.47	2754.84	835043.38
61	0.17	2812.31	57.28	2755.03	832288.35
62	0.17	2812.31	57.09	2755.22	829533.13
63	0.18	2812.31	56.90	2755.41	826777.72
64	0.18	2812.31	56.71	2755.60	824022.12
65	0.18	2812.31	56.52	2755.79	821266.33
66	0.18	2812.31	56.33	2755.98	818510.35
67	0.19	2812.31	56.14	2756.17	815754.18
68	0.19	2812.31	55.95	2756.36	812997.83
69	0.19	2812.31	55.77	2756.54	810241.28
70	0.19	2812.31	55.58	2756.73	807484.55
71	0.20	2812.31	55.39	2756.92	804727.63
72	0.20	2812.31	55.20	2757.11	801970.52
73	0.20	2812.31	55.01	2757.30	799213.21
74	0.21	2812.31	54.82	2757.49	796455.72
75	0.21	2812.31	54.63	2757.68	793698.04
76	0.21	2812.31	54.44	2757.87	790940.18
77	0.21	2812.31	54.25	2758.06	788182.12
78	0.22	2812.31	54.06	2758.25	785423.87
79	0.22	2812.31	53.87	2758.44	782665.44
80	0.22	2812.31	53.69	2758.63	779906.81
81	0.22	2812.31	53.50	2758.81	777148.00
82	0.23	2812.31	53.31	2759.00	774388.99
83	0.23	2812.31	53.12	2759.19	771629.80
84	0.23	2812.31	52.93	2759.38	768870.42
85	0.24	2812.31	52.74	2759.57	766110.84
86	0.24	2812.31	52.55	2759.76	763351.08
87	0.24	2812.31	52.36	2759.95	760591.13
88	0.24	2812.31	52.17	2760.14	757830.99
89	0.25	2812.31	51.98	2760.33	755070.67
90	0.25	2812.31	51.79	2760.52	752310.15
91	0.25	2812.31	51.60	2760.71	749549.44
92	0.26	2812.31	51.41	2760.90	746788.54
93	0.26	2812.31	51.22	2761.09	744027.46
94	0.26	2812.31	51.04	2761.28	741266.18

95	0.26	2812.31	50.85	2761.47	738504.72
96	0.27	2812.31	50.66	2761.65	735743.06
97	0.27	2812.31	50.47	2761.84	732981.22
98	0.27	2812.31	50.28	2762.03	730219.18
99	0.28	2812.31	50.09	2762.22	727456.96
100	0.28	2812.31	49.90	2762.41	724694.55
101	0.28	2812.31	49.71	2762.60	721931.95
102	0.28	2812.31	49.52	2762.79	719169.16
103	0.29	2812.31	49.33	2762.98	716406.18
104	0.29	2812.31	49.14	2763.17	713643.01
105	0.29	2812.31	48.95	2763.36	710879.65
106	0.29	2812.31	48.76	2763.55	708116.10
107	0.30	2812.31	48.57	2763.74	705352.36
108	0.30	2812.31	48.38	2763.93	702588.43
109	0.30	2812.31	48.19	2764.12	699824.31
110	0.31	2812.31	48.00	2764.31	697060.00
111	0.31	2812.31	47.81	2764.50	694295.51
112	0.31	2812.31	47.62	2764.69	691530.82
113	0.31	2812.31	47.43	2764.88	688765.94
114	0.32	2812.31	47.24	2765.07	686000.88
115	0.32	2812.31	47.05	2765.26	683235.62
116	0.32	2812.31	46.87	2765.45	680470.17
117	0.32	2812.31	46.68	2765.64	677704.54
118	0.33	2812.31	46.49	2765.82	674938.71
119	0.33	2812.31	46.30	2766.01	672172.70
120	0.33	2812.31	46.11	2766.20	669406.50
121	0.34	2812.31	45.92	2766.39	666640.10
122	0.34	2812.31	45.73	2766.58	663873.52
123	0.34	2812.31	45.54	2766.77	661106.74
124	0.34	2812.31	45.35	2766.96	658339.78
125	0.35	2812.31	45.16	2767.15	655572.63
126	0.35	2812.31	44.97	2767.34	652805.28
127	0.35	2812.31	44.78	2767.53	650037.75
128	0.36	2812.31	44.59	2767.72	647270.03
129	0.36	2812.31	44.40	2767.91	644502.12
130	0.36	2812.31	44.21	2768.10	641734.01
131	0.36	2812.31	44.02	2768.29	638965.72
132	0.37	2812.31	43.83	2768.48	636197.24
133	0.37	2812.31	43.64	2768.67	633428.57
134	0.37	2812.31	43.45	2768.86	630659.71
135	0.38	2812.31	43.26	2769.05	627890.65
136	0.38	2812.31	43.07	2769.24	625121.41
137	0.38	2812.31	42.88	2769.43	622351.98

138	0.38	2812.31	42.69	2769.62	619582.36
139	0.39	2812.31	42.50	2769.81	616812.55
140	0.39	2812.31	42.31	2770.00	614042.55
141	0.39	2812.31	42.12	2770.19	611272.35
142	0.39	2812.31	41.93	2770.38	608501.97
143	0.40	2812.31	41.74	2770.57	605731.40
144	0.40	2812.31	41.55	2770.76	602960.64
145	0.40	2812.31	41.36	2770.95	600189.69
146	0.41	2812.31	41.17	2771.14	597418.54
147	0.41	2812.31	40.98	2771.33	594647.21
148	0.41	2812.31	40.79	2771.52	591875.69
149	0.41	2812.31	40.60	2771.71	589103.98
150	0.42	2812.31	40.41	2771.90	586332.08
151	0.42	2812.31	40.22	2772.09	583559.98
152	0.42	2812.31	40.03	2772.28	580787.70
153	0.42	2812.31	39.84	2772.47	578015.23
154	0.43	2812.31	39.65	2772.66	575242.57
155	0.43	2812.31	39.46	2772.85	572469.71
156	0.43	2812.31	39.27	2773.04	569696.67
157	0.44	2812.31	39.08	2773.23	566923.44
158	0.44	2812.31	38.89	2773.42	564150.01
159	0.44	2812.31	38.70	2773.61	561376.40
160	0.44	2812.31	38.51	2773.80	558602.59
161	0.45	2812.31	38.32	2773.99	555828.60
162	0.45	2812.31	38.13	2774.18	553054.41
163	0.45	2812.31	37.94	2774.38	550280.04
164	0.46	2812.31	37.75	2774.57	547505.47
165	0.46	2812.31	37.56	2774.76	544730.72
166	0.46	2812.31	37.36	2774.95	541955.77
167	0.46	2812.31	37.17	2775.14	539180.64
168	0.47	2812.31	36.98	2775.33	536405.31
169	0.47	2812.31	36.79	2775.52	533629.79
170	0.47	2812.31	36.60	2775.71	530854.09
171	0.48	2812.31	36.41	2775.90	528078.19
172	0.48	2812.31	36.22	2776.09	525302.10
173	0.48	2812.31	36.03	2776.28	522525.82
174	0.48	2812.31	35.84	2776.47	519749.35
175	0.49	2812.31	35.65	2776.66	516972.69
176	0.49	2812.31	35.46	2776.85	514195.84
177	0.49	2812.31	35.27	2777.04	511418.80
178	0.49	2812.31	35.08	2777.23	508641.57
179	0.50	2812.31	34.89	2777.42	505864.15
180	0.50	2812.31	34.70	2777.61	503086.54

181	0.50	2812.31	34.51	2777.80	500308.73
182	0.51	2812.31	34.32	2777.99	497530.74
183	0.51	2812.31	34.13	2778.18	494752.56
184	0.51	2812.31	33.94	2778.37	491974.18
185	0.51	2812.31	33.75	2778.56	489195.62
186	0.52	2812.31	33.56	2778.76	486416.86
187	0.52	2812.31	33.36	2778.95	483637.92
188	0.52	2812.31	33.17	2779.14	480858.78
189	0.52	2812.31	32.98	2779.33	478079.45
190	0.53	2812.31	32.79	2779.52	475299.94
191	0.53	2812.31	32.60	2779.71	472520.23
192	0.53	2812.31	32.41	2779.90	469740.33
193	0.54	2812.31	32.22	2780.09	466960.24
194	0.54	2812.31	32.03	2780.28	464179.96
195	0.54	2812.31	31.84	2780.47	461399.49
196	0.54	2812.31	31.65	2780.66	458618.83
197	0.55	2812.31	31.46	2780.85	455837.97
198	0.55	2812.31	31.27	2781.04	453056.93
199	0.55	2812.31	31.08	2781.23	450275.70
200	0.56	2812.31	30.89	2781.42	447494.27
201	0.56	2812.31	30.69	2781.62	444712.66
202	0.56	2812.31	30.50	2781.81	441930.85
203	0.56	2812.31	30.31	2782.00	439148.85
204	0.57	2812.31	30.12	2782.19	436366.66
205	0.57	2812.31	29.93	2782.38	433584.28
206	0.57	2812.31	29.74	2782.57	430801.71
207	0.57	2812.31	29.55	2782.76	428018.95
208	0.58	2812.31	29.36	2782.95	425236.00
209	0.58	2812.31	29.17	2783.14	422452.86
210	0.58	2812.31	28.98	2783.33	419669.53
211	0.59	2812.31	28.79	2783.52	416886.00
212	0.59	2812.31	28.60	2783.72	414102.29
213	0.59	2812.31	28.40	2783.91	411318.38
214	0.59	2812.31	28.21	2784.10	408534.28
215	0.60	2812.31	28.02	2784.29	405750.00
216	0.60	2812.31	27.83	2784.48	402965.52
217	0.60	2812.31	27.64	2784.67	400180.85
218	0.61	2812.31	27.45	2784.86	397395.99
219	0.61	2812.31	27.26	2785.05	394610.93
220	0.61	2812.31	27.07	2785.24	391825.69
221	0.61	2812.31	26.88	2785.43	389040.26
222	0.62	2812.31	26.69	2785.63	386254.63
223	0.62	2812.31	26.49	2785.82	383468.81

224	0.62	2812.31	26.30	2786.01	380682.81
225	0.62	2812.31	26.11	2786.20	377896.61
226	0.63	2812.31	25.92	2786.39	375110.22
227	0.63	2812.31	25.73	2786.58	372323.64
228	0.63	2812.31	25.54	2786.77	369536.87
229	0.64	2812.31	25.35	2786.96	366749.90
230	0.64	2812.31	25.16	2787.15	363962.75
231	0.64	2812.31	24.97	2787.35	361175.40
232	0.64	2812.31	24.77	2787.54	358387.87
233	0.65	2812.31	24.58	2787.73	355600.14
234	0.65	2812.31	24.39	2787.92	352812.22
235	0.65	2812.31	24.20	2788.11	350024.11
236	0.66	2812.31	24.01	2788.30	347235.81
237	0.66	2812.31	23.82	2788.49	344447.32
238	0.66	2812.31	23.63	2788.68	341658.63
239	0.66	2812.31	23.44	2788.88	338869.76
240	0.67	2812.31	23.24	2789.07	336080.69
241	0.67	2812.31	23.05	2789.26	333291.43
242	0.67	2812.31	22.86	2789.45	330501.98
243	0.68	2812.31	22.67	2789.64	327712.34
244	0.68	2812.31	22.48	2789.83	324922.51
245	0.68	2812.31	22.29	2790.02	322132.49
246	0.68	2812.31	22.10	2790.21	319342.27
247	0.69	2812.31	21.90	2790.41	316551.87
248	0.69	2812.31	21.71	2790.60	313761.27
249	0.69	2812.31	21.52	2790.79	310970.48
250	0.69	2812.31	21.33	2790.98	308179.50
251	0.70	2812.31	21.14	2791.17	305388.33
252	0.70	2812.31	20.95	2791.36	302596.96
253	0.70	2812.31	20.76	2791.55	299805.41
254	0.71	2812.31	20.56	2791.75	297013.66
255	0.71	2812.31	20.37	2791.94	294221.73
256	0.71	2812.31	20.18	2792.13	291429.60
257	0.71	2812.31	19.99	2792.32	288637.28
258	0.72	2812.31	19.80	2792.51	285844.76
259	0.72	2812.31	19.61	2792.70	283052.06
260	0.72	2812.31	19.42	2792.90	280259.17
261	0.72	2812.31	19.22	2793.09	277466.08
262	0.73	2812.31	19.03	2793.28	274672.80
263	0.73	2812.31	18.84	2793.47	271879.33
264	0.73	2812.31	18.65	2793.66	269085.67
265	0.74	2812.31	18.46	2793.85	266291.82
266	0.74	2812.31	18.27	2794.04	263497.77

267	0.74	2812.31	18.07	2794.24	260703.53
268	0.74	2812.31	17.88	2794.43	257909.11
269	0.75	2812.31	17.69	2794.62	255114.49
270	0.75	2812.31	17.50	2794.81	252319.67
271	0.75	2812.31	17.31	2795.00	249524.67
272	0.76	2812.31	17.12	2795.20	246729.48
273	0.76	2812.31	16.92	2795.39	243934.09
274	0.76	2812.31	16.73	2795.58	241138.51
275	0.76	2812.31	16.54	2795.77	238342.74
276	0.77	2812.31	16.35	2795.96	235546.78
277	0.77	2812.31	16.16	2796.15	232750.62
278	0.77	2812.31	15.97	2796.35	229954.28
279	0.78	2812.31	15.77	2796.54	227157.74
280	0.78	2812.31	15.58	2796.73	224361.01
281	0.78	2812.31	15.39	2796.92	221564.09
282	0.78	2812.31	15.20	2797.11	218766.98
283	0.79	2812.31	15.01	2797.30	215969.67
284	0.79	2812.31	14.81	2797.50	213172.18
285	0.79	2812.31	14.62	2797.69	210374.49
286	0.79	2812.31	14.43	2797.88	207576.61
287	0.80	2812.31	14.24	2798.07	204778.53
288	0.80	2812.31	14.05	2798.26	201980.27
289	0.80	2812.31	13.85	2798.46	199181.81
290	0.81	2812.31	13.66	2798.65	196383.17
291	0.81	2812.31	13.47	2798.84	193584.33
292	0.81	2812.31	13.28	2799.03	190785.29
293	0.81	2812.31	13.09	2799.22	187986.07
294	0.82	2812.31	12.89	2799.42	185186.65
295	0.82	2812.31	12.70	2799.61	182387.05
296	0.82	2812.31	12.51	2799.80	179587.25
297	0.82	2812.31	12.32	2799.99	176787.25
298	0.83	2812.31	12.13	2800.18	173987.07
299	0.83	2812.31	11.93	2800.38	171186.69
300	0.83	2812.31	11.74	2800.57	168386.12
301	0.84	2812.31	11.55	2800.76	165585.36
302	0.84	2812.31	11.36	2800.95	162784.41
303	0.84	2812.31	11.17	2801.14	159983.27
304	0.84	2812.31	10.97	2801.34	157181.93
305	0.85	2812.31	10.78	2801.53	154380.40
306	0.85	2812.31	10.59	2801.72	151578.68
307	0.85	2812.31	10.40	2801.91	148776.77
308	0.86	2812.31	10.21	2802.11	145974.66

309	0.86	2812.31	10.01	2802.30	143172.36
310	0.86	2812.31	9.82	2802.49	140369.87
311	0.86	2812.31	9.63	2802.68	137567.19
312	0.87	2812.31	9.44	2802.87	134764.31
313	0.87	2812.31	9.24	2803.07	131961.25
314	0.87	2812.31	9.05	2803.26	129157.99
315	0.88	2812.31	8.86	2803.45	126354.54
316	0.88	2812.31	8.67	2803.64	123550.89
317	0.88	2812.31	8.47	2803.84	120747.06
318	0.88	2812.31	8.28	2804.03	117943.03
319	0.89	2812.31	8.09	2804.22	115138.81
320	0.89	2812.31	7.90	2804.41	112334.40
321	0.89	2812.31	7.71	2804.61	109529.79
322	0.89	2812.31	7.51	2804.80	106724.99
323	0.90	2812.31	7.32	2804.99	103920.00
324	0.90	2812.31	7.13	2805.18	101114.82
325	0.90	2812.31	6.94	2805.37	98309.45
326	0.91	2812.31	6.74	2805.57	95503.88
327	0.91	2812.31	6.55	2805.76	92698.12
328	0.91	2812.31	6.36	2805.95	89892.17
329	0.91	2812.31	6.17	2806.14	87086.02
330	0.92	2812.31	5.97	2806.34	84279.68
331	0.92	2812.31	5.78	2806.53	81473.16
332	0.92	2812.31	5.59	2806.72	78666.43
333	0.92	2812.31	5.40	2806.91	75859.52
334	0.93	2812.31	5.20	2807.11	73052.41
335	0.93	2812.31	5.01	2807.30	70245.11
336	0.93	2812.31	4.82	2807.49	67437.62
337	0.94	2812.31	4.63	2807.68	64629.93
338	0.94	2812.31	4.43	2807.88	61822.06
339	0.94	2812.31	4.24	2808.07	59013.99
340	0.94	2812.31	4.05	2808.26	56205.72
341	0.95	2812.31	3.86	2808.46	53397.27
342	0.95	2812.31	3.66	2808.65	50588.62
343	0.95	2812.31	3.47	2808.84	47779.78
344	0.96	2812.31	3.28	2809.03	44970.75
345	0.96	2812.31	3.08	2809.23	42161.52
346	0.96	2812.31	2.89	2809.42	39352.10
347	0.96	2812.31	2.70	2809.61	36542.49
348	0.97	2812.31	2.51	2809.80	33732.69
349	0.97	2812.31	2.31	2810.00	30922.69
350	0.97	2812.31	2.12	2810.19	28112.50

```
351  0.98  2812.31              1.93        2810.38     25302.12
352  0.98  2812.31              1.74        2810.58     22491.54
353  0.98  2812.31              1.54        2810.77     19680.77
354  0.98  2812.31              1.35        2810.96     16869.81
355  0.99  2812.31              1.16        2811.15     14058.66
356  0.99  2812.31              0.96        2811.35     11247.31
357  0.99  2812.31              0.77        2811.54      8435.77
358  0.99  2812.31              0.58        2811.73      5624.04
359  1.00  2812.31              0.39        2811.92      2812.12
360  1.00  2812.31              0.19        2812.12         0.00

$Other
                     Details
Loan             1.000000e+06
Total Paid       1.012432e+06
Total Interest   1.243184e+04
Eff Rate         2.500000e-02
i^(360)          2.469346e-02

>   # Outputting: Figure 4.18b
```

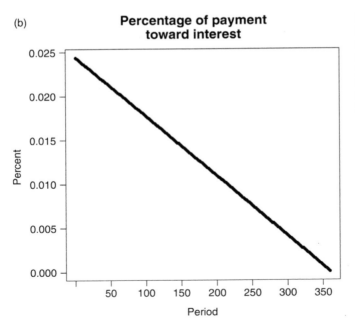

(b) **Percentage of payment toward interest**

Figure 4.18(b)
$1Million loan repaying by 360 equal monthly payments for loan at fixed annual mortgage rate @ 2.5%.

(B) How to Reduce Paying Total Interest and Save Money
Overall!

Part (A) shows that by making 360 equal monthly payments of
about $2,812, the total interest of the loan (after 3 years)
amounted to $1.243184e+04 or $12,431.84. By paying
$5,000.00, namely, ($5000.00 − $2,812.00) = $2,188 more per
monthly payment, the following computation in R shows the
concomitant benefit!

```
> amort.table(Loan=1000000, pmt=5000,
+                  i=0.025, pf=12, plot=TRUE)
> # Outputting:
$Schedule
```

Year	Payment	Interest	Paid	Principal	Paid	Balance
1	0.08	5000.00	2059.84		2940.16	997059.84
2	0.17	5000.00	2053.78		2946.22	994113.62
3	0.25	5000.00	2047.71		2952.29	991161.33
4	0.33	5000.00	2041.63		2958.37	988202.96
5	0.42	5000.00	2035.54		2964.46	985238.49
6	0.50	5000.00	2029.43		2970.57	982267.92
7	0.58	5000.00	2023.31		2976.69	979291.23
8	0.67	5000.00	2017.18		2982.82	976308.41
9	0.75	5000.00	2011.04		2988.96	973319.45
10	0.83	5000.00	2004.88		2995.12	970324.33
11	0.92	5000.00	1998.71		3001.29	967323.04
12	1.00	5000.00	1992.53		3007.47	964315.57
13	1.08	5000.00	1986.33		3013.67	961301.90
14	1.17	5000.00	1980.12		3019.88	958282.02
15	1.25	5000.00	1973.90		3026.10	955255.93
16	1.33	5000.00	1967.67		3032.33	952223.60
17	1.42	5000.00	1961.42		3038.58	949185.02
18	1.50	5000.00	1955.17		3044.83	946140.19
19	1.58	5000.00	1948.89		3051.11	943089.08
20	1.67	5000.00	1942.61		3057.39	940031.69
21	1.75	5000.00	1936.31		3063.69	936968.00
22	1.83	5000.00	1930.00		3070.00	933898.00
23	1.92	5000.00	1923.68		3076.32	930821.68
24	2.00	5000.00	1917.34		3082.66	927739.02
25	2.08	5000.00	1910.99		3089.01	924650.01
26	2.17	5000.00	1904.63		3095.37	921554.64
27	2.25	5000.00	1898.25		3101.75	918452.89
28	2.33	5000.00	1891.86		3108.14	915344.75
29	2.42	5000.00	1885.46		3114.54	912230.21
30	2.50	5000.00	1879.04		3120.96	909109.26

31	2.58	5000.00	1872.62	3127.38 905981.87
32	2.67	5000.00	1866.17	3133.83 902848.05
33	2.75	5000.00	1859.72	3140.28 899707.77
34	2.83	5000.00	1853.25	3146.75 896561.02
35	2.92	5000.00	1846.77	3153.23 893407.79
36	3.00	5000.00	1840.27	3159.73 890248.06
37	3.08	5000.00	1833.77	3166.23 887081.83
38	3.17	5000.00	1827.24	3172.76 883909.07
39	3.25	5000.00	1820.71	3179.29 880729.78
40	3.33	5000.00	1814.16	3185.84 877543.94
41	3.42	5000.00	1807.60	3192.40 874351.53
42	3.50	5000.00	1801.02	3198.98 871152.55
43	3.58	5000.00	1794.43	3205.57 867946.99
44	3.67	5000.00	1787.83	3212.17 864734.81
45	3.75	5000.00	1781.21	3218.79 861516.03
46	3.83	5000.00	1774.58	3225.42 858290.61
47	3.92	5000.00	1767.94	3232.06 855058.55
48	4.00	5000.00	1761.28	3238.72 851819.83
49	4.08	5000.00	1754.61	3245.39 848574.44
50	4.17	5000.00	1747.92	3252.08 845322.36
51	4.25	5000.00	1741.23	3258.77 842063.59
52	4.33	5000.00	1734.51	3265.49 838798.10
53	4.42	5000.00	1727.79	3272.21 835525.89
54	4.50	5000.00	1721.05	3278.95 832246.93
55	4.58	5000.00	1714.29	3285.71 828961.22
56	4.67	5000.00	1707.52	3292.48 825668.75
57	4.75	5000.00	1700.74	3299.26 822369.49
58	4.83	5000.00	1693.95	3306.05 819063.44
59	4.92	5000.00	1687.14	3312.86 815750.57
60	5.00	5000.00	1680.31	3319.69 812430.89
61	5.08	5000.00	1673.47	3326.53 809104.36
62	5.17	5000.00	1666.62	3333.38 805770.98
63	5.25	5000.00	1659.76	3340.24 802430.74
64	5.33	5000.00	1652.88	3347.12 799083.62
65	5.42	5000.00	1645.98	3354.02 795729.60
66	5.50	5000.00	1639.07	3360.93 792368.67
67	5.58	5000.00	1632.15	3367.85 789000.82
68	5.67	5000.00	1625.21	3374.79 785626.03
69	5.75	5000.00	1618.26	3381.74 782244.29
70	5.83	5000.00	1611.30	3388.70 778855.59
71	5.92	5000.00	1604.31	3395.69 775459.90
72	6.00	5000.00	1597.32	3402.68 772057.22
73	6.08	5000.00	1590.31	3409.69 768647.54
74	6.17	5000.00	1583.29	3416.71 765230.82

75	6.25	5000.00	1576.25	3423.75	761807.07
76	6.33	5000.00	1569.20	3430.80	758376.27
77	6.42	5000.00	1562.13	3437.87	754938.40
78	6.50	5000.00	1555.05	3444.95	751493.45
79	6.58	5000.00	1547.95	3452.05	748041.41
80	6.67	5000.00	1540.84	3459.16	744582.25
81	6.75	5000.00	1533.72	3466.28	741115.97
82	6.83	5000.00	1526.58	3473.42	737642.54
83	6.92	5000.00	1519.42	3480.58	734161.97
84	7.00	5000.00	1512.25	3487.75	730674.22
85	7.08	5000.00	1505.07	3494.93	727179.29
86	7.17	5000.00	1497.87	3502.13	723677.16
87	7.25	5000.00	1490.66	3509.34	720167.82
88	7.33	5000.00	1483.43	3516.57	716651.24
89	7.42	5000.00	1476.18	3523.82	713127.43
90	7.50	5000.00	1468.93	3531.07	709596.35
91	7.58	5000.00	1461.65	3538.35	706058.01
92	7.67	5000.00	1454.36	3545.64	702512.37
93	7.75	5000.00	1447.06	3552.94	698959.43
94	7.83	5000.00	1439.74	3560.26	695399.17
95	7.92	5000.00	1432.41	3567.59	691831.58
96	8.00	5000.00	1425.06	3574.94	688256.64
97	8.08	5000.00	1417.70	3582.30	684674.34
98	8.17	5000.00	1410.32	3589.68	681084.65
99	8.25	5000.00	1402.92	3597.08	677487.58
100	8.33	5000.00	1395.51	3604.49	673883.09
101	8.42	5000.00	1388.09	3611.91	670271.18
102	8.50	5000.00	1380.65	3619.35	666651.83
103	8.58	5000.00	1373.19	3626.81	663025.02
104	8.67	5000.00	1365.72	3634.28	659390.74
105	8.75	5000.00	1358.24	3641.76	655748.98
106	8.83	5000.00	1350.74	3649.26	652099.72
107	8.92	5000.00	1343.22	3656.78	648442.94
108	9.00	5000.00	1335.69	3664.31	644778.62
109	9.08	5000.00	1328.14	3671.86	641106.76
110	9.17	5000.00	1320.57	3679.43	637427.34
111	9.25	5000.00	1313.00	3687.00	633740.33
112	9.33	5000.00	1305.40	3694.60	630045.73
113	9.42	5000.00	1297.79	3702.21	626343.52
114	9.50	5000.00	1290.17	3709.83	622633.69
115	9.58	5000.00	1282.52	3717.48	618916.21
116	9.67	5000.00	1274.87	3725.13	615191.08
117	9.75	5000.00	1267.19	3732.81	611458.27
118	9.83	5000.00	1259.50	3740.50	607717.78

119	9.92	5000.00	1251.80	3748.20	603969.57
120	10.00	5000.00	1244.08	3755.92	600213.65
121	10.08	5000.00	1236.34	3763.66	596449.99
122	10.17	5000.00	1228.59	3771.41	592678.58
123	10.25	5000.00	1220.82	3779.18	588899.40
124	10.33	5000.00	1213.04	3786.96	585112.44
125	10.42	5000.00	1205.24	3794.76	581317.68
126	10.50	5000.00	1197.42	3802.58	577515.10
127	10.58	5000.00	1189.59	3810.41	573704.68
128	10.67	5000.00	1181.74	3818.26	569886.42
129	10.75	5000.00	1173.87	3826.13	566060.29
130	10.83	5000.00	1165.99	3834.01	562226.28
131	10.92	5000.00	1158.09	3841.91	558384.38
132	11.00	5000.00	1150.18	3849.82	554534.56
133	11.08	5000.00	1142.25	3857.75	550676.81
134	11.17	5000.00	1134.30	3865.70	546811.11
135	11.25	5000.00	1126.34	3873.66	542937.46
136	11.33	5000.00	1118.36	3881.64	539055.82
137	11.42	5000.00	1110.37	3889.63	535166.18
138	11.50	5000.00	1102.35	3897.65	531268.54
139	11.58	5000.00	1094.33	3905.67	527362.86
140	11.67	5000.00	1086.28	3913.72	523449.15
141	11.75	5000.00	1078.22	3921.78	519527.37
142	11.83	5000.00	1070.14	3929.86	515597.51
143	11.92	5000.00	1062.05	3937.95	511659.55
144	12.00	5000.00	1053.93	3946.07	507713.49
145	12.08	5000.00	1045.81	3954.19	503759.29
146	12.17	5000.00	1037.66	3962.34	499796.96
147	12.25	5000.00	1029.50	3970.50	495826.46
148	12.33	5000.00	1021.32	3978.68	491847.78
149	12.42	5000.00	1013.13	3986.87	487860.90
150	12.50	5000.00	1004.91	3995.09	483865.82
151	12.58	5000.00	996.68	4003.32	479862.50
152	12.67	5000.00	988.44	4011.56	475850.94
153	12.75	5000.00	980.18	4019.82	471831.11
154	12.83	5000.00	971.89	4028.11	467803.01
155	12.92	5000.00	963.60	4036.40	463766.61
156	13.00	5000.00	955.28	4044.72	459721.89
157	13.08	5000.00	946.95	4053.05	455668.84
158	13.17	5000.00	938.60	4061.40	451607.45
159	13.25	5000.00	930.24	4069.76	447537.68
160	13.33	5000.00	921.85	4078.15	443459.54
161	13.42	5000.00	913.45	4086.55	439372.99
162	13.50	5000.00	905.04	4094.96	435278.03

163	13.58 5000.00	896.60	4103.40 431174.63
164	13.67 5000.00	888.15	4111.85 427062.78
165	13.75 5000.00	879.68	4120.32 422942.46
166	13.83 5000.00	871.19	4128.81 418813.65
167	13.92 5000.00	862.69	4137.31 414676.34
168	14.00 5000.00	854.17	4145.83 410530.50
169	14.08 5000.00	845.63	4154.37 406376.13
170	14.17 5000.00	837.07	4162.93 402213.20
171	14.25 5000.00	828.49	4171.51 398041.69
172	14.33 5000.00	819.90	4180.10 393861.59
173	14.42 5000.00	811.29	4188.71 389672.88
174	14.50 5000.00	802.66	4197.34 385475.54
175	14.58 5000.00	794.02	4205.98 381269.56
176	14.67 5000.00	785.35	4214.65 377054.91
177	14.75 5000.00	776.67	4223.33 372831.58
178	14.83 5000.00	767.97	4232.03 368599.56
179	14.92 5000.00	759.25	4240.75 364358.81
180	15.00 5000.00	750.52	4249.48 360109.33
181	15.08 5000.00	741.77	4258.23 355851.10
182	15.17 5000.00	732.99	4267.01 351584.09
183	15.25 5000.00	724.21	4275.79 347308.30
184	15.33 5000.00	715.40	4284.60 343023.70
185	15.42 5000.00	706.57	4293.43 338730.27
186	15.50 5000.00	697.73	4302.27 334428.00
187	15.58 5000.00	688.87	4311.13 330116.86
188	15.67 5000.00	679.99	4320.01 325796.85
189	15.75 5000.00	671.09	4328.91 321467.94
190	15.83 5000.00	662.17	4337.83 317130.11
191	15.92 5000.00	653.24	4346.76 312783.35
192	16.00 5000.00	644.28	4355.72 308427.63
193	16.08 5000.00	635.31	4364.69 304062.94
194	16.17 5000.00	626.32	4373.68 299689.26
195	16.25 5000.00	617.31	4382.69 295306.57
196	16.33 5000.00	608.28	4391.72 290914.85
197	16.42 5000.00	599.24	4400.76 286514.09
198	16.50 5000.00	590.17	4409.83 282104.26
199	16.58 5000.00	581.09	4418.91 277685.35
200	16.67 5000.00	571.99	4428.01 273257.34
201	16.75 5000.00	562.87	4437.13 268820.20
202	16.83 5000.00	553.73	4446.27 264373.93
203	16.92 5000.00	544.57	4455.43 259918.50
204	17.00 5000.00	535.39	4464.61 255453.88
205	17.08 5000.00	526.19	4473.81 250980.08
206	17.17 5000.00	516.98	4483.02 246497.06

207	17.25	5000.00	507.74	4492.26	242004.80
208	17.33	5000.00	498.49	4501.51	237503.29
209	17.42	5000.00	489.22	4510.78	232992.51
210	17.50	5000.00	479.93	4520.07	228472.43
211	17.58	5000.00	470.62	4529.38	223943.05
212	17.67	5000.00	461.29	4538.71	219404.34
213	17.75	5000.00	451.94	4548.06	214856.27
214	17.83	5000.00	442.57	4557.43	210298.84
215	17.92	5000.00	433.18	4566.82	205732.02
216	18.00	5000.00	423.77	4576.23	201155.80
217	18.08	5000.00	414.35	4585.65	196570.15
218	18.17	5000.00	404.90	4595.10	191975.05
219	18.25	5000.00	395.44	4604.56	187370.48
220	18.33	5000.00	385.95	4614.05	182756.44
221	18.42	5000.00	376.45	4623.55	178132.89
222	18.50	5000.00	366.92	4633.08	173499.81
223	18.58	5000.00	357.38	4642.62	168857.19
224	18.67	5000.00	347.82	4652.18	164205.01
225	18.75	5000.00	338.24	4661.76	159543.24
226	18.83	5000.00	328.63	4671.37	154871.88
227	18.92	5000.00	319.01	4680.99	150190.89
228	19.00	5000.00	309.37	4690.63	145500.26
229	19.08	5000.00	299.71	4700.29	140799.96
230	19.17	5000.00	290.02	4709.98	136089.99
231	19.25	5000.00	280.32	4719.68	131370.31
232	19.33	5000.00	270.60	4729.40	126640.91
233	19.42	5000.00	260.86	4739.14	121901.77
234	19.50	5000.00	251.10	4748.90	117152.87
235	19.58	5000.00	241.32	4758.68	112394.19
236	19.67	5000.00	231.51	4768.49	107625.70
237	19.75	5000.00	221.69	4778.31	102847.39
238	19.83	5000.00	211.85	4788.15	98059.24
239	19.92	5000.00	201.99	4798.01	93261.23
240	20.00	5000.00	192.10	4807.90	88453.33
241	20.08	5000.00	182.20	4817.80	83635.53
242	20.17	5000.00	172.28	4827.72	78807.80
243	20.25	5000.00	162.33	4837.67	73970.13
244	20.33	5000.00	152.37	4847.63	69122.50
245	20.42	5000.00	142.38	4857.62	64264.88
246	20.50	5000.00	132.38	4867.62	59397.26
247	20.58	5000.00	122.35	4877.65	54519.61
248	20.67	5000.00	112.30	4887.70	49631.91
249	20.75	5000.00	102.23	4897.77	44734.14

250	20.83	5000.00	92.15	4907.85	39826.29
251	20.92	5000.00	82.04	4917.96	34908.32
252	21.00	5000.00	71.91	4928.09	29980.23
253	21.08	5000.00	61.75	4938.25	25041.98
254	21.17	5000.00	51.58	4948.42	20093.56
255	21.25	5000.00	41.39	4958.61	15134.95
256	21.33	5000.00	31.18	4968.82	10166.13
257	21.42	5000.00	20.94	4979.06	5187.07
258	21.50	5000.00	10.68	4989.32	197.75
258.04	21.50	197.77	0.02	197.75	0.00

```
$Other
                          Details
Loan                 1.000000e+06
Total Paid           1.290198e+06
Total Interest       2.901978e+05
Balloon PMT          5.197754e+03
Drop PMT             1.981611e+02
Last Regular PMT     1.977699e+02
Eff Rate             2.500000e-02
i^(12)               2.471804e-02
>
>   # Outputting: Figure 4.19
```

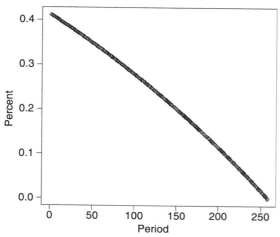

Percentage of payment toward interest

Figure 4.19 $1 Million loan repaying by equal monthly payments, of $5,000 each, for a loan at fixed annual mortgage rate @ 2.5%.

Remark: In this case, it took only 258.04 months to pay off the loan. Hence, the overall saving in interest payments, for the $1M loan, is

$$
\begin{aligned}
\$(1.290198e + 06 - 1.012432e + 06) &= \$(1,290,198 - 1,012,432) \\
&= \$277,766
\end{aligned}
$$

(Seems attractive?)

4.4.2 The Function list() and the Construction of data.frame() in R

The Function list()

A list in R consists of an ordered collection of objects – its components, which may be of any modes or types. For examples, a list may consist of a matrix, a numeric vector, a complex vector, a logical value, a character array, a function, and so on. Thus, some simple way to create a list would be

```
Example 1:   It is as easy as "1, 2, 3"!
> x <- 1
> y <- 2
> z <- 3
> list1 <- list(x, y, z) #  Forming a simple list
> list1 #  Outputting:
[[1]]
[1] 1
[[2]]
[1] 2
[[3]]
[1] 3
```

Moreover, the components are always numbered, and may be referred to as such. Thus, if my.special.list is the name of a list with four components, they may be referred to, individually, as

```
my.special.list[[1]], my.special.list[[2]],
my.special.list[[3]], and my.special.list[[4]].
If one defines my.special.list as:
> my.special.list <- list(name="John", wife="Mary",
+    number.of.children=3, children.age=c(2, 4, 6))
then
> my.special.list[[1]] #  Outputting:
[1] "John"
> my.special.list[[2]]
[1] "Mary"
> my.special.list[[3]]
```

```
[1]  3
> my.special.list[[4]]
[1]  2 4 6
```

The Number of Components in a List   The number of (top-level) components in a list may be found by the function length(). Thus

```
> length(my.special.list)
[1]  4
viz., the list my.special.list has 4 components.
```

To combine a set of objects into a larger composite collection for more efficient processing, the list function may be used to construct a list from its components.

As an example, consider

```
> odds <- c(1, 3, 5, 7, 9, 11,13,15,17,19)
> evens <- c(2, 4, 6, 8, 10, 12, 14, 16, 18, 20)
> mylist <- list(before=odds, after=evens)
> mylist
$before
 [1]  1  3  5  7  9 11 13 15 17 19
$after
 [1]  2  4  6  8 10 12 14 16 18 20
> mylist$before
 [1]  1  3  5  7  9 11 13 15 17 19
> mylist$after
 [1]  2  4  6  8 10 12 14 16 18 20
```

Components of a List   Components of a list may be named. In such a case, the component may be referred to either

1) by giving the component name as a character string in place of the number in double square brackets or
2) by giving an expression of the form

```
> name$component_name
        for the same object.
Example 2:  A family affair -
> my.special.list <- list(name="John", wife="Mary",
+    number.of.children=3, children.age=c(2, 4, 6))
> my.special.list # Outputting:
$name
```

```
[1] "John"
$wife
[1] "Mary"
$number.of.children
[1] 3
$children.age
[1] 2 4 6
Thus, for this list:
> my.special.list[[1]]
[1] "John"
> my.special.list$name
>  # This is the same as my.special.list[[1]]
[1] "John"
> my.special.list[[2]]
[1] "Mary"
> my.special.list$wife
> # This is the same as my.special.list[[2]]
[1] "Mary"
> my.special.list[[2]]
[1] "Mary"
> my.special.list[[3]]
[1] 3
> my.special.list$number.of.children
> # This is the same as my.special.list[[3]]
[1] 3
> my.special.list[[4]]
[1] 2 4 6
> my.special.list$children.age
> # This is the same as my.special.list[[4]]
[1] 2 4 6
```

Extraction of a Variable from a List

To extract the name of a component stored in another variable, one may use the names of the list components in double square brackets, viz., list1[["name"]]. The following R code segment may be used:

```
> x <- "name"; my.special.list[[John]]
[1] "John"
Constructing, Modifying, and Concatenating Lists:
New lists may be constructed from existing objects by the
function
list().
Thus,  the form
```

```
> new.list <- list(name_1=object_1,... name-
+                       n=object_n)
```
will set up s list, list1, of n components using object_1, . . . ,

object_n for the components and giving then names as specified.

4.4.3 Stock Market Risk Analysis: ES (Expected Shortfall) in the Black–Scholes Model

Package 'Dowd'

March 11, 2016

Type Package

Title Functions Ported from 'MMR2' Toolbox Offered in Kevin Dowd's

Book Measuring Market Risk

Version 0.12

Date 2015-08-20

Author Dinesh Acharya <dines.acharya@gmail.com>

Maintainer Dinesh Acharya <dines.acharya@gmail.com>

Description 'Kevin Dowd's' book Measuring Market Risk is a widely read book

in the area of risk measurement by students and

practitioners alike. As he claims, 'MATLAB' indeed might have been the most

suitable language when he originally wrote the functions, but, with growing popularity of R it is not entirely

valid. As 'Dowd's' code was not intended to be error free and were mainly for reference, some functions in this package have inherited those errors. An attempt will be made in future releases to identify and correct them. 'Dowd's' original code can be downloaded from www .kevindowd.org/measuring-marketrisk/.

It should be noted that 'Dowd' offers both 'MMR2'

and 'MMR1' toolboxes. Only 'MMR2' was ported to

R. 'MMR2' is more recent version of 'MMR1' toolbox and they both have mostly similar function. The toolbox mainly contains different parametric and non-parametric methods for measurement of market risk as well as backtesting risk measurement methods.

Depends R (>= 3.0.0), bootstrap, MASS, forecast

Suggests PerformanceAnalytics, testthat

License GPL

NeedsCompilation no
Repository CRAN
Date/Publication 2016-03-11 00:45:03
BlackScholesCallESSim ES of Black-
Scholes call using Monte Carlo Simulation
Description
Estimates ES of Black-
Scholes call Option using Monte Carlo simulation
Usage
BlackScholesCallESSim
(amountInvested, stockPrice, strike, r, mu, sigma,
maturity, numberTrials, cl, hp)
Arguments
amountInvested Total amount paid for the Call Option and is
  positive (negative) if the option
position is long (short)
stockPrice Stock price of underlying stock
strike Strike price of the option
r Risk-free rate
mu Expected rate of return on the underlying asset and is
  in annualised term
sigma Volatility of the underlying stock and is in
  annualised term
maturity The term to maturity of the option in days
numberTrials The number of interactions in the Monte Carlo
  simulation exercise
cl Confidence level for which ES is computed and is scalar
hp Holding period of the option in days and is scalar
Value
ES
Author(s)
Dinesh Acharya
References
Dowd, Kevin. Measuring Market Risk, Wiley, 2007.
Lyuu, Yuh-
Dauh. Financial Engineering & Computation: Principles,
  Mathematics, Algorithms,
Cambridge University Press, 2002.
Examples
Market Risk of American call with given parameters.
BlackScholesCallESSim
(0.20, 27.2, 25, .16, .2, .05, 60, 30, .95, 30)
> In the R domain

```
>
> install.packages("Dowd")
Installing package into 'C:/Users/Bert/Documents/R/win-
library/3.2'
(as 'lib' is unspecified)
--- Please select a CRAN mirror for use in this session ---
# A CRAN mirror is selected
The downloaded binary packages are in
      C:\Users\Bert\AppData\Local\Temp\RtmpuYe2Ox
\downloaded_packages
> library(4.4.3        Stock Market Risk Analysis:)
> # ES (Expected Shortfall) in the Black-Scholes Model
> library(Dowd)
> ls("package:Dowd")  # Outputting:
  [1] "AdjustedNormalESHotspots"
  [2] "AdjustedNormalVaRHotspots"
  [3] "AdjustedVarianceCovarianceES"
  [4] "AdjustedVarianceCovarianceVaR"
  [5] "ADTestStat"
  [6] "AmericanPutESBinomial"
  [7] "AmericanPutESSim"
  [8] "AmericanPutPriceBinomial"
  [9] "AmericanPutVaRBinomial"
 [10] "BinomialBacktest"
 [11] "BlackScholesCallESSim"
 [12] "BlackScholesCallPrice"
 [13] "BlackScholesPutESSim"
 [14] "BlackScholesPutPrice"
 [15] "BlancoIhleBacktest"
-----------------------------------------------
 [145] "tVaRPlot3D"
[146] "VarianceCovarianceES"
[147] "VarianceCovarianceVaR"
>
>
> # Market Risk of American call with given
+   # parameters.
> BlackScholesCallESSim(0.20, 27.2, 25, .16, .2, .05,
+                                   60, 30, .95, 30)
> # Outputting the Black-Scholes Call ES (Expected Shortfall):
[1] 0.001294227
```

```
> #   viz., according to the Black-Scholes Model, for this Call,
> #   the ES (Expected Shortfall) is predicted to be at the 0.1%
> #   level, or, very unlikely indeed!
```

Some Remarks on Expected Shortfall

- **Expected shortfall (ES)** is a risk measure – a concept used in the field of financial risk measurement to evaluate the market risk, or credit risk, of a portfolio.

  The "expected shortfall at q% level" is the expected return on the portfolio in the worst {\displaystyle q}qqq% of cases. ES is an alternative to Value at Risk that is more sensitive to the shape of the tail of the loss distribution.
- Expected shortfall is also called
  a) Conditional Value at Risk (CVaR),
  b) Average Value at Risk (AVaR), and
  c) Expected Tail Loss (ETL).
- ES estimates the risk of an investment in a conservative way, focusing on the less profitable outcomes. A value of {\displaystyle q}ES often used in practice is 5%.
- Expected shortfall is a coherent, as well as a spectral, measure of financial portfolio risk. It requires a quantile-level, {\displaystyle q}and is defined to be the expected loss of portfolio value.

Review Questions for Section 4.4

1 To use R in data analysis, the data to be processed must first be entered into the R environment; discuss 7 ways of data entry, giving examples.

2 How can the function list() be used to enter data into the R environment? Provide an example.

3 Use the function data.frame() to enter data into the R environment, giving an example.

4 Use the following functions to input data into the R environment, giving an example of each: c(), matrix(), and array().

5 Use the function source() to enter data into the R environment, giving an example.

6 When using the functions data.entry() and edit() for data entry, what are the limitations?

7   Show that the function list() may be used to combine several components to form a new list. Give an example.

8   Write a code segment in R to extract the name of a component stored in another variable, giving an example.

9   With an example, use the concatenation function c() with given list arguments, obtain a list whose components are those of the argument list joined together sequentially, in the form of

```
> list.ABC <- c(list.A, list.B, list.C)
```

10 Look up the software Epidata from its web site http://www.epidata.dk, and suggest an efficient method of data entry in R.

4.5   Univariate, Bivariate, and Multivariate Data Analysis

A univariate data set has only one variable:{x}, for example, {patient name}
A bivariate data set has two variables:{$x1, x2$}, or {x, y}, for example, {patient name, gender}
A multivariate data set has more than two, or many, variables: {$x1, x2, x3, \ldots, xn$},
for example, {investor name, gender, age, capital, preferred minimum return on investment, periodic payout, . . . }

4.5.1   Univariate Data Analysis

As an example, enter the following code segments:

```
> x <- rexp(100); x
> #  Outputting 100 exponentially-distributed
> # random numbers:
 [1]  0.39136880 0.66948212 1.48543076 0.34692128 0.71533079 0.12897216
 [7]  1.08455419 0.07858231 1.01995665 0.81232737 0.78253619 4.27512555
[13]  2.11839466 0.47024886 0.62351482 1.02834522 2.17253419 0.37622879
[19]  0.16456926 1.81590741 0.16007371 0.95078524 1.26048607 5.92621325
[25]  0.21727112 0.07086311 0.83858727 1.01375231 1.49042968 0.53331210
[31]  0.21069467 0.37559212 0.10733795 2.84094906 0.17899040 1.34612473
[37]  0.00290699 1.77078060 1.79505318 0.09763821 1.96568170 0.15911043
[43]  4.36726420 0.33652419 0.01196883 0.35657882 0.72797670 0.91958975
[49]  0.68777857 0.29100399 0.22553560 1.56909742 0.20617517 0.37169621
[55]  0.53173534 0.26034316 0.21965356 2.94355695 1.88392667 1.13933083
[61]  0.31663107 0.23899975 0.01544856 1.30674088 0.53674598 1.72018758
```

```
[67]  0.31035278  0.81074737  0.09104104  1.52426229  1.35520172  0.27969075
[73]  1.36320488  0.56317216  0.85022837  0.49031656  0.17158651  0.31015165
[79]  2.07315953  1.29566872  1.28955269  0.33487343  0.20902716  2.84732652
[85]  0.58873236  1.54868210  2.93994181  0.46520037  0.73687959  0.50062507
[91]  0.20275282  0.49697531  0.58578119  0.49747575  1.53430435  4.56340237
[97]  0.90547787  0.72972219  2.60686316  0.33908320
```

Note: The function rexp() is defined as follows:

rexp(n, rate = 1)

with arguments:

x   vector

n   number of observations. If length(n) > 1, the length is taken to be the number required.

The exponential distribution with rate λ has density

f(x) = λ e $^{-\lambda x}$,   for x ≥ 0.

If the rate λ is not specified, it assumes the default value of 1.

Remark: The function rexp() is one of the functions in R under exponential in the CRAN package stats.

To undertake a biostatistical analysis of this set of univariate data, one may call up the function univax(), in the package epibasix, using the following code segments:

```
> library(epibasix)
> univar(x)  #  Outputting:
Univariate Summary
Sample Size: 100
Sample Mean: 1.005
Sample Median: 0.646
Sample Standard Deviation: 1.067
>
```

Thus, for this sample size of 100 elements, the mean, median and standard deviation have been computed.

For data analysis of univariate data sets, the R package epibasix may be used.

This CRAN package covers many elementary financial functions for statistics and econometrics. It contains elementary tools for analysis of common financial problems, ranging from sample size estimation, through 2 x 2 contingency table analysis, and basic measures of agreement (kappa, sensitivity/specificity).

Appropriate print and summary statements are also written to facilitate interpretation wherever possible. This work is appropriate for graduate financial engineering courses.

This package is a work in progress.

To start, enter the R environment and use the code segment:

```
> install.packages("epibasix")
Installing package(s) into 'C:/Users/bertchan/Documents/R/
win-library/2.14
(as 'lib' is unspecified)
--- Please select a CRAN mirror for use in this session ---
> #  Select CA1
trying URL
'http://cran.cnr.Berkeley.edu/bin/windows/contrib/2.14/
epibasix_1.1.zip'
Content type 'application/zip' length 57888 bytes (56 Kb)
opened URL
downloaded 56 Kb
package 'epibasix' successfully unpacked and MD5
sums checked
The downloaded packages are in
C:\Users\bertchan\AppData\Local\Temp\RtmpMFOrEn
\downloaded_packages
With epibasix loaded into the R environment, to learn more
about this package,
follow these steps:
1.  Go to the CRAN website: http://cran.r-project.org/
2.  Select (single-click) Packages, on the left-hand column
3.  On the page: select E (for epibasix)
        Available CRAN Packages By Name
        A B C D E F G H I J K L M N O P Q R S T U V W X Y Z
4.  Scroll down list of packages whose name starts with "E"
or "e", and select:
epibasix
5. When the epibasix page opens up, select:   Reference manual:
epibasix.pdf
6.  The information is now on displayed, as follows:
Package       'epibasix'
              January 2, 2012
Version       1.1
Date          2009-05-13
Author        Michael A Rotondi <mrotondi@uwo.ca>
Maintainer    Michael A Rotondi mrotondi@uwo.ca
Depends R (>= 2.01)
```

For another example, consider the same analysis on the first
one hundred Natural
Numbers, using the following R code segments:

```
> x <-1:100; x #  Consider, and then output, the first 100
>                 #  Natural Numbers
[1]    1    2    3    4    5    6    7    8    9   10   11   12   13
14   15   16   17   18
[19]   19   20   21   22   23   24   25   26   27   28   29   30   31
32   33   34   35   36
[37]   37   38   39   40   41   42   43   44   45   46   47   48   49
50   51   52   53   54
[55]   55   56   57   58   59   60   61   62   63   64   65   66   67
68   69   70   71   72
[73]   73   74   75   76   77   78   79   80   81   82   83   84   85
86   87   88   89   90
[91]   91   92   93   94   95   96   97   98   99 100
> #  ANOVA Tables: Summarized in the following tables,
> # ANOVA is used for
> # two different purposes:
> library(epibasix)
> univar(x) #  Performing a univariate data analysis
>            #  on the vector x, and Outputting:
Univariate Summary
Sample Size: 100
Sample Mean: 50.5
Sample Median: 50.5
Sample Standard Deviation: 29.011
And that's it!
```

4.5.2 Bivariate and Multivariate Data Analysis

When there are two variables, (X, Y), one need to consider the following two
cases:

Case I: In the classical regression model, only Y called the dependent variable, is
required to be random. X is defined as a fixed, nonrandom, variable and is the
independent variable. Under this model, observations are obtained by
preselecting values of X and determining the corresponding value of Y.

Case II: If both X and Y are random variables, it is called the correlation model –
under which sample observations are obtained by selecting a random sample
of the units of association (such as persons, characteristics (age, gender,
locations, points of time, specific events/actions/ . . . ,) or elements on which
the two measurements are based) and by recording a measurement of X and

of Y. In this case, values of X are not preselected but occurring at random, depending on the unit of association selected in the sample.

Regression Analysis:

Case I: Correlation analysis cannot be meaningfully under this model.
Case II: Regression analysis can be performed under the correlation model.

Correlation for two variables implies a corelationship between the variables, and does not distinguish between them as to which is the dependent or the independent variable. Thus, one may fit a straight line to the data either by minimizing $\sum(x_i - x)^2$ or by minimizing $\sum(y_i - y)^2$. The fitted regression line will, in general, be different in the two cases – and a logical question arises as to which line to fit.

Two situation do exit, and should be considered:

1) If the objective is to obtain a measure of strength of the relationship between the two variables, it does not matter which line is fitted – the measure calculated will be the same in either case.
2) If one needs to use the equation describing the relationship between the two variables for the dependency of one upon the other, it does matter which line is to be fitted. The variable for which one wishes to estimate means or to make predictions should be treated as the depending variable. That is, this variable should be regressed with respect to the other variable.

Available R Packages for Bivariate Data Analysis Among the R packages for bivariate data analysis, a notable one available for sample size calculations for bivariate random intercept regression model is the bivariate power.

An Example in Bivariate Data Analysis As an example, this package may be used to calculate necessary sample size to achieve 80% power at 5% alpha level for null and alternative hypotheses that correlate between RI 0 and 0.2, respectively, across six time points. Other covariance parameter are set as follows:

```
Correlation between residuals = 0;
Standard deviations: 1st RI = 1,   2nd RI = 2,  1st residual
= 0.5,   2nd residual = 0.75
The following R code segment may be used:
> library(bivarRIpower)
> bivarcalcn(power=.80,powerfor='RI',timepts=6,
+ d1=1,d2=2,  p=0,p1=.2,s1=.5,s2=.75,r=0,r1=.1)
#  Outputting:
Variance parameters
Clusters                            = 209.2
```

```
Repeated measurements          = 6
Standard deviations
  1st random intercept          = 1
  2nd random intercept          = 2
  1st residual term             = 0.5
  2nd residaul term             = 0.75
Correlations
  RI under H_o                  = 0
  RI under H_a                  = 0.2
  Residual under H_o            = 0
  Residual under H_a            = 0.1
  Con obs    under H_o          = 0
  Con obs    under H_a          = 0.1831984
  Lag obs    under H_o          = 0
  Lag obs    under H_a          = 0.1674957
Correlation variances under H_o
-------------------------------------------------------------
Random intercept               = 0.005096138
Residual                       = 0.0009558759
Concurrent observations        = 0.00358999
Lagged observations            = 0.003574277
Power (%) for correlations
-------------------------------------------------------------
Random intercept               = 80%
Residual                       = 89.9%
Concurrent observations        = 86.4%
Lagged observations            = 80%
>
```

Bivariate Normal Distribution   Under the correlation model, the bivariates X and Y vary together in a joint distribution, which, if this joint distribution is a normal distribution, is called a bivariate normal distribution, from which inferences may be made based on the results of sampling properly from the population. If the joint distribution is known to be non-normal, or if the form is unknown, inferential procedures are invalid. The following assumptions must hold for inferences about the population to be valid when sampling from a bivariate distribution:

i) For each value of X, there is a normally distributed subpopulation of Y values.
ii) For each value of Y, there is a normally distributed subpopulation of X values.
iii) The joint distribution of X and Y is a normal distribution called the bivariate normal distribution.

iv) The subpopulation of Y values have the same variance.
v) The sub-population of X values have the same variance.

Two random variables X and Y are said to be jointly normal if they can be expressed in the form

$$X = aU + bV \qquad (4.1)$$

$$Y = cU + dV \qquad (4.2)$$

where U and V are independent normal random variables.

If X and Y are jointly normal, then any linear combination

$$Z = s_1 X + s_2 Y \qquad (4.3)$$

has a normal distribution. The reason is that if one has $X = aU + bV$ and $Y = cU + dV$ for some independent normal random variables U and V, then

$$
\begin{aligned}
Z &= s_1(aU + bV) + s_2(cU + dV) \\
&= (as_1 + cs_2)U + (bs_1 + ds_2)V
\end{aligned} \qquad (4.4)
$$

Thus, Z is the sum of the independent normal random variables $(as_1 + cs_2)U$ and $(bs_1 + ds_2)V$, and is therefore normal.

A very important property of jointly normal random variables is that zero correlation implies independence.

Zero Correlation Implies Independence   If two random variables X and Y are jointly normal and are uncorrelated, then they are independent.

(This property can be verified using multivariate transforms)

Multivariate Data Analysis   The following are the two similar, but distinct, approaches used for multivariate data analysis:

1) *The Multiple Linear Regression Analysis*: Assuming that a linear relationship exists between some variable Y, call the dependent variable, and n independent variables, $X_1, X_2, X_3, \ldots, X_n$, which are called explanatory or predictor variables because of their use.

The following are the assumptions underlying multiple regression model analysis:

a) The X_i are nonrandom fixed variables indicating that any inferences drawn from sample data apply only to the set of X values observed, but not to larger collections of X. Under this regression model, correlation analysis is not meaningful.

b) For each set of X_i values, there is a subpopulation of Y values. Usually, one assumes that these Y values are normally distributed.

c) The variances of Y are all equal.

d) The Y values are independent of the different selected set of X values.

For multiple linear regression, the model equation is

$$y_j = \beta_0 + \beta_1 x_{1j} + \beta_2 x_{2j} + \beta_3 x_{3j} + \cdots + \beta_n x_{nj} + e_j \ldots \tag{4.5}$$

where y_j is a typical value from one of the subpopulations of Y values, and the β_i values are the regression coefficients.

$x_{1j}, x_{2j}, x_{3j}, \ldots, x_{nj}$ are, respectively, particular values of the independent variables $X_1, X_2, X_3, \ldots, X_n$, and e_j is a random variable with mean 0 and variance σ_2, the common variance of the subpopulation of Y values. Generally, e_j is assumed normal and independently distributed.

When Equation (4.1) consists of one dependent variable and two independent variables, the model becomes

$$y_j = \beta_0 + \beta_1 x_{1j} + \beta_2 x_{2j} + e_j \ldots \tag{4.6}$$

A plane in three-dimensional space may be fitted to the data points. For models containing more than two variables, it is a hyperplane.

Let $f_j = \beta_0 + \beta_1 x_{1j} + \beta_2 x_{2j}$
Now, $e_j = y_j - f_j$, formimgavectore.

If y is the mean of the observed data,

namely, $\underline{y} = (1/n)\Sigma y_i,$ for $i = 1, 2, 3, \ldots, n$

then the variability of the data set may be measured using three sums of squares (proportional to the variance of the data):

1) The Total Sum of Squares (proportional to the variance of the data):

$$SS_{total} = \Sigma(y_i - y_i)^2 \tag{4.7}$$

2) The Regression Sum of Squares:

$$SS_{reg} = \Sigma(f_i - y_i)^2 \tag{4.8}$$

3) The Sum of Squares of Residuals:

$$SS_{res} = \Sigma_1(y_i - f_i)^2 = \Sigma_i e_2^i \tag{4.9}$$

The most general definition of the coefficient of multiple determination is

$$R_{y,12\ldots n}^2 \equiv 1 - (SS_{res}/SS_{total}) \tag{4.10}$$

The parameter of interest in this model is the coefficient of multiple determination, $R_{y|1,2,3,\ldots n}^2$, obtained by dividing the explained sum of squares by the total sum of squares:

$$R_{y,12\ldots n}^2 = \Sigma(y_i - f_i)^2 / \Sigma(y_i - y)^2 = SSR/SSE \tag{4.11}$$

where:

$\sum (y_i - fi)^2$ = the explained variation

= the original observed values from the calculated Y values

= the sum of squared deviation of the calculated values from the mean of the observed Y values, or

= the sum of squares due to regression (SSR)

$\sum (y_i - y)^2$ = the unexplained variation

= the sum of squared deviations of the original observations from the calculated values

= the sum of squares about regression, or

= the error sum of squares (SSE)

The total variation is the sum of squared deviations of each observation of Y from the mean of the observations:

$$\Sigma(y_j - y)^2 = \Sigma(y_i - y)^2 + \Sigma(y_i - y)^2 \ldots \tag{4.12}$$

namely,

$$\text{SST} = \text{SSR} + \text{SSE} \tag{4.13}$$

or

total sum of squares = explained(regression)sum of squares
+ unexplained(error)sum squares $+ \cdots$ (4.14)

The Multiple Correlation Model Analysis – The object of this approach is to gain insight into the strength of the relationship between variables.

The multiple regression correlation model analysis equation is

$$y_j = \beta_0 + \beta_1 x_{1j} + \beta_2 x_{2j} + \beta_3 x_{3j} + \cdots + \beta_n x_{nj} + e_j + \cdots \tag{4.15}$$

where y_j is a typical value from one of the subpopulations of Y values, the β_i are the regression coefficients, $x_{1j}, x_{2j}, x_{3j}, \ldots, x_{nj}$ are, respectively, particular known values of the random variables $X_1, X_2, X_3, \ldots, X_n$, and e_j is a random variable with mean 0 and variance σ^2, the common variance of the subpopulation of Y values. Generally, e_j is assumed normal and independently distributed.

This model is similar to model Equation (4.5), with the following important distinction:

in Equation (4.5), the x_i are nonrandom variables, but
in Equation (4.9), the x_i are random variables.

That is, in the correlation model, Equation (4.9), there is a joint distribution of Y and the X_i that is called a multivariate distribution.

Under this model, the variables are no longer considered as being dependent or independent, because logically they are interchangeable, and either of the X_i may play the role of Y.

2) The Correlation Model Analysis

The Multiple Correlation Coefficient: To analyze the relationships among the variables, consider the multiple correlation coefficient, which is the square root of the coefficient of multiple determination, and hence the sample value may be computed by taking the square root of Equation (4.12), namely,

$$R_{y,12\ldots n}^2 = \sqrt{R_{y,12\ldots n}^2} = \sqrt{\{\Sigma(y_i - y)^2 / \Sigma(y_i - y)^2\}}$$
$$= \sqrt{(SSR/SSE)} \tag{4.16}$$

3) Analysis of Variance (ANOVA)

In statistics, ANalysis Of VAriance (ANOVA) is a collection of statistical models in which the observed variance in a particular variable is partitioned into components from different sources of variation. ANOVA provides a statistical test of whether or not the means of several groups are all equal, and therefore, generalizes t-test to more than two groups. Doing multiple two-sample t-tests would result in an increased chance of committing a Type I error. For this reason, ANOVAs are useful in comparing two, three, or multiple means.

ANOVA Tables: Summarized in the following tables, ANOVA is used for two different purposes:

1) To estimate and test hypotheses for simple linear regression about population variances
2) To estimate and test hypotheses about population means

ANOVA table for testing hypotheses about simple linear regression

Source	DF	Sum of squares	Mean squares
F-value	P-value	$MSG/MSE = F_{1,n-2}$	$Pr(F > F_{1,n-2})$
Model	1	$\Sigma(y_i - y)^2 = SSModel$	$SSM = MSM$
Residual	$n - 2$	$\Sigma e_i^2 = SSResidual$	$SSR/(n-2) = MSE$
Total	$n - 1$	$\Sigma(y_i - y)^2 = SSTotal$	$SST/(n-1) = MST$

Residuals are often called errors since they are the part of the variation that the line could *not* explain, so $MSR = MSE = $ sum of squared residuals/df $= \sigma = $ estimate for variance of the population regression line $SSTot/(n-1) = MSTOT = s_y^2 = $ the total variance of the y, $F = t^2$ for simple linear regression.

The larger the F (the smaller the p-value) the more the y's variation in the line explained and so the less likely that H_0 is true. One rejects a hypothesis when the p-value $< \alpha$.

R^2 = proportion of the total variation of y explained by the regression line

= SSM/SST

= $1 - $ SSResidual/SST

ANOVA table for testing hypotheses about population means

Source	DF	Sum of squares	Mean squares	F-value P-value	p-value
Group (between)	$k-1$	$\Sigma n_i(\bar{x}_i\bar{x})^2 = $ SSG	SSG/ $(k-1) = $ MSG	MSG/MSE $\Pr = F_{k-1,}$ $N-k = (F > Fk-1,N-k)$	$\Pr(F > F_{k-1,N-k})$
Error (within)	$N-k$	$\Sigma(n_i - 1)$ $s_i^2 = $ SSE	SSE/ $(N-k) = $ MSE		
Total	$N-1$	$\Sigma(x_{ij}-)^2 = $ SSTot	SSTot/ $(N-1) = $ MST		

$N = $ total number of observations $= \Sigma n_i$, where $n_i = $ number of observations for group i.

The F test statistic has two different degrees of freedom: the numerator $= k - 1$, and the denominator $= N - k \cdot Fk-1,N-k$

Note: SSE/$(N-k) = $ MSE $= sp^2 = $ (pooled sample variance)

$$\frac{(n_1 - 1)s_1^2 + \cdots + (n_k - 1)s_k^2}{(n_1 - 1) + \cdots + (n_k - 1)}$$

$$= \hat{\sigma}^2$$

= estimate for assumed equal variance

(This is the "average" variance for each group.) SSTot/$(N-1) = $ MSTOT $= $ s2 $= $ the total variance of the data (assuming NO groups)

$F \approx$ variance of the (between) sample means divided by the $\sim$average variance of the data, the larger the F (the smaller the p-value) the more varied the means are, so the less likely H_0 is true. It is rejected when the p-value $< \alpha$.

R2 $= $ proportion of the total variation explained by the difference in means $= \frac{SSG}{SSTot}$

Example 1
Package 'fAssets'

February 19, 2015
Title Rmetrics - Analysing and Modelling Financial Assets
Date 2014-10-30
Version 3011.83
Author Rmetrics Core Team,
Diethelm Wuertz [aut],
Tobias Setz [cre],
Yohan Chalabi [ctb]
Maintainer Tobias Setz <tobias.setz@rmetrics.org>
Description Environment for teaching
ˋFinancial Engineering and Computational Finance''.
Depends R (>= 2.15.1), timeDate, timeSeries, fBasics
Imports fMultivar, robustbase, MASS, sn,
ecodist, mvnormtest, energy
Suggests methods, mnormt, RUnit
Note SEVERAL PARTS ARE STILL PRELIMINARY AND
MAY BE CHANGED IN THE FUTURE. THIS TYPICALLY INCLUDES
FUNCTION AND ARGUMENT NAMES, AS WELL AS DEFAULTS FOR
ARGUMENTS AND RETURN VALUES.
LazyData yes
License GPL (>= 2)
URL https://www.rmetrics.org
NeedsCompilation no
Repository CRAN
Date/Publication 2014-10-30 13:38:28
R topics documented:

```
>
> install.packages("fAssets")
```

```
Installing package into 'C:/Users/Bert/Documents/R/win-
library/3.2'
> library(fAssets)
https://www.rmetrics.org --- Mail to: info@rmetrics.org
Warning messages:
1: package 'fAssets' was built under R version 3.2.5
2: package 'timeDate' was built under R version 3.2.5
3: package 'timeSeries' was built under R version 3.2.5
4: package 'fBasics' was built under R version 3.2.5
> ls("package:fAssets")
 [1] "abcArrange"                "assetsArrange"
 [3] "assetsBasicStatsPlot"      "assetsBoxPercentilePlot"
 [5] "assetsBoxPlot"             "assetsBoxStatsPlot"
 [7] "assetsCorEigenPlot"        "assetsCorgramPlot"
 [9] "assetsCorImagePlot"        "assetsCorTestPlot"
[11] "assetsCumulatedPlot"       "assetsDendrogramPlot"
[13] "assetsDist"                "assetsFit"
[15] "assetsHistPairsPlot"       "assetsHistPlot"
[17] "assetsLogDensityPlot"      "assetsLPM"
[19] "assetsMeanCov"             "assetsMomentsPlot"
[21] "assetsNIGFitPlot"          "assetsNIGShapeTrianglePlot"
[23] "assetsOutliers"            "assetsPairsPlot"
[25] "assetsQQNormPlot"          "assetsReturnPlot"
[27] "assetsRiskReturnPlot"      "assetsSelect"
[29] "assetsSeriesPlot"          "assetsSim"
[31] "assetsSLPM"                "assetsStarsPlot"
[33] "assetsTest"                "assetsTreePlot"
[35] "binaryDist"                "braycurtisDist"
[37] "canberraDist"              "corDist"
[39] "covEllipsesPlot"           "euclideanDist"
[41] "getCenterRob"              "getCovRob"
[43] "hclustArrange"             "jaccardDist"
[45] "kendallDist"               "mahalanobisDist"
[47] "manhattanDist"             "maximumDist"
[49] "minkowskiDist"             "mutinfoDist"
[51] "mvenergyTest"              "mvshapiroTest"
[53] "orderArrange"              "pcaArrange"
[55] "sampleArrange"             "sorensenDist"
[57] "spearmanDist"              "statsArrange"
```

```
assets-selection Selecting Assets from Multivariate
Asset Sets
```

Description
Select assets from Multivariate Asset Sets based on clustering.

Usage

```
assetsSelect(x, method = c
("hclust", "kmeans"), control = NULL, ...)
```

Arguments

x   any rectangular time series object which can be
            converted by the function as.matrix() into a matrix
            object, e.g. like an object of class timeSeries, data.
            frame, or mts.

method   a character string, which clustering method should be
            used? Either hclust for hierarchical clustering of
            dissimilarities, or kmeans for k-means clustering.
            control a character string with two entries controlling
            the parameters used in the underlying cluster
            algorithms. If set to NULL, then default settings are
            taken: For hierarchical clustering this is method=c
            (measure="euclidean", method="complete"), and for
            kmeans clustering this is method=c(centers=3,
            algorithm="Hartigan-Wong").

. . .   optional arguments to be passed. Note, for the k-means
            algorithm the number of centers has to be specified!

Details
The function assetsSelect calls the functions hclust or kmeans from R's "stats" package.

```
hclust performs a hierarchical cluster analysis on the set
of dissimilarities hclust(dist(t(x)))
```

and kmeans performs a k-means clustering on the data matrix itself. Note, the hierarchical clustering method has in addition a plot method.

Value

```
if use="hclust" was selected then the function returns
a S3 object of class "hclust", otherwise if
use="kmeans" was selected then the function returns an
object of class "kmeans".
```

For details we refer to the help pages of hclust and kmeans.
Author(s)
Diethelm Wuertz for the Rmetrics port.
References
Wuertz, D., Chalabi, Y., ChenW., Ellis A. (2009); Portfolio Optimization with
R/Rmetrics, Rmetrics
eBook, Rmetrics Association and Finance Online, Zurich.
16 assets-testing

Examples

```
## LPP -
# Load Swiss Pension Fund Data:
LPP <- LPP2005REC
colnames(LPP)
## assetsSelect -
# Hierarchical Clustering:
hclust <- assetsSelect(LPP, "hclust")
plot(hclust)
## assetsSelect -
# kmeans Clustering:
assetsSelect(LPP, "kmeans", control =
c(centers = 3, algorithm = "Hartigan-Wong"))
assets-
> ## LPP -
> # Load Swiss Pension Fund Data:
> LPP <- LPP2005REC
> colnames(LPP)
[1] "SBI"    "SPI"    "SII"    "LMI"    "MPI"
     "ALT"    "LPP25"  "LPP40"  "LPP60"
>
> ## assetsSelect -
> # Hierarchical Clustering:
> hclust <- assetsSelect(LPP, "hclust")
> plot(hclust)
```

Cluster dendrogram

Figure 4.20 Output for plot(hclust)

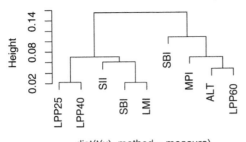

dist(t(x), method = measure)
hclust (*, "complete")

```
> # Outputting: Figure 4.20
>
> ## assetsSelect -
> # kmeans Clustering:
> assetsSelect(LPP, "kmeans", control =
+ c(centers = 3, algorithm = "Hartigan-Wong"))
K-means clustering with 3 clusters of sizes 3, 1, 5
```

Cluster Means

	2005-11-01	2005-11-02	2005-11-03	2005-11-04	2005-11-07
1	-7.174567e-05	-0.0004228193	0.0070806633	0.0086669900	0.003481978
2	8.414595e-03	0.0025193420	0.0127072920	-0.0007027570	0.006205226
3	-9.685162e-04	-0.0021474822	-0.0004991982	-0.0005400452	0.000752058

	2005-11-08	2005-11-09	2005-11-10	2005-11-11	2005-11-14
1	0.0006761683	0.001840756	0.001091617	0.008291539	-0.0006469167
2	0.0003292600	-0.002378200	0.000922087	0.013334906	-0.0046930640
3	0.0014039486	-0.001694911	0.001119758	0.001929053	-0.0003708446

	2005-11-15	2005-11-16	2005-11-17	2005-11-18	2005-11-21
1	-0.0005408223	0.001686769	0.0048643273	0.005051224	0.001901881
2	0.0012668650	-0.007187498	0.0076581030	0.012527202	0.002659666
3	-0.0008278810	0.001879324	0.0003510514	-0.001005688	0.001511155

	2005-11-22	2005-11-23	2005-11-24	2005-11-25	2005-11-28
1	0.0037938340	0.0024438233	-0.000394091	0.0022790387	-0.006316902
2	0.0021424940	0.0035671280	-0.002559544	0.0033748180	-0.009816739
3	0.0003204556	0.0008044326	0.001046522	0.0007771762	-0.001717004

	2005-11-29	2005-11-30	2005-12-01	2005-12-02	2005-12-05
1	0.0031941750	-0.0037985373	0.011412029	0.0030362450	-0.003450339
2	0.0018864440	-0.0040370550	0.015977559	0.0070552900	-0.000268694
3	0.0001304264	-0.0001264086	0.002280999	0.0004551984	-0.001906796

	2005-12-06	2005-12-07	2005-12-08	2005-12-09	2005-12-12

1	-0.001325485	0.0007240550	-0.0066049723	0.0031775860	-0.001708432
2	0.002864672	-0.0026064460	-0.0034829450	0.0007472420	-0.000278182
3	0.002119992	-0.0007569246	0.0008520178	0.0006191402	0.000879002
	2005-12-13	2005-12-14	2005-12-15	2005-12-16	2005-12-19
1	0.0046313440	-0.004764346	0.0029376953	0.001213580	0.0004901883
2	0.0013586830	-0.007283576	-0.0073441080	0.004292528	0.0049806720
3	0.0008959768	-0.001095387	0.0007868516	0.001616794	0.0023614496
	2005-12-20	2005-12-21	2005-12-22	2005-12-23	2005-12-26
1	0.0071593677	0.007886516	0.0001159380	0.001659755	0.001861593
2	-0.0016334200	0.004319502	-0.0036299400	-0.002234004	0.000000000
3	0.0009404428	0.002617884	-0.0002855498	0.000507245	0.000246350
	2005-12-27	2005-12-28	2005-12-29	2005-12-30	2006-01-02
1	-0.001653545	-0.0005308263	0.004244265	-0.0020539910	0.0007404967
2	0.006571546	0.0016811710	0.007701353	-0.0039764730	0.0000000000
3	0.001191907	0.0015647450	0.001259883	-0.0007382012	0.0000960024
	2006-01-03	2006-01-04	2006-01-05	2006-01-06	2006-01-09
1	-0.0013538470	0.001985783	-0.001291249	0.0015069157	0.006899025
2	0.0065956320	0.012163609	-0.002285923	0.0022278130	-0.000174350
3	-0.0001111704	0.001657156	-0.000790521	0.0005815596	0.000574890
	2006-01-10	2006-01-11	2006-01-12	2006-01-13	2006-01-16
1	-0.001533005	0.0040728800	0.004428087	-0.001791852	-0.0005044543
2	-0.006484459	0.0109940950	0.005887780	0.000629145	0.0041673760
3	-0.000713456	-0.0008200066	0.001761519	-0.002024939	-0.0000711238
	2006-01-17	2006-01-18	2006-01-19	2006-01-20	2006-01-23
1	-0.0064264427	-0.009959086	0.008732922	-0.005498331	-0.012300008
2	-0.0060610800	-0.007104074	0.004901245	-0.008712876	0.000080500
3	-0.0004887026	-0.000363750	0.001841324	-0.002770640	-0.000594795

	2006-01-24	2006-01-25	2006-01-26	2006-01-27	2006-01-30	2006-01-31
1	0.0043293403	0.003928724	0.007596804	0.013018524	0.004510001	-0.000363935
2	-0.0009696750	-0.001245166	0.008348556	0.004847171	0.001426750	0.002736095
3	-0.0007514866	-0.001468543	0.001332284	0.002198388	0.000585831	0.000676559

	2006-02-01	2006-02-02	2006-02-03	2006-02-06	2006-02-07
1	0.0025679990	-0.0033918090	0.001392298	0.003829826	-0.0016356553
2	0.0033390250	-0.0052315370	0.006229563	0.000010100	-0.0003661590
3	-0.0001937426	-0.0009799974	0.001561223	0.000789857	-0.0005261388
	2006-02-08	2006-02-09	2006-02-10	2006-02-13	2006-02-14
1	-0.0034102587	0.004724740	-0.0002670790	-0.0043036747	0.0071958650
2	-0.0007797950	0.005323768	-0.0010290070	0.0030521410	-0.0005124970
3	0.0002705448	0.001218430	-0.0001116748	-0.0002136692	0.0009441608
	2006-02-15	2006-02-16	2006-02-17	2006-02-20	2006-02-21
1	-0.0012863397	0.0045143087	-0.0008902317	-0.002237808	0.0050509303
2	-0.0099343840	0.0129171990	0.0012844190	0.004250758	0.0063638350
3	0.0002048986	0.0009410282	0.0012602358	0.001315374	0.0009478938
	2006-02-22	2006-02-23	2006-02-24	2006-02-27	2006-02-28
1	0.007771075	0.0008127150	0.0050935397	0.0050390380	-0.0108695373
2	0.004433658	-0.0060669620	-0.0027457400	0.0053722110	-0.0119569840
3	0.002287663	0.0003148286	-0.0003851528	0.0009570004	-0.0006375348
	2006-03-01	2006-03-02	2006-03-03	2006-03-06	2006-03-07
1	0.0038673107	-0.003960535	-0.004273008	0.0005488833	-0.001580974
2	0.0105047210	-0.004177747	0.001755361	0.0004602100	-0.007028455

3	0.0001534274	-0.003340033	-0.001016466	0.0001020288	-0.001403624
	2006-03-08	2006-03-09	2006-03-10	2006-03-13	2006-03-14
1	-0.004631919	0.0054281930	0.0077513163	0.0026288817	-0.0011077267
2	-0.004230567	0.0035344700	0.0128746620	0.0082543960	-0.0005305960
3	-0.000989756	0.0001691144	-0.0002940006	0.0006133858	0.0006779516
	2006-03-15	2006-03-16	2006-03-17	2006-03-20	2006-03-21
1	0.0039076300	-0.002874349	0.0035407443	-0.0002971253	0.0017513113
2	-0.0017679700	0.000666818	0.0043129660	0.0021624820	-0.0002753190
3	0.0001087012	0.000484471	-0.0001434976	0.0006474828	-0.0008360246
	2006-03-22	2006-03-23	2006-03-24	2006-03-27	2006-03-28
1	0.0034417023	0.0042167183	-0.0005082437	-0.002477860	-0.005123098
2	0.0013273100	-0.0043185900	0.0027725900	-0.006945127	-0.000727865
3	0.0006235516	0.0002566278	0.0001903118	-0.000559792	-0.001748221
	2006-03-29	2006-03-30	2006-03-31	2006-04-03	2006-04-04
1	0.0083719110	-0.0023536467	0.0032189740	0.002929918	-0.0053531483
2	0.0009252440	0.0062064890	-0.0009292510	0.006944399	-0.0026184930
3	-0.0004704764	0.0003112066	-0.0003227242	-0.001220841	-0.0005425296
	2006-04-05	2006-04-06	2006-04-07	2006-04-10	2006-04-11
1	0.001672089	0.0035391920	0.0006688370	-0.0018577213	-0.0078612010
2	0.004752739	0.0015471340	0.0005119710	0.0000160000	-0.0095747430
3	0.000371034	0.0001308954	-0.0001606024	0.0004964256	0.0008645472
	2006-04-12	2006-04-13	2006-04-14	2006-04-17	2006-04-18
1	-0.001794527	-6.576767e-05	0.0003451073	-0.01162293	0.0072958027
2	-0.002978612	3.927437e-03	0.0000000000	0.00000000	-0.0050312550
3	-0.001221078	-2.154967e-03	-0.0000179728	-0.00127895	0.0008801984
	2006-04-19	2006-04-20	2006-04-21	2006-04-24	2006-04-25
1	0.005262631	0.004346695	-0.0001275733	-0.0055535810	-0.0008195537
2	0.010893639	0.004201724	0.0048380810	-0.0016947400	-0.0041756790
3	0.001021676	0.001559365	-0.0000182248	-0.0001482958	-0.0017556356
	2006-04-26	2006-04-27	2006-04-28	2006-05-01	2006-05-02
1	0.001310372	-0.003174862	-0.0081970307	-0.0042204567	0.005113142
2	0.004699408	0.000883969	-0.0029437410	0.0000000000	0.003189327
3	-0.000656350	0.000053330	0.0002421188	-0.0007466388	0.001444051
	2006-05-03	2006-05-04	2006-05-05	2006-05-08	2006-05-09
1	-0.001263695	-0.0006502133	0.006613750	0.0063954017	-0.003357079
2	-0.011943802	0.0030146350	0.011235363	0.0061756300	0.002517020
3	-0.001037496	-0.0000032026	0.002150292	0.0007181488	-0.001401839
	2006-05-10	2006-05-11	2006-05-12	2006-05-15	2006-05-16
1	-0.0025826107	-0.008643966	-0.016678837	-0.006493649	-0.002878691
2	-0.0013872090	-0.001096609	-0.017989732	-0.013399941	0.003285150
3	-0.0005785598	-0.003397807	-0.004047129	-0.000274325	0.001044719
	2006-05-17	2006-05-18	2006-05-19	2006-05-22	2006-05-23
1	-0.010999777	-0.0085399823	0.005933202	-0.021077161	0.007398834
2	-0.028406916	-0.0097114170	0.001848007	-0.025997761	0.018970677
3	-0.003777801	0.0001129756	0.002052023	-0.001940486	0.002266990
	2006-05-24	2006-05-25	2006-05-26	2006-05-29	2006-05-30
1	-0.0004267907	0.0061746340	0.016697015	-0.0011805273	-0.017512962
2	-0.0111155890	0.0000000000	0.025842125	0.0017720070	-0.019842413
3	-0.0011767074	0.0006506848	0.002053482	0.0005608306	-0.003672446
	2006-05-31	2006-06-01	2006-06-02	2006-06-05	2006-06-06

```
1   0.004247769    0.0048529913    0.002548541    -0.006215087   -0.007154035
2   0.009323326    0.0015364830    0.006993059     0.000000000   -0.022326545
3   0.001607127   -0.0008596914    0.002280503    -0.001082275   -0.001973220
    2006-06-07     2006-06-08      2006-06-09      2006-06-12     2006-06-13
1  -0.0015405703  -0.008894756     0.004967781    -0.004911457   -0.017072989
2   0.0056383270  -0.027379310     0.012429170    -0.013895865   -0.023992295
3  -0.0003462346  -0.001317298     0.002478040    -0.001395419   -0.003113317
    2006-06-14     2006-06-15      2006-06-16      2006-06-19     2006-06-20
1  -0.0008148080   0.0180758310    0.0023954587    0.001269003   -0.002988990
2   0.0022722680   0.0215693860   -0.0029262460    0.008192221    0.003859327
3  -0.0006585664   0.0004955194   -0.0000012182   -0.001387469   -0.001853536
    2006-06-21     2006-06-22      2006-06-23      2006-06-26     2006-06-27
1   0.0019448093   0.005720239     0.0003236553    0.0022263750  -0.004784092
2   0.0033826850   0.007397751     0.0005459380   -0.0024934820  -0.006852493
3  -0.0002515638  -0.000396063    -0.0009850436   -0.0002276698  -0.001709434
    2006-06-28     2006-06-29      2006-06-30      2006-07-03     2006-07-04    2006-07-05
1   0.002418113    0.013946565     0.0007082933    0.0051894390   0.0036357180  -0.003018747
2   0.002764602    0.014041765     0.0144738140    0.0087848690   0.0019620600  -0.008807793
3   0.000155729    0.003295715     0.0007134052    0.0007825544  -0.0005997304  -0.002343919
    2006-07-06     2006-07-07      2006-07-10      2006-07-11     2006-07-12
1   0.0006826483  -0.0053623837    0.006527871    -0.0026637643  -0.000460215
2   0.0055633270  -0.0057979700    0.005982938    -0.0038346560   0.004470950
3   0.0008409888   0.0001182382    0.001124756     0.0001879154  -0.000366662
    2006-07-13     2006-07-14      2006-07-17      2006-07-18     2006-07-19
1  -0.013003389   -0.0051940770   -0.0004672143    0.0002666863   0.010286439
2  -0.015294320   -0.0103727210   -0.0022662700   -0.0056064500   0.018419580
3  -0.002107831    0.0007863122    0.0008104944   -0.0013472198   0.001975698
    2006-07-20     2006-07-21      2006-07-24      2006-07-25     2006-07-26    2006-07-27
1   0.001826095   -0.0080939067    0.013847674     0.0046651393   0.0005912200  -0.000995908
2   0.008023110   -0.0053387720    0.019239808     0.0006067870   0.0043343300   0.005560211
3   0.001140619   -0.0006294194    0.003391497     0.0009658202   0.0007748634   0.001358731
    2006-07-28     2006-07-31      2006-08-01      2006-08-02     2006-08-03
1   0.0054530087   0.0009621367   -0.0016914860    0.001702776    3.388133e-05
2   0.0102982490   0.0002672360    0.0000000000   -0.003903872   -1.101953e-02
3   0.0008377254   0.0012288024   -0.0002328868   -0.000336470   -1.946893e-03
    2006-08-04     2006-08-07      2006-08-08      2006-08-09     2006-08-10
1   0.0004074303  -0.005170347     0.001632508     0.0004397503   0.005330442
2   0.0095373900  -0.012098329     0.000553560     0.0107776200  -0.005122863
3   0.0025471316  -0.001348114     0.001391077     0.0003048744   0.000744234
    2006-08-11     2006-08-14      2006-08-15      2006-08-16     2006-08-17
1   0.0003234277   0.0042420490    0.006281944     0.002505853    0.0022312940
2   0.0016952790   0.0080151810    0.014854679     0.004466055    0.0029039360
3  -0.0003461608  -0.0005048628    0.002816991     0.001751071    0.0003821246
    2006-08-18     2006-08-21      2006-08-22      2006-08-23     2006-08-24
1   0.001322035   -0.0061511060    0.007819312    -0.0024164317  -0.0006088927
2  -0.003261951   -0.0030139970    0.003136533    -0.0002148440   0.0020208950
3   0.001275974   -0.0002004374    0.002229195    -0.0002547412  -0.0001443246
    2006-08-25     2006-08-28      2006-08-29      2006-08-30     2006-08-31
1   0.001879964    0.0001150750    0.0039964717   -0.0013533323   0.003702581
```

2	0.000665318	0.0018015970	0.0060754350	0.0027585440	-0.001466397
3	0.000606633	0.0008119902	-0.0000166692	0.0008165558	0.002074951

	2006-09-01	2006-09-04	2006-09-05	2006-09-06	2006-09-07	
1	0.003825121	0.0022519797	0.002990688	-0.005576266	-0.0046239360	
2	0.003126659	0.0045617030	-0.000424630	-0.005904139	-0.0061051260	
3	0.001543151	-0.0000862626	0.000150264	-0.002215612	-0.0008681144	

	2006-09-08	2006-09-11	2006-09-12	2006-09-13	2006-09-14	2006-09-15
1	0.005100734	-0.006268633	0.0079822300	0.004665103	-0.0020439300	0.007261748
2	0.005038040	-0.008630266	0.0115443790	0.003047893	-0.0032925020	0.005426908
3	0.002150122	-0.002463850	0.0007386444	0.001504165	-0.0003697208	0.001212241

	2006-09-18	2006-09-19	2006-09-20	2006-09-21	2006-09-22	
1	-0.002313630	-0.0022077103	0.002879550	0.0017740240	-0.0120944313	
2	0.003786049	-0.0025686540	0.012243951	0.0039715930	-0.0091896050	
3	-0.001438723	0.0004212428	0.001333139	0.0009339414	0.0000184756	

	2006-09-25	2006-09-26	2006-09-27	2006-09-28	2006-09-29	2006-10-02
1	0.0015608587	0.007422705	0.0054981810	0.003993309	0.002248728	-0.004799167
2	-0.0027169680	0.012816618	0.0036014040	0.000706224	0.001409431	-0.005008288
3	0.0006968214	0.002279966	-0.0000253728	0.000284538	0.000532484	-0.001994171

	2006-10-03	2006-10-04	2006-10-05	2006-10-06	2006-10-09	
1	0.0011308367	0.007532294	0.007702274	0.0013383343	-1.842333e-06	
2	0.0006241600	0.006657881	0.007166048	0.0010145910	3.046625e-03	
3	-0.0000141456	0.001957197	0.001609076	-0.0009962644	2.434124e-04	

	2006-10-10	2006-10-11	2006-10-12	2006-10-13	2006-10-16	
1	0.0058490050	-0.001032232	0.005763076	0.0057859347	0.0016738733	
2	0.0089863210	0.001066884	0.004239901	-0.0020605700	-0.0018373360	
3	-0.0004574944	0.000480683	0.001357373	0.0004827066	-0.0004780952	

	2006-10-17	2006-10-18	2006-10-19	2006-10-20	2006-10-23	
1	-0.0072083463	0.0065625757	-0.004172731	0.0013755277	0.0067030360	
2	-0.0105018630	0.0091540220	0.001017744	0.0026080710	0.0055130300	
3	-0.0002134596	0.0004904236	-0.001158779	-0.0000334576	-0.0001258452	

	2006-10-24	2006-10-25	2006-10-26	2006-10-27	2006-10-30	
1	0.0006909480	0.0007045297	0.0004360173	-0.006801505	-0.0013724143	
2	-0.0035125790	0.0028470680	-0.0006172110	0.002649247	-0.0047902870	
3	0.0001992902	0.0006721280	0.0005680150	0.000502184	0.0003469408	

	2006-10-31	2006-11-01	2006-11-02	2006-11-03	2006-11-06	
1	-0.001518782	1.051133e-05	-0.0013112907	0.0043942913	0.006986784	
2	-0.008238610	4.875862e-03	0.0034278750	0.0059986060	0.010778959	
3	0.001557833	1.065105e-03	-0.0000054624	-0.0000061736	0.002111045	

	2006-11-07	2006-11-08	2006-11-09	2006-11-10	2006-11-13	
1	-0.000492798	0.0010355810	-0.0042464430	-0.0022569473	0.0017332780	
2	0.003869904	-0.0059257000	-0.0005821010	-0.0026741380	0.0015220670	
3	0.002714624	0.0004917422	0.0007154908	-0.0007612248	-0.0003075366	

	2006-11-14	2006-11-15	2006-11-16	2006-11-17	2006-11-20	
1	0.005042735	0.0039623527	0.000571091	-0.0041334430	-0.0008046817	
2	-0.001419646	0.0067483290	0.000088600	-0.0046228550	0.0012852990	
3	0.001633393	-0.0003903638	-0.000064044	-0.0004950332	0.0008012360	

	2006-11-21	2006-11-22	2006-11-23	2006-11-24	2006-11-27	
1	0.0035024650	-0.001423817	-0.0022464847	-0.0074566157	-0.008884057	
2	0.0015966630	0.000878824	-0.0028429630	-0.0120525280	-0.014190477	
3	0.0000896774	-0.000362714	-0.0009733886	-0.0007358458	-0.001985401	

	2006-11-28	2006-11-29	2006-11-30	2006-12-01	2006-12-04
1	-0.0005720070	0.010011509	-0.0028207700	-0.0045097767	0.008035140
2	-0.0064998550	0.012901202	-0.0086468080	-0.0064254010	0.006552949
3	0.0004750374	0.001182122	0.0003041304	-0.0001392252	0.003841311

	2006-12-05	2006-12-06	2006-12-07	2006-12-08	2006-12-11
1	0.0006129080	0.0012128337	0.0014398583	-0.001522731	0.006351745
2	-0.0028113180	0.0071741430	0.0074404230	-0.002915097	0.006216417
3	-0.0002819442	-0.0002501772	-0.0002762668	0.000266215	0.000853780

	2006-12-12	2006-12-13	2006-12-14	2006-12-15	2006-12-18
1	0.0003822723	0.0032613870	0.0085999303	0.004852111	0.0007975687
2	0.0077836640	0.0016147920	0.0100852550	0.001294973	0.0045667980
3	0.0014112930	-0.0003111648	0.0006167068	0.002972998	-0.0018724352

	2006-12-19	2006-12-20	2006-12-21	2006-12-22	2006-12-25
1	-0.007259400	0.0039956217	-0.001713398	-0.001404090	-0.0004215033
2	-0.006190852	0.0015531190	0.000940550	-0.005018284	0.0000000000
3	-0.001259345	0.0002665752	0.002535426	-0.002011179	-0.0000406544

	2006-12-26	2006-12-27	2006-12-28	2006-12-29	2007-01-01
1	0.0026915907	0.009066652	0.0004164007	-0.001200784	-9.820667e-06
2	0.0000000000	0.010333820	-0.0020348850	-0.001101976	0.000000e+00
3	0.0004873476	0.002159424	-0.0010962306	-0.000159042	-1.140000e-05

	2007-01-02	2007-01-03	2007-01-04	2007-01-05	2007-01-08
1	0.0016820867	0.004437189	-0.0002060123	-0.002683635	-0.001690284
2	0.0000000000	0.014834908	0.0001279650	-0.002580941	-0.004541686
3	0.0004246192	0.001879946	0.0006782908	-0.001927514	-0.001011925

	2007-01-09	2007-01-10	2007-01-11	2007-01-12	2007-01-15
1	0.0044103520	-0.0006727953	0.007703170	0.003633612	0.005514158
2	0.0032808260	-0.0014696800	0.013481449	0.005427095	0.008135203
3	0.0001659808	-0.0012173928	0.001547265	-0.000644134	0.001385737

	2007-01-16	2007-01-17	2007-01-18	2007-01-19	2007-01-22
1	-0.0002334037	-0.0009348027	0.0018915933	0.005485118	-0.001422125
2	-0.0023385150	0.0036009040	-0.0000306000	0.004899725	-0.004463529
3	0.0019940186	-0.0003052906	-0.0002513524	0.001870075	0.000533480

	2007-01-23	2007-01-24	2007-01-26	2007-01-26	2007-01-29
1	-0.0024795987	0.010078597	-0.005617319	0.0003207103	0.0026204570
2	-0.0005070610	0.006027774	-0.001732155	-0.0095941300	0.0074250550
3	-0.0000391516	0.001921089	-0.001420198	0.0000189644	0.0006710216

	2007-01-30	2007-01-31	2007-02-01	2007-02-02	2007-02-05	2007-02-06
1	0.002703326	-0.0017867277	0.005665622	0.005261889	1.452767e-05	0.0016964537
2	0.004207812	-0.0000649000	0.009002711	0.004474549	2.320000e-05	0.0009123640
3	0.002512929	0.0001274554	0.001459012	0.002004385	1.534503e-03	-0.0001289984

	2007-02-07	2007-02-08	2007-02-09	2007-02-12	2007-02-13
1	0.001088440	0.0005737837	-0.0000244410	-0.003474908	0.0039971257
2	0.002937055	-0.0060861880	0.0065013820	-0.004102506	-0.0023852220
3	-0.001036504	0.0008835816	-0.0007572888	-0.001322240	0.0002060698

	2007-02-14	2007-02-15	2007-02-16	2007-02-19	2007-02-20
1	0.003429143	0.0003378953	-0.0010496707	0.001237031	0.002481922
2	0.006839015	0.0014878740	0.0030582220	0.000894606	-0.003270610
3	0.001787848	0.0004864332	-0.0005123012	-0.002126213	0.001312554

	2007-02-21	2007-02-22	2007-02-23	2007-02-26	2007-02-27

1	-0.0003544737	0.004133925	-0.003687659	-0.0006432653	-0.028181932	
2	-0.0099202370	0.003907766	-0.000693145	-0.0035030230	-0.035746244	
3	0.0008831304	-0.000198173	0.000224204	0.0003787182	-0.003557775	
	2007-02-28	2007-03-01	2007-03-02	2007-03-05	2007-03-06	2007-03-07
1	-0.006545974	-0.004681529	-0.007204097	-0.016965306	0.0131994283	0.001848668
2	-0.011946524	-0.001959064	0.002364749	-0.014462125	0.0116318040	0.011790712
3	-0.001276448	-0.000431924	-0.001183209	-0.000934475	0.0006908982	0.000013807
	2007-03-08	2007-03-09	2007-03-12	2007-03-13	2007-03-14	2007-03-15
1	0.01238125	0.0054817963	-0.0019347513	-0.011489525	-0.016127064	0.010698674
2	0.01192565	0.0003297920	-0.0031733170	-0.007777588	-0.028205491	0.014756610
3	0.00211282	0.0003630802	0.0003176836	-0.001476343	-0.003026142	0.003330527
	2007-03-16	2007-03-19	2007-03-20	2007-03-21	2007-03-22	2007-03-23
1	-0.0040807663	0.011023104	0.006051645	0.008317470	0.006371441	0.003735611
2	0.0011294700	0.014488916	0.004157206	0.007430739	0.013969587	0.001698465
3	-0.0004977044	0.001539199	0.001811865	0.002705776	0.002193543	-0.000066116
	2007-03-26	2007-03-27	2007-03-28	2007-03-29	2007-03-30	
1	-0.0038874940	-0.0024984800	-0.007625166	0.008954610	0.0028986417	
2	-0.0075840220	-0.0047538740	-0.009870398	0.011153959	0.0004970940	
3	-0.0006071596	-0.0008749648	-0.001164874	0.001407261	-0.0002237774	
	2007-04-02	2007-04-03	2007-04-04	2007-04-05	2007-04-06	
1	-0.0013403807	0.008656804	0.0031424653	-0.0014326240	-0.0004997497	
2	-0.0014515930	0.010335207	0.0013071590	0.0050004920	0.0000000000	
3	0.0000685862	0.002129822	0.0004029036	-0.0001666756	-0.0000500682	
	2007-04-09	2007-04-10	2007-04-11			
1	0.006564997	0.0004269093	-0.001272110			
2	0.000000000	0.0063294250	-0.001044170			
3	0.000570819	-0.0000510372	-0.000411593			

Clustering Vector

SBI	SPI	SII	LMI	MPI	ALT	LPP25	LPP40	LPP60
3	2	3	3	1	1	3	3	1

Within cluster sum of squares by cluster:

```
[1] 0.003806242 0.000000000 0.005432037
 (between_SS / total_SS =  75.3 %)
```

Available components:

```
[1] "cluster"    "centers"    "totss"     "withinss"   "tot.withinss"
[6] "betweenss"  "size"       "iter"      "ifault"
>
++++++++++++++++++++++++++++++++++++++++++++
```

Example 1 in Multivariate Data Analysis – (see the illustrative example in Section 4.2.1)

ANOVA Analysis

ANOVA (Analysis of Variance) may be achieved using the R function anova(), which is in the CRAN package stats. The standard usage form of this function is

```
> anova(object, . . . )
with arguments:
object≡ an object for model fitting (e.g., lm or glm).
 . . . ≡ additional objects of the same type.
```

Example 2 in Multivariate Data Analysis

Consider some real econometric data as illustrated in the CRAN package for financial engineering analysis:

Package:	'quantmod':
Type:	Package
Title:	Quantitative Financial Modelling Framework
Version:	0.4-7
Date:	2016-10-24
Depends:	xts(>= 0.9-0), zoo, TTR(>= 0.2), methods
Suggests:	DBI,RMySQL,RSQLite,timeSeries,its,XML, downloader,jsonlite(>=1.1)
Description:	Specify, build, trade, and analyze quantitative financial trading strategies.
LazyLoad:	yes
License:	GPL-3
URL:	http://www.quantmod.com
	https://github.com/joshuaulrich/quantmod
BugReports:	https://github.com/joshuaulrich/quantmod/issues
NeedsCompilation:	yes
Authors:	Jeffrey A. Ryan [aut, cph],
	Joshua M. Ulrich [cre, ctb],
	Wouter Thielen [ctb]
Maintainer:	Joshua M. Ulrich <josh.m.ulrich@gmail.com>
Repository:	CRAN
Date/Publication:	2016-10-24 23:30:16 adjustOHLC

Adjust Open, High, Low, Close Prices For Splits and Dividends

Description
Adjust all columns of an OHLC object for split and dividend.

Usage

```
adjustOHLC(x,
                adjust = c("split","dividend"),
                        use.Adjusted = FALSE,
                        ratio = NULL,
        symbol.name=deparse(substitute(x)))
```

Arguments

x	An OHLC object
adjust	adjust by split, dividend, or both (default)
use.Adjusted	use the 'Adjusted' column in Yahoo! data to adjust
ratio	ratio to adjust with, bypassing internal calculations
symbol.name	used if x is not named the same as the symbol adjusting

Details
This function calculates the adjusted Open, High, Low, and Close prices according to split and dividend information.

There are three methods available to calculate the new OHLC object prices.

By default, getSplits and getDividends are called to retrieve the respective information. These may dispatch to custom methods following the "." methodology used by quantmod dispatch. See getSymbols for information related to extending quantmod. This information is passed to adjRatios from the TTR package, and the resulting ratio calculations are used to adjust to observed historical prices. *This is the most precise way to adjust a series.*

The second method works only on standard Yahoo! data containing an explicit Adjusted column.

A final method allows for one to pass a ratio into the function directly.

All methods proceed as follows:

1) New columns are derived by taking the ratio of adjusted value to original Close, and multiplying by the difference of the respective column and the original Close.
2) This is then added to the modified Close column to arrive at the remaining "adjusted" Open, High, Low column values. If no adjustment is needed, the function returns the original data unaltered.

Value

An object of the original class, with prices adjusted for splits and dividends.

Warning

Using use.Adjusted = TRUE will be less precise than the method that employs actual split and dividend information. This is due to loss of precision from Yahoo! using Adjusted columns of only two decimal places. The advantage is that this can be run offline, and for short series or those with few adjustments the loss of precision will be small.

The resulting precision loss will be from row observation to row observation, as the calculation will be exact for intraday values.

Author(s)

Jeffrey A. Ryan

References

Yahoo Finance http://finance.yahoo.com

See Also

getSymbols.yahoo getSplits getDividends

Examples

```
getSymbols("AAPL", from="1990-01-01",
                  src="yahoo")
head(AAPL)
head(AAPL.a <- adjustOHLC(AAPL))
head(AAPL.uA <- adjustOHLC(AAPL,
       use.Adjusted=TRUE))
# intraday adjustments are precise across all
# methods
# an example with Open to Close (OpCl)
head(cbind(OpCl(AAPL),OpCl(AAPL.a),
OpCl(AAPL.uA)))
# Close to Close changes may lose precision
head(cbind(ClCl(AAPL),ClCl(AAPL.a),
       ClCl(AAPL.uA)))
## End
```

In the R domain:

```
>
> install.packages("quantmod")
Installing package into 'C:/Users/Bert/Documents/R/win-library/3.2'
```

```
(as 'lib' is unspecified)
--- Please select a CRAN mirror for use in this session
A CRAN mirror is selected.
---
also installing the dependencies 'xts', 'TTR'
sums checked
The downloaded binary packages are in
        C:\Users\Bert\AppData\Local\Temp\Rtmp2jzxqk
\downloaded_packages
> library(quantmod)
> ls("package:quantmod")
  [1]  "Ad"                 "add_axis"          "add_BBands"
  [4]  "add_DEMA"           "add_EMA"           "add_EVWMA"
  [7]  "add_GMMA"           "add_MACD"          "add_RSI"
 [10]  "add_Series"         "add_SMA"           "add_SMI"
 [13]  "add_TA"             "add_VMA"           "add_Vo"
 [16]  "add_VWAP"           "add_WMA"           "addADX"
 [19]  "addAroon"           "addAroonOsc"       "addATR"
 [22]  "addBBands"          "addCCI"            "addChAD"
 [25]  "addChVol"           "addCLV"            "addCMF"
 [28]  "addCMO"             "addDEMA"           "addDPO"
 [31]  "addEMA"             "addEMV"            "addEnvelope"
 [34]  "addEVWMA"           "addExpiry"         "addKST"
 [37]  "addLines"           "addMACD"           "addMFI"
 [40]  "addMomentum"        "addOBV"            "addPoints"
 [43]  "addROC"             "addRSI"            "addSAR"
 [46]  "addShading"         "addSMA"            "addSMI"
 [49]  "addTA"              "addTDI"            "addTRIX"
 [52]  "addVo"              "addVolatility"     "addWMA"
 [55]  "addWPR"             "addZigZag"         "addZLEMA"
 [58]  "adjustOHLC"         "allReturns"        "annualReturn"
 [61]  "as.quantmod.OHLC"   "attachSymbols"     "axTicksByTime2"
 [64]  "axTicksByValue"     "barChart"          "buildData"
 [67]  "buildModel"         "candleChart"       "chart_pars"
 [70]  "chart_Series"       "chart_theme"       "chartSeries"
 [73]  "chartShading"       "chartTA"           "chartTheme"
 [76]  "Cl"                 "ClCl"              "current.chob"
 [79]  "dailyReturn"        "Delt"              "dropTA"
 [82]  "findPeaks"          "findValleys"       "fittedModel"
 [85]  "fittedModel<-"      "flushSymbols"      "futures.expiry"
 [88]  "getDefaults"        "getDividends"      "getFin"
 [91]  "getFinancials"      "getFX"             "getMetals"
 [94]  "getModelData"       "getOptionChain"    "getPrice"
 [97]  "getQuote"           "getSplits"         "getSymbolLookup"
[100]  "getSymbols"         "getSymbols.csv"    "getSymbols.FRED"
```

```
[103]  "getSymbols.google"     "getSymbols.mysql"      "getSymbols.MySQL"
[106]  "getSymbols.oanda"      "getSymbols.rda"        "getSymbols.RData"
[109]  "getSymbols.SQLite"     "getSymbols.yahoo"      "getSymbols.yahooj"
[112]  "has.Ad"                "has.Ask"               "has.Bid"
[115]  "has.Cl"                "has.Hi"                "has.HLC"
[118]  "has.Lo"                "has.OHLC"              "has.OHLCV"
[121]  "has.Op"                "has.Price"             "has.Qty"
[124]  "has.Trade"             "has.Vo"                "Hi"
[127]  "HiCl"                  "HLC"                   "importDefaults"
[130]  "is.BBO"                "is.HLC"                "is.OHLC"
[133]  "is.OHLCV"              "is.quantmod"           "is.quantmodResults"
[136]  "is.TBBO"               "Lag"                   "lineChart"
[139]  "listTA"                "Lo"                    "loadSymbolLookup"
[142]  "loadSymbols"           "LoCl"                  "LoHi"
[145]  "matchChart"            "modelData"             "modelSignal"
[148]  "monthlyReturn"         "moveTA"                "new.replot"
[151]  "newTA"                 "Next"                  "oanda.currencies"
[154]  "OHLC"                  "OHLCV"                 "Op"
[157]  "OpCl"                  "OpHi"                  "OpLo"
[160]  "OpOp"                  "options.expiry"        "peak"
[163]  "periodReturn"          "quantmodenv"           "quarterlyReturn"
[166]  "reChart"               "removeSymbols"         "saveChart"
[169]  "saveSymbolLookup"      "saveSymbols"           "seriesAccel"
[172]  "seriesDecel"           "seriesDecr"            "seriesHi"
[175]  "seriesIncr"            "seriesLo"              "setDefaults"
[178]  "setSymbolLookup"       "setTA"                 "show"
[181]  "showSymbols"           "specifyModel"          "standardQuote"
[184]  "summary"               "swapTA"                "tradeModel"
[187]  "unsetDefaults"         "unsetTA"               "valley"
[190]  "viewFin"               "viewFinancials"        "Vo"
[193]  "weeklyReturn"          "yahooQF"               "yahooQuote.EOD"
[196]  "yearlyReturn"          "zoom_Chart"            "zoomChart"
[199]  "zooom"
> adjustOHLC
function (x, adjust = c("split", "dividend"), use.Adjusted = FALSE,
    ratio = NULL, symbol.name = deparse(substitute(x)))
{
    if (is.null(ratio)) {
        if (use.Adjusted) {
            if (!has.Ad(x))
                stop("no Adjusted column in 'x'")
            ratio <- Ad(x)/Cl(x)
        }
```

```
        else {
            div <- getDividends(symbol.name, from = "1900-01-01")
            splits <- getSplits(symbol.name, from = "1900-01-01")
            if (is.xts(splits) && is.xts(div) && nrow(splits) >
                0 && nrow(div) > 0)
                div <- div * 1/adjRatios(splits = merge(splits,
                    index(div)))[, 1]
            ratios <- adjRatios(splits, div, Cl(x))
            if (length(adjust) == 1 && adjust == "split") {
                ratio <- ratios[, 1]
            }
            else if (length(adjust) == 1 && adjust == "dividend") {
                ratio <- ratios[, 2]
            }
            else ratio <- ratios[, 1] * ratios[, 2]
        }
    }
    Adjusted <- Cl(x) * ratio
    structure(cbind((ratio * (Op(x) - Cl(x)) + Adjusted), (ratio *
        (Hi(x) - Cl(x)) + Adjusted), (ratio * (Lo(x) - Cl(x)) +
        Adjusted), Adjusted, if (has.Vo(x))
        Vo(x)
    else NULL, if (has.Ad(x))
        Ad(x)
    else NULL), .Dimnames = list(NULL, colnames(x)))
}
<environment: namespace:quantmod>
> ## Not run:
> getSymbols("AAPL", from="1990-01-01", src="yahoo")
    As of 0.4-0, 'getSymbols' uses env=parent.frame() and
 auto.assign=TRUE by default.
```

This behavior will be phased out in 0.5-0 when the call will default to use auto.assign=FALSE. getOption("getSymbols.env") and getOptions("getSymbols.auto.assign") are now checked for alternate defaults.

This message is shown once per session and may be disabled by setting options("getSymbols.warning4.0"=FALSE). See ?getSymbols for more details.

```
[1] "AAPL"
Warning message:
In download.file(paste(yahoo.URL, "s=", Symbols.name, "&a=", from.m,   :
  downloaded length 452489 != reported length 200
> head(AAPL)
```

	AAPL. Open	AAPL High	AAPL. Low	AAPL. Close	AAPL. Volume	AAPL. Adjusted
1990-01-02	35.25	37.50	35.00	37.250	45799600	1.132075
1990-01-03	38.00	38.00	37.50	37.500	51998800	1.139673
1990-01-04	38.25	38.75	37.25	37.625	55378400	1.143471
1990-01-05	37.75	38.25	37.00	37.75	30828000	1.147270
1990-01-08	37.50	38.00	37.00	38.000	25393200	1.154868
1990-01-09	38.00	38.00	37.00	37.625	21534800	1.143471

```
> head(AAPL.a <- adjustOHLC(AAPL))
```

	AAPL. Open	AAPL. High	AAPL. Low	AAPL. Close	AAPL. Volume	AAPL. Adjusted
1990-01-02	1.071292	1.139673	1.063694	1.132075	45799600	1.132075
1990-01-03	1.154868	1.154868	1.139673	1.139673	51998800	1.139673
1990-01-04	1.162466	1.177662	1.132075	1.143471	55378400	1.143471
1990-01-05	1.147270	1.162466	1.124477	1.147270	30828000	1.147270
1990-01-08	1.139673	1.154868	1.124477	1.154868	25393200	1.154868
1990-01-09	1.154868	1.154868	1.124477	1.143471	21534800	1.143471

```
> head(AAPL.uA <- adjustOHLC(AAPL,
+         use.Adjusted=TRUE))
```

	AAPL. Open	AAPL. High	AAPL. Low	AAPL. Close	AAPL. Volume	AAPL. Adjusted
1990-01-02	1.071292	1.139673	1.063695	1.132075	45799600	1.132075
1990-01-03	1.154869	1.154869	1.139673	1.139673	51998800	1.139673
1990-01-04	1.162466	1.177661	1.132074	1.143471	55378400	1.143471
1990-01-05	1.147270	1.162466	1.124477	1.147270	30828000	1.147270
1990-01-08	1.139672	1.154868	1.124477	1.154868	25393200	1.154868
1990-01-09	1.154868	1.154868	1.124476	1.143471	21534800	1.143471

```
>
> # intraday adjustments are precise across all methods
> # an example with Open to Close (OpCl)
> head(cbind(OpCl(AAPL),OpCl(AAPL.a),
+         OpCl(AAPL.uA)))
```

	OpCl.AAPL	OpCl.AAPL.a	OpCl.AAPL.uA
1990-01-02	0.056737647	0.056737647	0.056737647
1990-01-03	-0.013157869	-0.013157869	-0.013157869
1990-01-04	-0.016339895	-0.016339895	-0.016339895
1990-01-05	0.000000000	0.000000000	0.000000000
1990-01-08	0.013333307	0.013333307	0.013333307
1990-01-09	-0.009868395	-0.009868395	-0.009868395

```
> # Close to Close changes may lose precision
> head(cbind(ClCl(AAPL),ClCl(AAPL.a),
+     ClCl(AAPL.uA)))
```

	ClCl.AAPL	ClCl.AAPL.a	ClCl.AAPL.uA

1990-01-02	NA	NA	NA
1990-01-03	0.006711382	0.006711382	0.006711569
1990-01-04	0.003333333	0.003333333	0.003332535
1990-01-05	0.003322259	0.003322259	0.003322340
1990-01-08	0.006622490	0.006622490	0.006622678
1990-01-09	-0.009868395	-0.009868395	-0.009868660

```
>
```

chartSeries **Create Financial Charts**

Description

Charting tool to create standard financial charts gives a time series like object. Serves as the base function for future technical analysis additions. Possible chart styles include candles, matches (1 pixel candles), bars, and lines. Chart may have white or black background.

reChart allows for dynamic changes to the chart without having to respecify the full chart parameters.

Usage

```
chartSeries(x, type = c("auto",
    "candlesticks", "matchsticks",
    "bars","line"), subset = NULL,
    show.grid = TRUE, name = NULL,
    time.scale = NULL,
    log.scale = FALSE, TA = 'addVo()',
    TAsep=';', line.type = "l",
    bar.type = "ohlc",
    theme = chartTheme("black"),
    layout = NA,
    major.ticks='auto', minor.ticks=TRUE,
    yrange=NULL,
    plot=TRUE,
    up.col,dn.col,color.vol
    = TRUE, multi.col = FALSE, ...)
reChart(type = c("auto", "candlesticks",
        "matchsticks", "bars","line"),
        subset = NULL,
        show.grid = TRUE,
        chartSeries 23
        name = NULL,
        time.scale = NULL,
        line.type = "l",
        bar.type = "ohlc",
```

```
theme = chartTheme("black"),
major.ticks='auto', minor.ticks=TRUE,
yrange=NULL,
up.col,dn.col,color.vol =
TRUE, multi.col = FALSE,
...)
```

Arguments

x	an OHLC object – see details
type	style of chart to draw
subset xts	style date subsetting argument
show.grid	display price grid lines?
name	name of chart
time.scale	what is the timescale? automatically deduced (broken)
log.scale	should the y-axis be log-scaled?
TA	a vector of technical indicators and params, or character strings
TAsep	TA delimiter for TA strings
line.type	type of line in line chart
bar.type	type of barchart - ohlc or hlc
theme	a chart.theme object
layout	if NULL bypass internal layout
major.ticks	where should major ticks be drawn
minor.ticks	should minor ticks be drawn?
yrange	override y-scale
plot	should plot be drawn
up.col	up bar/candle color
dn.col	down bar/candle color
color.vol	color code volume?
multi.col	4 color candle pattern
. . .	additional parameters

Details

Currently, chart displays standard style OHLC charts familiar in financial applications or line charts not passing OHLC data. Works with objects having explicit time-series properties.

Line charts are created with close data, or from single column time series.

The subset argument can be used to specify a particular area of the series to view. The underlying series is left intact to allow for TA functions to use the full data set. Additionally, it is possible to use syntax borrowed from the first and last functions, for example, "last 4 months."

TA allows for the inclusion of a variety of chart overlays and technical indicators. A full list is available from addTA. The default TA argument is addVo() – which adds volume, if available, to the chart being drawn.

theme requires an object of class chart.theme, created by a call to chartTheme. This function can be used to modify the look of the resulting chart. See chart.theme for details. line.type and bar.type allow further fine tuning of chart styles to user tastes. multi.col implements a color coding scheme used in some charting applications, and follows the following rules:

- grey => Op[t] < Cl[t] and Op[t] < Cl[t-1]
- white => Op[t] < Cl[t] and Op[t] > Cl[t-1]
- red => Op[t] > Cl[t] and Op[t] < Cl[t-1]
- black => Op[t] > Cl[t] and Op[t] > Cl[t-1]

reChart takes any number of arguments from the original chart call and redraws the chart with the updated parameters. One item of note: if multiple color bars/candles are desired, it is necessary to respecify the theme argument. Additionally, it is not possible to change TA parameters at present. This must be done with addTA/dropTA/swapTA/moveTA commands.

Value
Returns a standard chart plus volume, if available, suitably scaled.

If plot=FALSE a chob object will be returned.

Note
Most details can be fine-tuned within the function, though the code does a reasonable job of scaling and labeling axes for the user. The current implementation maintains a record of actions carried out for any particular chart. This is used to recreate the original when adding new indicator. A list of applied TA actions is available with a call to listTA. This list can be assigned to a variable and used in new chart calls to recreate a set of technical indicators. It is also possible to force all future charts to use the same indicators by calling setTA.

Additional motivation to add outlined candles to allow for scaling and advanced color coding is owed to Josh Ulrich, as are the base functions (from TTR) for the yet to be released technical analysis charting code.

Many improvements in the current version were the result of conversations with Gabor Grothendieck. Many thanks to him.

Author(s)

Jeffrey A. Ryan

References

Josh Ulrich - TTR package and multi.col coding

See Also

getSymbols, addTA, setTA, chartTheme

Examples

```
## Not run:
getSymbols("YHOO")
chartSeries(YHOO)
chartSeries(YHOO, subset='last 4 months')
chartSeries(YHOO, subset='2007::2008-01')
chartSeries(YHOO,theme=chartTheme('white'))
chartSeries(YHOO,TA=NULL) #no volume
chartSeries(YHOO,TA=c(addVo(),addBBands()))
 #add volume and Bollinger Bands from TTR
addMACD() # add MACD indicator to current chart
setTA()
chartSeries(YHOO)
# draws chart again, this time will all indicators present
## End(Not run)
```

chartTheme **Create A Chart Theme**

Description

Charting tool to create standard financial charts gives a time series like object. Serves as the base function for future technical analysis additions. Possible chart styles include candles, matches (1 pixel candles), bars, and lines. Chart may have white or black background. reChart allows for dynamic changes to the chart without having to respecify the full chart parameters.

Usage

```
chartSeries(x,
type = c("auto", "candlesticks", "matchsticks", "bars",
"line"),
subset = NULL,
show.grid = TRUE,
name = NULL,
time.scale = NULL,
log.scale = FALSE,
```

```
TA = 'addVo()',
TAsep=';',
line.type = "l",
bar.type = "ohlc",
theme = chartTheme("black"),
layout = NA,
major.ticks='auto', minor.ticks=TRUE,
yrange=NULL,
plot=TRUE,
up.col,dn.col,color.vol = TRUE, multi.col = FALSE,
...)
reChart(type = c("auto", "candlesticks", "matchsticks",
"bars", "line"),
subset = NULL,
show.grid = TRUE,
chartSeries 23
name = NULL,
time.scale = NULL,
line.type = "l",
bar.type = "ohlc",
theme = chartTheme("black"),
major.ticks='auto', minor.ticks=TRUE,
yrange=NULL,
up.col,dn.col,color.vol = TRUE, multi.col = FALSE,
...)
```

Arguments

x	an OHLC object – see details
type	style of chart to draw
subset xts	style date subsetting argument
show.grid	display price grid lines?
name	name of chart
time.scale	what is the timescale? automatically deduced (broken)
log.scale	should the y-axis be log-scaled?
TA	a vector of technical indicators and params, or character strings
TAsep	TA delimiter for TA strings

`line.type`	type of line in line chart
`bar.type`	type of barchart – ohlc or hlc
`theme`	a chart.theme object
`layout`	if NULL bypass internal layout
`major.ticks`	where should major ticks be drawn
`minor.ticks`	should minor ticks be drawn?
`yrange`	override y-scale
`plot`	should plot be drawn
`up.col`	up bar/candle color
`dn.col`	down bar/candle color
`color.vol`	color code volume?
`multi.col`	4 color candle pattern
`. . .`	additional parameters

Details

Currently, charts displays standard style OHLC charts familiar in financial applications or line charts not passing OHLC data. Works with objects having explicit time-series properties.

Line charts are created with close data, or from single column time series.

The subset argument can be used to specify a particular area of the series to view. The underlying series is left intact to allow for TA functions to use the full data set. Additionally, it is possible to use syntax borrowed from the first and last functions, for example, "last 4 months."

TA allows for the inclusion of a variety of chart overlays and technical indicators. A full list is available from addTA. The default TA argument is addVo() – which adds volume, if available, to the chart being drawn. theme requires an object of class chart.theme, created by a call to chartTheme. This function can be used to modify the look of the resulting chart. See chart.theme for details. line.type and bar.type allow further fine tuning of chart styles to user tastes. multi.col implements a color coding scheme used in some charting applications, and follows the following rules:

- grey => Op[t] < Cl[t] and Op[t] < Cl[t-1]
- white => Op[t] < Cl[t] and Op[t] > Cl[t-1]
- red => Op[t] > Cl[t] and Op[t] < Cl[t-1]
- black => Op[t] > Cl[t] and Op[t] > Cl[t-1]

reChart takes any number of arguments from the original chart call and redraws the chart with the updated parameters. One item of note: If multiple color bars/candles are desired, it is necessary to respecify the theme argument. Additionally, it is not possible to change TA parameters at present.

This must be done with `addTA/dropTA/swapTA/moveTA` commands.

Value

Returns a standard chart plus volume, if available, suitably scaled.

If `plot=FALSE`, a chob object will be returned.

Note

Most details can be fine-tuned within the function, though the code does a reasonable job of scaling and labeling axes for the user.

The current implementation maintains a record of actions carried out for any particular chart. This is used to recreate the original when adding new indicator. A list of applied TA actions is available with a call to `listTA`.

This list can be assigned to a variable and used in new chart calls to recreate a set of technical indicators. It is also possible to force all future charts to use the same indicators by calling `setTA`.

Additional motivation to add outlined candles to allow for scaling and advanced color coding is owed to Josh Ulrich, as are the base functions (from TTR) for the yet to be released technical analysis charting code.

Many improvements in the current version were the result of conversations with Gabor Grothendieck. Many thanks to him.

Author(s)

Jeffrey A. Ryan

ReferencesJosh Ulrich - TTR package and multi.col coding

See Also

`getSymbols, addTA, setTA, chartTheme`

Examples

```
## Not run:
getSymbols("YHOO")
chartSeries(YHOO)
chartSeries(YHOO, subset='last 4 months')
chartSeries(YHOO, subset='2007::2008-01')
chartSeries(YHOO,theme=chartTheme('white'))
chartSeries(YHOO,TA=NULL) #no volume
chartSeries(YHOO,TA=c(addVo(),addBBands()))
 #add volume and Bollinger Bands from TTR
addMACD() # add MACD indicator to current chart
setTA()
chartSeries(YHOO)
 #draws chart again, this time will all indicators present
## End(Not run)
```

```
>
> getSymbols("YHOO")
[1] "YHOO"
Warning message:
In download.file(paste(yahoo.URL, "s=", Symbols.
name, "&a=", from.m,  :
  downloaded length 147128 != reported length 200
>
> getSymbols("YHOO")
[1] "YHOO"
Warning message:
In download.file(paste(yahoo.URL, "s=", Symbols.
name, "&a=", from.m,  :
  downloaded length 147128 != reported length 200
> chartSeries(YHOO)
> utils:::menuShowCRAN()
>
>
>
>
>
>
>
> chartSeries(YHOO)

> # Outputting: Figure 4.21
```

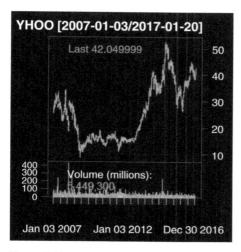

Figure 4.21

```
>
> chartSeries(YHOO)
> chartSeries(YHOO,
                     subset='last 4 months')
> # Outputting: Figure 4.22
```

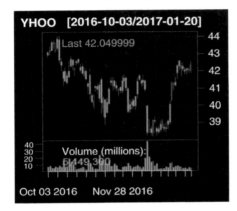

Figure 4.22

```
>
> chartSeries(YHOO, subset='last 4 months')
> chartSeries(YHOO,
                     subset='2007::2008-01')

> # Outputting: Figure 4.23
>
> chartSeries(YHOO, subset='2007::2008-01')
```

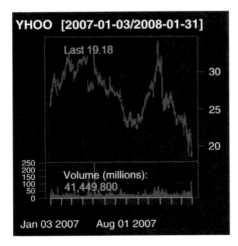

Figure 4.23

```
>
> chartSeries(YHOO,theme=chartTheme
                        ('white'))

> # Outputting: Figure 4.24
```

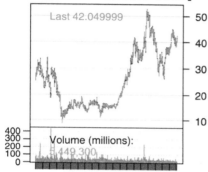

Figure 4.24

```
>
> chartSeries(YHOO,theme=chartTheme
                        ('white'))
>
> chartSeries(YHOO,TA=NULL) #no volume

> # Outputting: Figure 4.25
```

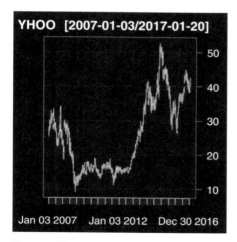

Figure 4.25

```
chartSeries(YHOO,TA=NULL) #no volume
>
> chartSeries(YHOO,TA=c(addVo(),
    addBBands()))
#add volume and Bollinger

> # Outputting: Figure 4.26
```

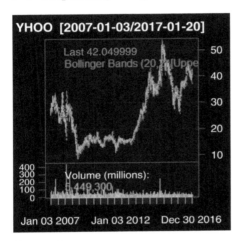

Figure 4.26

```
> chartSeries
+                   (YHOO,TA=c(addVo(),
+                     addBBands()))
>           # add volume and Bollinger
>
> addMACD() # add MACD indicator to
+           # current chart

> # Outputting: Figure 4.27
> addMACD() # add MACD indicator to
+           current chart
>
> setTA()

> # Outputting: Figure 4.28
>
> setTA()
>
> chartSeries(YHOO) # draws chart again,
+              # this time will all indicators present
```

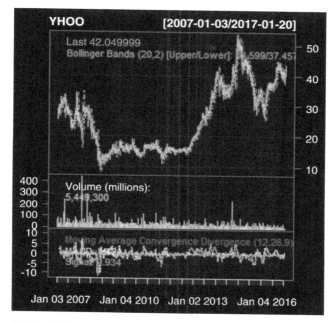

Figure 4.27

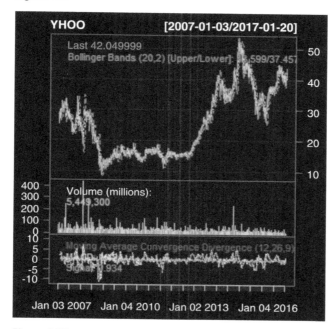

Figure 4.28

Figure 4.29

```
> # Outputting: Figure 4.29
>
> chartSeries(YHOO)
            + # draws chart again, this time
            + # with all indicators present
>
```

Review Questions for Section 4.5

1 Define: univariate, bivariate, and multivariate data analyses, giving an example of each.

2 **a** How are these analyses carried out in the R environment?
   b Give examples of the R code segments for these analyses.

3 **a** What is meant by regression analysis?
   b How is regression analysis used in data analysis?

4 **a** How is regression analysis carried in the R environment?
   b Provide examples of the R functions used for regression analysis.

5 **a** Summarize the two uses of the ANOVA Table in data analysis.
   b For data analysis, suggest an applicable R code segments.

Exercise for Section 4.5

Using the R code segment below,

a create a 50-vector x of 50 random numbers from the standard normal distribution,

b output x, and

c perform a univariate data analysis on x.

```
> x <- rnorm(1:50)
> x
> install.packages("epibasix")
> library(epibasix)
> univar(x)
> x
```

Appendix 1: Documentation for the Plot Function

```
plot {graphics}
R Documentation
Generic X-Y Plotting
```

Description

Generic function for plotting of R objects. For more details about the graphical parameter arguments, see *par*.

For simple scatter plots, *plot.default* will be used. However, there are plot methods for many R objects, including *functions, data.frames, density* objects, and so on. Use methods(plot) and the documentation for these.

Usage

```
plot(x, y, . . . )
```

Arguments

x: the coordinates of points in the plot. Alternatively, a single plotting structure, function, or any R object with a plot method can be provided.

Y: the y coordinates of points in the plot, optional if x is an appropriate structure.

. . . Arguments to be passed to methods, such as *graphical parameters* (see *par*). Many methods will accept the following arguments:

type

what type of plot should be drawn. Possible types are

- "p" for points,
- "l" for lines,

- "b" for both,
- "c" for the lines part alone of "b",
- "o" for both "overplotted,"
- "h" for "histogram" like (or "high-density") vertical lines,
- "s" for stair steps,
- "S" for other steps, see "Details" below,
- "n" for no plotting.

All other types give a warning or an error; using, for example, type = "punkte" being equivalent to type = "p" for S compatibility. Note that some methods, for example, *plot.factor,* do not accept this.

Main an overall title for the plot: see *title.*

Sub a sub title for the plot: see *title.*

Xlab a title for the *x*-axis: see *title.*

Ylab a title for the *y*-axis: see *title.*

Asp the y/x aspect ratio: see *plot.window.*

Details

The two step types differ in their *x*–*y* preference: Going from $(x1,y1)$ to $(x2,y2)$ with $x1 < x2$, type = "s" moves first horizontal, then vertical, whereas type = "S" moves the other way around.

See Also

`plot.default, plot.formula` and other methods; `points, lines, par.`
For X-Y-Z plotting see `contour, persp,` and `image.`

Examples

```
require(stats)
plot(cars)
lines(lowess(cars))
plot(sin, -pi, 2*pi) # see ?plot.function
## Discrete Distribution Plot:
plot(table(rpois(100,5)), type = "h", col = "red", lwd=10,
     main=" rpois(100,lambda=5)")
## Simple quantiles/ECDF, see ecdf() {library(stats)}
 for a better one:
plot(x <- sort(rnorm(47)), type = "s", main = "plot
(x, type = \"s\")")
points(x, cex = .5, col = "dark red")
```

5

Assets Allocation Using R

5.1 Risk Aversion and the Assets Allocation Process

The main objectives for asset allocation include the maximizing of returns, with/ without periodic incomes derived from the returns. During these processes, the investor may freely use assets allocation to achieve such goals. Moreover, it certainly should not escape one's attention that the process of assets allocation is laden with the state of risk aversion on the part of the investors.

A growing body of research investigates whether investor risk aversion varies over time. Determining risk aversion is becoming increasingly important both in the United States and internationally for compliance and suitability purposes among financial advisors who are building portfolios for their clients. It appears that significant evidence has been established that risk aversion is time varying and that *changes in risk aversion are primarily related to changes in investor expectations instead of historical market returns.* Time-varying risk aversion carries important implications for the demand for risky assets if investors reduce demand for stocks when valuations are most attractive. It is noted that a statistically significant relation exists between time-varying risk aversion and net equity mutual fund flows, or a variable risk preference bias. It is found that net equity flows for more sophisticated investors, such as those who use index (versus active) mutual funds and purchase institutional equities.

Using a unique dataset with daily responses to a risk tolerance questionnaire from participants in a defined contribution plan, it is found that risk aversion is time-varying. It is also noted that investor expectations tend to be a *better* predictor of time-varying risk aversion than historical performance. So it appears that *risk aversion is driven more by the unknown future than the known past*! While investor expectations influence risk aversion over time, other investor attributes such as age, equity allocation, and salary appear to play an even greater role in shaping risk aversion. Time-varying risk aversion has important implications for the demand for risky assets since investors tend to

Applied Probabilistic Calculus for Financial Engineering: An Introduction Using R, First Edition.
Bertram K. C. Chan.
© 2017 John Wiley & Sons, Inc. Published 2017 by John Wiley & Sons, Inc.
Companion website: www.wiley.com/go/chan/appliedprobabilisticcalculus

shy away from risky assets when valuations are most attractive and when traditional portfolio theory would predict greater demand for equities to rebalance a portfolio. This effect is especially noteworthy given the increasing use of *risk tolerance questionnaires* (RTQs) by advisors when recommending an optimal client portfolio allocation. This is significant because the primary reason that individual investors underperform institutional investors is *their tendency to sell equity mutual funds during a bear market*!

Sophisticated investors, such as those who use index (versus active) mutual funds and buy institutional share classes (i.e., have more wealth and, therefore, more implied investing human capital), exhibit lower time-varying risk aversion. It is also noted that a consistent negative relation between time-varying risk aversion and net equity mutual fund flows may be established. Trail compensation, which provides the same incentive whether the client buys new funds to match their changing risk preferences, is associated with a lower variable risk preference bias.

5.2 Classical Assets Allocation Approaches

Asset allocation is the undertaking of an investment strategy in order to balance risk versus reward by adjusting the percentage of each asset in an investment portfolio according to the investor's risk tolerance, goals, and investment time frame.

Figure 5.1 is a typical asset allocation pie chart.

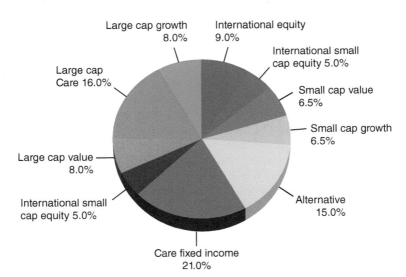

Figure 5.1 An asset investment pie chart – with a diverse portfolio.

It is not easy to measure the overall potential benefits from efficient financial planning: For a given portfolio, a specific investment decision may be analyzed in terms of two fundamental components, generally called alpha α and beta β:

- α *is the luck/skill-based residual component associated with the various approaches of active financial management – such as security selection, tactical asset allocation, and so on*
- β *is the systematic risk exposures of the portfolio – normally achieved via asset allocation.*

α and β are at the heart of traditional performance analysis. However, they are only two of many important financial planning decisions, such as savings and withdrawal strategies, which can have a substantial impact on the retirement outcome for an investor.

5.2.1 Going Beyond α and β

The notions of α and β (in particular α) have long challenged financial advisors and their clients:

- α allows a financial advisor to demonstrate (and potentially quantify) the excess returns generated, which can help justify fees.
- On the other hand, β (systematic risk exposures) helps to explain the risk factors of a portfolio to the market, that is, the asset allocation.

However, another concept gamma γ is designed to measure the value added achieved by an investor from making more sophisticated financial planning decisions. This value may be measured by calculating the certainty equivalent utility-adjusted retirement income across different scenarios. The Greek letter gamma, γ, is a widely used parameter in some areas of finance, such as in the trading of derivatives and in risk management.

5.2.2 γ Factors

The potential value, or *gamma*, γ, may be obtained from making five different "intelligent" financial planning decisions during retirement. A potential retiree has a number of risks, some of which are unique to retirement planning and are not serious concerns during accumulation. These five different factors are as follows:

1) *Asset Location and Withdrawal Sourcing:* Tax-efficient investing for a retiree may be considered in terms of both "asset location" and intelligent withdrawal sequencing from accounts that differ in tax status. Asset location is defined as locating assets in the most tax-efficient account type. For example, it generally makes sense to place less tax-efficient assets (i.e., the

majority of total return comes from dividends that are taxed as ordinary income), such as bonds, in retirement accounts (e.g., IRAs or 401(k)s) and more tax-efficient assets (i.e., the majority of total return comes from capital gains taxed at rates less than ordinary income), such as stocks, in taxable accounts. When thinking about withdrawal sequencing, it is prudent to withdraw monies from taxable accounts first and more tax-efficient accounts (e.g., IRAs or 401(k)s) later.

2) *Total Wealth Asset Allocation:* Most techniques used for determining the asset allocation for a client are often relatively subjective and focus primarily on risk preference (i.e., an investor's aversion to risk) and ignore risk capacity (i.e., an investor's ability to assume risk). In practice, however, it is often believed that asset allocation should be based on a combination of risk preference and risk capacity, although primarily risk capacity. One determines an investor's risk capacity by evaluating the total wealth of the client, which is a combination of human capital (an investor's future potential savings) and financial capital. One may either use the market portfolio as the target aggregate asset allocation for each investor (as suggested by the Capital Asset Pricing Model) or build an investor-specific asset allocation that incorporates an investor's risk preferences. In both approaches, the financial assets are invested, subject to certain constraints, in order to achieve an optimal asset allocation that takes both human and financial capital into account.

3) *Annuity Allocation:* Among retirees, one of the greatest fears is outliving one's savings! For example, a study by Allianz Life has shown that more retirees feared outliving their resources (61%) versus death (39%). Annuities allow a retiree to hedge against longevity risk and can, therefore, improve the overall efficiency of a retiree's portfolio. The contribution of an annuity within a total portfolio framework, (benefit, risk, and cost) should be considered before determining the appropriate amount and annuity type.

4) *Dynamic Withdrawal Strategy:* Most retirement research has focused on static withdrawal strategies where the annual withdrawal during retirement is based on the initial account balance at retirement, increased annually for inflation. For example, a "5% withdrawal rate" would really mean a retiree may take a 5% withdrawal of the initial portfolio value and continue withdrawing that amount each year, adjusted for inflation. If the initial portfolio value was $2 million, and the withdrawal rate was 5%, the retiree would be expected to generate 5% of $2 million, or $100,000, in the first year. If inflation during the first year was 3%, the actual cash flow amount in the second year (in nominal terms) would be $103,000 (= $100,000 × 1.03). Using this approach the withdraw amount is based entirely on the initial income target, and is not updated based on market performance or expected investor longevity.

5) *Liability-Relative Optimization:* Asset allocation methodologies commonly ignore the funding risks, like inflation and currency, associated with an investor's goals. By incorporating the liability into the portfolio optimization process it is possible to build portfolios that can better hedge the risks faced by a retiree. While these "liability-driven" portfolios may appear to be less efficient asset allocations when viewed from an asset-only perspective, one finds they are actually more efficient when it comes to achieving the sustainable retirement income.

On the other hand, each of these five gamma γ concepts may be thought of as actions and services provided by financial planners. One may take a practical function approach to quantify the benefit of different income-maximizing decisions. The goal of this approach is to provide some perspective, as well as quantify the potential benefits that can be realized by an investor (in particular a retiree) from using a gamma-optimized portfolio.

5.2.2.1 Measuring γ

One approach to measure the economic gain from making more intelligent financial planning decisions is to calculate the net present value of the additional income generated by the improved strategy, for example, to quantify the economic benefit that American investors can obtain from strategically timing the start of social security benefits (e.g., adjusting the initial claiming age). This approach does not require any explicit assumptions about subjective investor preferences.

5.2.2.2 How to Measure γ

For a single period, expected utility theory expects that an investor ranks alternative uncertain amounts of income by the expected utility of each. Letting I denote the random amount of income in the given period, then the expected utility of I is

$$\text{EU}[I] = E[u(I)] \tag{5.1}$$

where $u(\cdot)$ is an increasing concave utility function that reflects the risk tolerance of the investor.

Since values of $u(\cdot)$ are abstract undefined measures of "utility," one may define the expected utility to the utility-adjusted certainty equivalent level of income:

$$\text{CE}[I] = u^{-1}\{E[u(I)]\} \tag{5.2}$$

which means that the investor is indifferent to the random amount of income I and the certain amount of income $\text{CE}[I]$. γ measures how much additional utility-adjusted income a strategy increases over and above the utility-adjusted income from a set of base-case decisions.

In the financial literature, several parametric forms of $u(\cdot)$ are commonly used. The most common one is the constant relative risk aversion (CRRA) utility function, which may be expressed as:

$$u(x) = \{[\theta/(\theta - 1)]x^{[(\theta-1)/\theta]}\}, \theta > 0, \quad \theta \neq 1$$
$$\{\ln(x)\}, \qquad\qquad\qquad \theta = 1 \tag{5.3}$$

where θ is the risk tolerance parameter.

Owing to its analytical simplicity and its ability to represent a wide range of opinions toward risk with a single parameter, the CRRA utility function is used not only in single-period models, but also in multiperiod models.

Let I_t denote the random amount of income in period t, the expected utility of the sequence of incomes from periods 0 to T in these models is

$$EU = \sum_{t=0}^{T} d_t[\theta/(\theta - 1)] E[I_t^{[(\theta-1)/\theta]}] \tag{5.4}$$

where d_t is the discount factor for period t. However, the parameter θ does have a role in the calculation of utility even if there is no uncertainty – because, in a multiperiod context, θ plays the following two roles:

1) The investor's risk tolerance parameter
2) The investor's elasticity of intertemporal substitution (EOIS) preference parameter.

It was pointed out by Epstein and Zin that there is no reason that the risk tolerance parameter and the EOIS parameter are equal. The reason for setting them equal is mathematical expediency. By recursively nesting the certainty equivalence function inside of the intertemporal utility function, Epstein and Zin formulated expected utility that makes these distinct parameters:

$$V_t = \left[I_t^{\frac{\eta-1}{\eta}} + \frac{d_{t+1}}{d_t} CE_t[V_{t+1}]^{\frac{\eta-1}{\eta}} \right]^{\frac{\eta}{\eta-1}} \tag{5.5}$$

where

V_t is the utility of the stream of income, starting at time t (measured in the same unit as income), and
η is the investor's elasticity of intertemporal substitution preference parameter.

On the certainty equivalent operator, the t subscript denotes that it is conditional on what is known at time t.

The utility function generalizes the recursive expected utility maximization problem formulated by Lucas (1978), whereas gamma is derived from a measure of the utility of a given set of simulated income paths. Here, one may formulate a utility function with the same EOIS and risk parameters as the Epstein–Zin utility function that may be evaluated without recursion. This may be achieved by reversing the order of the nesting of intertemporal and risk components of utility. Thus, for each simulated income path, one may calculate its utility-equivalent constant income level based on the EOIS parameter, which is denoted as II. That is, for a given simulated income path, II is the constant amount of income with the same utility as the actual income path. This is given by

$$II = \left(\frac{\sum_{t=0}^{T} q_t (1+\rho)^{-t} I_t^{\frac{\eta-1}{\eta}}}{\sum_{t=0}^{T} q_t (1+\rho)^{-t}} \right)^{\frac{\eta}{\eta-1}} \tag{5.6}$$

where

I_t is the level of income in year t
q_t is the probability of surviving to at least year t
r is the last year for which $q_t > 0$

ρ is the investor's subjective discount rate (so that d_t in (5.5) is $q_t (1+\rho)^{-t}$ Note that while (5.6) contains two preference parameters (ρ and η) that describe how the investor feels about having income to consume at different points in time, it does not show how the investor feels about risk. As was discussed already, one treats the elasticity of intertemporal substitution as a parameter distinct from the risk tolerance parameter. One may introduce the risk tolerance parameter next by treating the entire path as unknown and evaluating expected utility.

One may measure expected utility using the CRRA utility function with its risk tolerance parameter θ that we introduced in (5.3).

$$EU = \sum_{i=1}^{M} p_i \frac{\theta}{\theta - 1} (II_i)^{\frac{\theta-1}{\theta}} \tag{5.7}$$

where

M is the number of paths
the subscript i denotes which of M paths is being referred to

p_i is the probability of path i occurring which is set to $1/M$. One may define Y as the constant value for II that yields this level of expected utility. This is the certainty equivalent of the stochastic utility-adjusted income II, and Y is given by

$$Y = \left[\sum_{i=1}^{M} p_i (II_i)^{\frac{\theta-1}{\theta}} \right]^{\frac{\theta}{\theta-1}} \tag{5.8}$$

One may now define the gamma γ of a given strategy or a set of decisions as

$$\text{Gamma(strategy)} = \frac{Y(\text{strategy}) - Y(\text{benchmark})}{Y(\text{benchmark})} \tag{5.9}$$

As a base case, one may use the following parameter values:

$$\rho = 2.5\%, \quad \eta = 0.5, \quad \text{and} \quad \theta = 0.33.$$

Later, one may perform sensitivity analysis to explore the impact of how gamma γ is affected by the choice values for these parameter values.

5.2.3 The Mortality Model

Mortality may be modeled using the "Gompertz law of mortality":

"The probability of a person dying increases at a relatively constant exponential rate as age increases."

Here, one may use the formulation of Gompertz law for mortality – for which the Milevsky–Robinson form of this law is the probability of survival to age t ≤ 115 conditional on a life at age a, is given by

$$q_t = \exp[\exp\{(a - m)/b\}[1 - \exp\{(t - a)/b\}]] \tag{5.10}$$

where

m is the modal lifespan
b is the dispersion coefficient.

One may use the Gompertz parameters that are fitted to the discrete, "Annuity 2000 Basic Table," mortality table using the following procedure: the probability of having at least one member of a married couple surviving to age t is

$$q_t = q_t^{\text{Male}} + q_t^{\text{Female}} - q_t^{\text{Male}} q_t^{\text{Female}} \tag{5.11}$$

Using this table, which contains the mortality rates per 1,000 persons (for males and females), ages 5–115, in Equation (5.12) – in which $\text{Mort}_t^{\text{Gender}}$ denotes the mortality rate for a person of the given age t and gender – one may compute the survival rates for persons of ages 65 to 115 for each gender, as follows:

$$q_t^{\text{Gender}} = \begin{cases} 1\}, t = 65 \\ \{q_{t-1}^{\text{Gender}}\{1 - (\text{Mort}_t^{\text{Gender}}/1000)\}, t > 65 \end{cases} \tag{5.12}$$

From this model, one may calculate the probability of a person, of a given gender, dying at age $t > 65$

$$p_t^{\text{Gender}} = q_{t-1}^{\text{Gender}} - q_t^{\text{Gender}} \tag{5.13}$$

For a given gender, the age which reaches its maximum value is the *modal age* for the given gender: m in (5.10). It has been shown that

for male	$m = 86$
for female	$m = 90$

For each gender, one may estimate the dispersion coefficient b in (5.10) by minimizing the sum of squared differences:

$$\text{SSD} = \sum_{i=65}^{i=115} (q_t \underset{\_}{q_t})^2 \tag{5.14}$$

where q_t is the survival probability given by the Gompertz model in (5.6). The SSD results are 10.48 for males and 8.63 for females.

For each gender, one may estimate the dispersion coefficient, b in (5.10), by minimizing the sum of squared differences, (5.14).

5.2.4 Sensitivity Analysis

To assess the impact of the asset allocation of the base case, one may recalculate Test 2 (say, bootstrapping by repeating each 10,000-trial Monte Carlo simulation 100 times) using different equity allocations.

One might note that for all equity allocations greater than the base case of 20%, the gamma estimate is almost within one standard error of the base case result. Hence, one may conclude that gamma varies little across base case asset allocations. To see the impact on gamma of varying the preference parameters for Test 2, one took the simulation from the initial bootstrap analysis that had the closest value to the average, which was 19.36%. Using this simulation, one varied the values of risk tolerance (θ), the subjective discount rate (ρ), and the elasticity of intertemporal substitution (η).

It has been found that varying θ *had little impact on gamma.*

5.2.4.1 The Elasticity of Intertemporal Substitution (EOIS)

To illustrate the role of the EOIS parameter, η, and its impact on gamma, consider a model in which a given amount of total wealth (financial assets plus human capital) is used to finance consumption over an infinite time horizon. Assume that the market offers a risk-free flat yield curve. If the single market interest rate equals the subjective discount factor (ρ), the optimal level of consumption is the same in every year regardless of the marginal rate of intertemporal substitution.

If the market interest rate is less than the subjective discount rate, the optimal consumption path is downward sloping:

1) If the EOIS is high ($\eta = 0.9$), the optimal level of consumption starts high and the path is steeply sloped. However, if the EOIS is low ($\eta = 0.1$), the optimal consumption path is nearly flat.

   This model of the sensitivity of gamma to the η parameter illustrates that gaging investors' willingness to reschedule income can be an important input to financial planning. While most financial planning questionnaires are designed to elicit information about an investor's investment horizon and risk tolerance, few seek to assess an investor's elasticity of intertemporal substitution.

5.3   Allocation with Time Varying Risk Aversion

5.3.1   Risk Aversion

In financial engineering, *risk aversion* is the behavior of humans (especially investors), when exposed to *uncertainty*, to try to *reduce* that uncertainty. For example, it is the reluctance of the investor to accept a bargain with an uncertain payoff rather than another bargain with a more certain, but possibly/usually lower, *expected* payoff. For example, a risk-averse investor may prefer to put the investment into a savings *bank* account with a low but guaranteed interest rate, rather than into a *stock* that may have much higher expected returns, but also involves a chance of losing value.

Such *psychological* characteristics are illustrated in Figure 5.2.

5.3.1.1   Example of a Risk-Averse/Neutral/Loving Investor

Suppose an investor is given the choice between the following two scenarios:

1) One with a guaranteed payoff
2) One without

In the guaranteed scenario (1), the investor receives $100,000 zillion.

In the uncertain scenario (2), a coin is flipped to decide whether the person receives $200,000 (that is, 2 × $100,000) or nothing. The expected payoff for both scenarios is $100,000, meaning that an individual who was insensitive to risk would *not* care whether they took the guaranteed payment or the gamble.

However, individuals may have different *risk attitudes*.

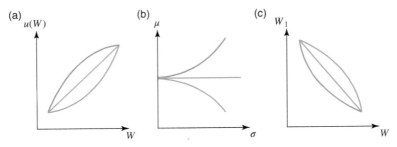

Figure 5.2 Utility function of a risk-seeking individual. Risk aversion (red) contrasted to risk neutrality (yellow) and risk loving (orange) in different settings.(a) A risk averse utility function is concave from below, while a risk loving utility function is convex from below.(b) In standard deviation-expected value space, risk averse indifference curves are upward sloped.(c) With fixed probabilities of two alternative states 1 and 2, risk averse indifference curves over pairs of state-contingent outcomes are convex.

An investor is said to be the following:

- *Risk-averse* (or *risk-avoiding*): If the investor would accept a certain payment (*certainty equivalent*) of less than $100,000 (e.g., $80,000), rather than taking the gamble and possibly receiving nothing.
- *Risk-neutral*: If the investor is indifferent between the bet and a certain $100,000 payment.
- *Risk-seeking*: If the investor would accept the bet even when the guaranteed payment is more than $100,000 (e.g., $120,000).

The average payoff of the gamble, called its *expected value*, is $100,000. The dollar amount that the individual would accept instead of the bet is called the *certainty equivalent*, and the difference between the expected value and the certainty equivalent is called the *risk premium*, that is,

$$\text{risk premium} = \text{expected value} - \text{certainty equivalent} \tag{5.15}$$

For risk-averse individuals, the risk premium is *positive*, for risk-neutral persons it is *zero*, and for risk-loving individuals their risk premium becomes *negative*.

5.3.1.2 Expected Utility Theory

In *expected utility* theory, a person (including an investor) has a utility function $u(x)$, where x represents the value that he might receive in money or goods (in the above example x could be 0 or 100).

Time does not come into this consideration, so inflation does not appear. (The utility function $u(x)$ is defined only up to positive linear affine transformation – in other words, a constant offset could be adjusted (added) to the value of $u(x)$ for all x, and/or $u(x)$ could be multiplied by a positive constant factor, without affecting the conclusions).

An agent possesses risk aversion if and only if the utility function is *concave*. For instance $u(0)$ could be 0, $u(100)$ might be 10, $u(60)$ might be 7, and for comparison $u(50)$ might be 6.

The expected utility of the above bet (with a 50% chance of receiving 100 and a 50% chance of receiving 0) is

$$E(u) = [u(0) + u(100)]/2 \tag{5.16}$$

and if the person has the utility function with $u(0) = 0$, $u(50) = 6$, and $u(100) = 10$, then the expected utility of the bet equals 5, which is the same as the known utility of the amount 40. Hence, the certainty equivalent is 40. The risk premium is ($50 – $40) = $10, or in proportional terms

$$(\$50 - \$40)/\$40 = \$10/\$40 = \tfrac{1}{4} \tag{5.17}$$

or 25%, where $50 is the expected value of the risky bet

$$(1/2)(0) + (1/2)(100) = 0 + 50 = 50$$

This risk premium means that the investor would be willing to sacrifice as much as $10 in expected value in order to achieve perfect certainty about how much money will be received. That is, the person would be indifferent between the bet and a guarantee of $40, and would prefer anything over $40 to the bet.

For a more wealthy investor, the risk of losing $100 would be less significant, and for such small amounts his utility function would be likely to be almost linear, for instance if $u(0) = 0$ and $u(100) = 10$, then $u(40)$ might be 4.0001 and u (50) might be 5.0001.

5.3.1.3 Utility Functions
The utility function for perceived gains has two key properties: an upward slope and concavity.

i) The upward slope implies that the investor feels that *more is better*: A larger amount received yields greater utility, and for risky bets the person would prefer a bet which is *first-order stochastically dominant* over an alternative bet (i.e., if the probability mass of the second bet is pushed to the right to form the first bet, then the first bet is preferred).

ii) The concavity of the utility function implies that the investor is risk averse: A sure amount would always be *preferred* over a risky bet having the same expected value; moreover, for risky bets the investor would prefer a bet that is a *mean-preserving contraction* of an alternative bet (that is, if some of the probability mass of the first bet is spread out without altering the mean to form the second bet, then the first bet is preferred) (Figures 5.3a, 5.3b, and 5.3c).

A person is given the choice between two scenarios, one with a guaranteed payoff and one without. In the guaranteed scenario, the person receives $50. In the uncertain scenario, a coin is flipped to decide whether the person receives

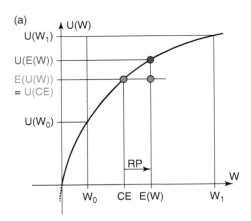

(a)

Figure 5.3 (a) Utility function of a risk-averse (risk-avoiding) investor. (b) Utility function of a risk-neutral investor. (c) Utility function of a risk-seeking individual. CE: certainty equivalent, E(U(W)): expected value of the utility (expected utility) of the uncertain payment, E(W): expected value of the uncertain payment, U(CE): utility of the certainty equivalent, U(E(W)): utility of the expected value of the uncertain payment, U(W0): utility of the minimal payment, U(W1): utility of the maximal payment, W0: minimal payment, W1: maximal payment, RP: risk premium.

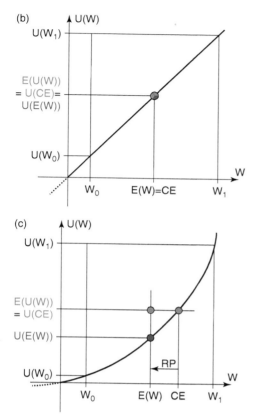

Figure 5.3 (*Continued*)

$100 or nothing. The expected payoff for both scenarios is $50, meaning that an individual who was insensitive to risk would not care whether they took the guaranteed payment or the gamble. However, individuals may have different *risk attitudes*.

5.3.2 Utility of Money

In *expected utility* theory, an investor has a utility function $u(x)$, where x represents the value that an investor might receive in money or goods (in the above example x could be 0 or 100).

Time does not come into this calculation, so inflation does not appear. (The utility function $u(x)$ is defined only up to positive linear *affine transformation* – namely, a *constant* offset could be added to the value of $u(x)$ for all x, and/or $u(x)$ could be multiplied by a positive constant factor, without affecting the conclusions).

An agent possesses risk aversion if and only if the utility function is *concave*. For instance $u(0)$ could be 0, $u(100)$ might be 20, $u(40)$ might be 10, and for comparison $u(50)$ might be 12.

The *expected utility* of the above bet (with a 50–50 chance of receiving 0 and a 50–50 chance of receiving 100) is

$$E(u) = [u(0) + u(100)]/2 \tag{5.18}$$

and if the person has the utility function with $u(0) = 0$, $u(40) = 10$, and $u(100) = 20$, then the expected utility of the bet equals

$$(0 + 20)/2 = 10$$

which is the same as the known utility of the amount 40: $u(40) = 10$.

Hence the certainty equivalent is 40.

The risk premium is ($50 − $40) = $10, or in proportional terms

$$(\$50 - \$40)/\$40 = \$10/\$40 = 10/40 = \tfrac{1}{4}$$

or 25%, where $50 is the expected value of the risky bet

$$[(\tfrac{1}{2})(0) + (\tfrac{1}{2})(100)] = 0 + 50 = 50$$

This risk premium means that the investor would be willing to sacrifice as much as $10 in expected value in order to achieve perfect certainty about how much money will be received. That is, the person would be indifferent between the bet and a guarantee of $40, and would prefer anything over $40 to the bet.

5.4 Variable Risk Preference Bias

It appears that there is a growing body of evidence of time-varying risk aversion.

Using a unique data set with daily responses to a risk tolerance questionnaire from participants in a defined contribution plan, one finds significant evidence that

a) *risk aversion is time-varying,* and that
b) *changes in risk aversion are primarily related to changes in investor expectations instead of historical market returns.*

Time-varying risk aversion has important implications for the demand for risky assets if investors reduce demand for stocks when valuations are most attractive.

- One may note a statistically significant relation between time-varying risk aversion and net equity mutual fund flows, or a variable risk preference bias.
- One may also find that net equity flows for more sophisticated investors, such as those who use index (versus active) mutual funds and purchase institutional share classes, exhibit lower time-varying risk aversion.

- One may also find a statistically significant negative relation between time-varying risk aversion and net equity mutual fund flows.

Hence, financial investment advisor compensation models focused on trail commissions may provide a valuable debiasing incentive that makes investors less susceptible to the variable risk preference bias.

5.4.1 Time-Varying Risk Aversion

Recent research in financial engineering provides significant evidences. S&P 500 index levels, from January 2000 through September 2010, showed the following:

a) The average investor risk aversion is not constant over time.
b) The maximum average monthly risk aversion score over the data series was 5.405, while the minimum average score was 4.525.
c) The overall monthly average risk aversion score was 4.949.
d) These changes represent a significant magnitude of variation because the potential scores are bounded by 3.000 and 9.000, where a score of 3.000 indicates a low level of risk aversion and a score of 9.000 indicates a high level of risk aversion.

5.4.1.1 The Rationale Behind Time-Varying Risk Aversion

While the average investor risk tolerance changes over time, there are a number of other investor and market characteristics that affect risk aversion. Two potential reasons of risk aversion are

i) historical experiences (i.e., past returns) and
ii) future expectations of stock market performance.

The research has shown that there is a significant relation between risk aversion score decile and net equity mutual fund flows, CAPE Ratio, and future one-year returns. The negative relation between risk aversion and net equity flows suggests the following:

i) Investors tend to favor equities when risk aversion is the lowest, which is also when markets have the least attractive valuations based on the CAPE Ratio and lowest future one-year returns.
ii) Investors appear to be less attracted to stock funds when valuations are most.

5.4.1.2 Risk Tolerance for Time-Varying Risk Aversion

Since risk aversion is time-varying, it is possible that the "run-of-the-mill" risk tolerance questionnaires (RTQs) may contribute to individual investor under-performance if the scores influence asset allocation selected by the plan providers. The more sophisticated investors may be less susceptible to risk preference bias.

One way to determine the relative investor sophistication is to look at the historical relation between net equity mutual fund flows for different types of mutual funds and risk aversion over time. The seasoned investor with constant risk preferences would rationally rebalance to equity mutual funds during a bear market. A biased investor may be less attracted to equity funds as valuations fall, increasing the portfolio share of safe assets when valuations of risky assets are most attractive. One may use three different methods to classify investors based on net equity mutual fund flows:

1) Sort by whether funds are actively or passively managed. Passive mutual funds are assumed to be those classified as "index" funds by Morningstar, Inc., while all other mutual funds are considered actively managed funds. There is evidence that investors in passive funds are more sophisticated than investors in active funds that are most often sold through the broker channel.
2) Classify mutual funds into four different fund groups: broker-sold, institutional, investor, or retirement.
3) One may further decompose the broker-sold category by the form of compensation paid to the financial advisor. Differences in the method of compensation provided through share class structure may influence whether the advisor gains from debiasing a client who is tempted to shift the portfolio to safety during an equity market decline.

5.5 A Unified Approach for Time Varying Risk Aversion

Although traditional finance theory assumes that risk aversion is not time-varying, there is a growing body of empirical research that suggests that risk preferences change with market conditions. Using a unique dataset with daily responses to a risk tolerance questionnaire from participants in a defined contribution plan, one may find significant evidence that risk aversion is time-varying. It is also noted that investor expectations tend to be a better predictor of time-varying risk aversion than historical performance. Thus, it seems that *risk aversion is driven more by the unknown future than the known past*.

While investor expectations influence risk aversion over time, other investor attributes (such as age, equity allocation, and salary, etc.) appear to play an even greater role in shaping risk aversion. *Time-varying risk aversion* has important implications for the demand for risky assets since investors tend to avoid risky assets when valuations are most attractive and when traditional portfolio theory would predict greater demand for equities to rebalance a portfolio. This effect is especially noteworthy given the increasing use of the "traditional" RTQs by investment advisors when recommending an optimal client portfolio allocation. This is important because the primary reason that individual investors under-perform institutional investors is their tendency to sell equity mutual funds during a bear market.

More sophisticated investors, such as those who use index (versus active) mutual funds and buy institutional share classes, exhibit lower time-varying risk aversion. It is also noted that a consistent negative relation between time-varying risk aversion and net equity mutual fund flows when sorted by fees. Trail compensation, which provides the same incentive whether the client buys new funds to match their changing risk preferences, is associated with a lower variable risk preference bias. Financial advisor compensation schemes that provide greater compensation for acceding to variable risk preference biases are associated with a failure to debias. This finding is consistent with research on perverse incentives from traditional commission.

5.6 Assets Allocation Worked Examples

5.6.1 Worked Example 1: Assets Allocation Using R

The Black–Litterman (B–L) Asset Allocation Model

Overview Combining information from *two* sources to create an estimate of expected returns:

Source 1: What does the current market tell us about the expected excess returns? Implied excess equilibrium returns?

Source 2: What views does the investment manager have about particular stocks, sectors, asset classes, or country?

The B–L model combines these different sources to produce estimates of excess returns.

Combining predicted and implied returns

$$E(r - r_j) = [(\tau S)^{-1} + P^T \Omega P]^{-1}[(\tau S)^{-1}\Pi + P^T \Omega^{-1}Q]$$

Basic Notation

$\tau = $ A scalar (assume $\tau = 1$).

$S = $ Variance–Covariance matrix for all assets under consideration

$\Omega = $ Uncertainty surrounding your views.

$\Pi = $ Implied excess returns

$Q = $ Views on expected excess returns for some or all assets

$P = $ Link matrix identifying which assets you have views about

Understanding the Formula Splitting into two sections for better understanding

Consider the second section first:

$$(\tau S)^{-1} \qquad \Pi \qquad + \qquad P^T \Omega^{-1} \qquad Q$$
$$\downarrow \qquad\qquad \downarrow \qquad\qquad\qquad \downarrow \qquad\qquad \downarrow$$

Measures of confidence	Implied excess returns	Confidence	Our views

We are combining implied excess returns with our own views on excess returns: A weighted average

What are the weights?

$$E(r - rj) = [(\tau S)^{-1} + P^T \Omega P]^{-1} [(\tau S)^{-1}\Pi + P^T \Omega^{-1} Q]$$

The investor about his/her views relative to the implied excess returns.

-How confident?

Ω will be *large* $\Rightarrow \Omega^{-1}$ will be *small*

Understanding the formula:

$$E(r - r_j) = [(\tau S)^{-1} + P^T \Omega P]^{-1}[(\tau S)^{-1}\Pi + P^T \Omega^{-1} Q]$$

The first part of this formula:

$$[(\tau S)^{-1} + P^T \Omega P]^{-1}$$

- τ implies excess returns and our views add up to 1.
- Namely, the formula is just a weighted average!

Consider a worked example:
Consider 1 stock: AMZN (Amazon.com)

- The implied excess returns are 0.74% per month and the variance is 2.018%
- We predict excess returns of 2% per month. The uncertainty surrounding this view is reflected by a variance of 0.50%

$$\Rightarrow \Pi = 0.74\%$$
$$S = 2.015\%$$
$$Q = 2\%$$
$$\Omega = 0.50\%$$

- Assume $\tau = 1$, $P = 1$

Substituting into the B–L formula:

$$E(r - r_j) = [(\tau S)^{-1} + P^T \Omega P]^{-1}[(\tau S)^{-1}\Pi + P^T \Omega^{-1}Q]$$

$$= [0.02015^{-1} + 0.005 - 1]^{-1} \times [0.02015^{-1} \times 0.0 \times [0.3674 + 4]/074$$

$$+ 0.005^{-1} \times 0.02]$$

$$= [49.63 + 200]^{-1} \times [49.63 \times 0.0074 + 200 \times 0.02]$$

$$= [249.64]^{-1} \times [0.3673 + 4]$$

$$= 4.3673/249.63$$

$$= 0.017495$$

$$= 1.749\% \text{ per month}$$

Does this estimate make sense?

Since we are confident about our views, now, try a more complicated example:

Incorporating Investor Views:

1) Analysts tell you that AEP has found a way to store electricity. Based on this breakthrough, they expect AEP to outperform XOM by 1% per month.

Stock[a]	Historical excess returns	CAPM excess returns	Implied excess returns
INTC[a]	0.0035	0.0135	0.0092
AEP	0.0026	0.0064	0.0032
AMZN	0.0326	0.0111	0.0074
MRK	0.0002	0.0057	0.0059
XOM	0.0111	0.0052	0.0064

a) Intel, American Electric Power Company, Inc., Amazon, Merck, Exxon Mobil

2) Given the current economic conditions, we think that INTC will outperform AMZN by 1.75% per month.
   - Relative views are common in reality.
   - Absolute views, such as AEP having returns of 2% per month, are much less common.

Views versus Implied Excess Returns:

View (1): AEP outperforms XOM by 1% per month
   - The difference in implied excess returns is −0.32% per month.
   - One would expect that incorporating these views would lead to an increase in one's holdings of AEP and a decrease in XOM:0.0092.

View (2): INTC will outperform AMZN by 1.75% per month.
- The difference in implied excess returns is 0.18% per month.
- We would expect that incorporating our view would increase our holdings of INTC and reduce our holdings of AMZN.

The views expressed above are relative views of assets.

Incorporating Our Views:

To link our views to implied excess returns, we need a link matrix, P. Matrix P is constructed in the following way:

Each row represents a view, each column represents a company

		INTC	AEP	AMZN	MRK	XOM	
View 1	[	0	1	0	0	−1	]
P =	[						]
View 2	[	1	0	−1	0	0	]
	[						]
View 3	[	0.5	0	0.5	−0.5	−0.5	]

The View Vector and Uncertainty

What do Q and Ω look like?

$$Q + \varepsilon = \begin{bmatrix} Q_1 \\ \vdots \\ Q_k \end{bmatrix} + \begin{bmatrix} \varepsilon_1 \\ \vdots \\ \varepsilon_k \end{bmatrix} \rightarrow Q + \varepsilon = \begin{matrix} \text{View 1} \\ \downarrow \\ \begin{bmatrix} 0.01 \\ 0.0175 \end{bmatrix} \\ \uparrow \\ \text{View 2} \end{matrix} + \begin{bmatrix} \varepsilon_1 \\ \varepsilon_2 \end{bmatrix}$$

where

$$\begin{bmatrix} \varepsilon_1 \\ \vdots \\ \varepsilon_k \end{bmatrix} \sim N \left(\begin{bmatrix} 0 \\ \vdots \\ 0 \end{bmatrix}, \underbrace{\begin{bmatrix} \omega_{11} & \cdots & \omega_{1k} \\ \vdots & \cdots & \vdots \\ \omega_{k1} & \cdots & \omega_{kk} \end{bmatrix}}_{\Omega} \right)$$

Calculating Ω

There is no best way to calculate Ω. It will depend on how confident you are of your predictions.

Black and Litterman recommend

$$\text{Base} \Rightarrow \Omega = \tau PSP^T \quad \text{, where } P = \text{Link martrix and}$$
$$S = \text{variance} - \text{covariance matrix}$$

We have assumed that $\tau = 1$, and so we can ignore it!

The well-known classical Black–Litterman models for asset allocation may be represented by the following two schematic illustrations:

Figures 5.4a and 5.5a.

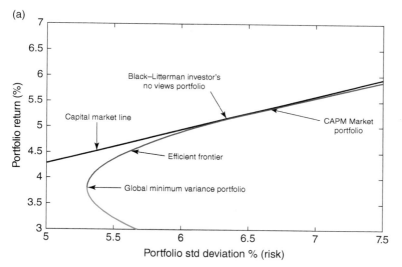

Figure 5.4a Classical Black–Litterman model I.

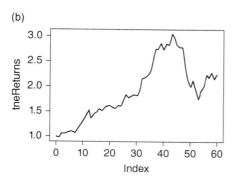

Figure 5.4b StockPortfolio-1.

5.6.2   Worked Example 2: Assets Allocation Using R, from CRAN

Harry Markowitz on Assets Allocation: R package `PortfolioAnalytics`
 In the R **domain**

```
> install.packages("PortfolioAnalytics")
> library(PortfolioAnalytics)
> ls("package:PortfolioAnalytics")
 [1] "ac.ranking"              "add.constraint"
 [3] "add.objective"          "add.objective_v1"
 [5] "add.objective_v2"       "add.sub.portfolio"
 >
> install.packages("stockPortfolio ")
Installing package into
'C:/Users/Bert/Documents/R/win-
library/3.2'
(as 'lib' is unspecified)
- - - Please select a CRAN mirror for use in this session - - -
```

A CRAN **mirror is selected.**

```
>
trying URL
'https://stat.ethz.ch/CRAN/bin/windows/contrib/3.2/
stockPortfolio
_1.2.zip'
Content type 'application/zip' length 115410 bytes (112 KB)
downloaded 112 KB
package 'stockPortfolio' successfully unpacked and MD5 sums
checked
The downloaded binary packages are in
    C:\Users\Bert\AppData\Local\Temp\RtmpeEU8Po\downloaded_
packages
> library(stockPortfolio)
> ls("package:stockPortfolio")
[1] "adjustBeta"   "getCorr"
"getReturns"   "optimalPort"
[5] "portCloud"    "portPossCurve"
"portReturn"   "stockModel"
 [9] "testPort"
```

```
>
> #1 stockPortfolio-package
> # Build and manage stock models and portfolios
>
> #===> two examples of downloading data <===#
> ## Not run: grEx1 <- getReturns(c('C','BAC'), start='2004-01-
01', end='2008-12-31')
> ## Not run: grEx2 <- getReturns(c('KEY', 'WFC', 'JPM', 'AMR',
'BIIB', 'AMGN'))
> #===> build four models <===#
> data(stock99)
> data(stock94Info)
> non <- stockModel(stock99, drop=25, model='none',
+ industry=stock94Info$industry)
> sim <- stockModel(stock99, model='SIM',
+ industry=stock94Info$industry, index=25)
> ccm <- stockModel(stock99, drop=25, model='CCM',
+ industry=stock94Info$industry)
> mgm <- stockModel(stock99, drop=25, model='MGM',
+ industry=stock94Info$industry)
> #===> build optimal portfolios <===#
> opNon <- optimalPort(non)
> opSim <- optimalPort(sim)
> opCcm <- optimalPort(ccm)
> opMgm <- optimalPort(mgm)
> #===> test portfolios on 2004-9 <===#
> data(stock04)
> tpNon <- testPort(stock04, opNon)
Warning message:
In testPort(stock04, opNon) : Allocation X was standardized
> tpSim <- testPort(stock04, opSim)
Warning message:
In testPort(stock04, opSim) : Allocation X was standardized
> tpCcm <- testPort(stock04, opCcm)
Warning message:
In testPort(stock04, opCcm) : Allocation X was standardized
> tpMgm <- testPort(stock04, opMgm)
Warning message:
In testPort(stock04, opMgm) : Allocation X was standardized
>
> #===> compare performances <===#
> plot(tpNon)
> # Figure 5.4b    StockPortfolio-1
```

(a)

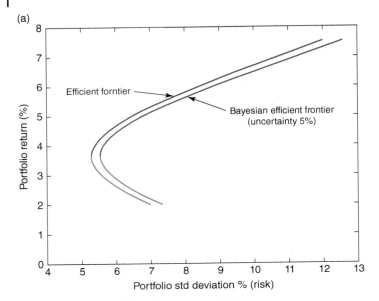

Figure 5.5a Classical Black–Litterman model II.

```
>
> lines(tpSim, col=2, lty=2)
> # Figure 5.5b    StockPortfolio-2
>
> lines(tpCcm, col=3, lty=3)
> # Figure 5.6    StockPortfolio-3
>
> lines(tpMgm, col=4, lty=4)
```

(b)

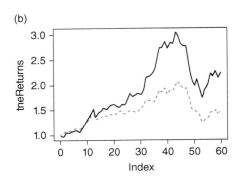

Figure 5.5b StockPortfolio-2.

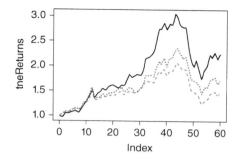

Figure 5.6 StockPortfolio-3.

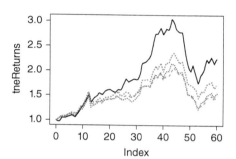

Figure 5.7 StockPortfolio-4.

```
> # Figure 5.7    StockPortfolio-4
>
> legend('topleft', col=1:4, lty=1:4, legend=c('none', 'SIM',
+       'CCM', 'MGM'))
> # Figure 5.8    StockPortfolio-5
```

5.6.3   Worked Example 3: The Black–Litterman Asset

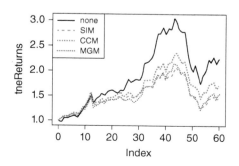

Figure 5.8 StockPortfolio-5. SIM (Grupo Simec SAB de CV) – Computing and Technology, CCM – Medical Services, and MGM – MGM Resorts International.

Review Questions and Exercises

1 At the heart of traditional performance analysis are two factors α and β. Briefly describe these two factors, giving some common examples.

2 **A** Describe and contrast investors who are
   a risk-averse
   b risk-neutral, and
   c risk-loving
   B Suggest some useful approaches suitable for each one of these three types of investors.

3 *Literature Survey:*
   A considerable amount of literature has been developed in the areas supporting the analytical approaches to assets allocation. Much of these materials are readily available in the Internet sources. A typical example is the following paper by Robert Merton (1973) available at conpapers.repec. org/article/ecmemetrp/default38.htm
   a *Study* this online paper and
   b *Comment* on the following issues:

   "We use the structure imposed by Merton's (1973) ICAPM to obtain monthly estimates of the market-level risk-return relationship from the cross-section of equity returns. Our econometric approach sidesteps the specification of time-series models for the conditional risk premium and volatility of the market portfolio. We show that the risk-return relation is mostly positive but varies considerably over time. It co-varies positively with counter-cyclical state variables. The relationship between the risk premium and hedge-related risk also exhibits strong time-variation, which supports the empirical evidence that aggregate risk aversion varies over time. Finally, the ICAPM's two components of the risk premium show distinctly different cyclical properties. The volatility component exhibits a counter-cyclical pattern whereas the hedging component is less related to the business cycle and falls below zero for extended periods. This suggests the market serves an important hedging role for long-term investors."

   These issues are found in a paper by
   Brandt, M. W. and Wang, L. P. (2010) Measuring the time-varying risk-return relation from the cross-section of equity returns.
   c *Do you agree, or disagree* with the conclusions in this paper? And *why*?

4   An exercise in financial modeling for assets allocation.
    The GARCH Process and the `CRAN` Package `tseries`
    *The **generalized** **autoregressive** **conditional** **heteroskedasticity** process*
    The generalized autoregressive conditional heteroskedasticity (GARCH)
    process is an econometric term developed in 1982 by Robert F. Engle, an
    economist and the 2003 winner of the Nobel Memorial Prize for Economics,
    to describe an approach to estimate volatility in financial markets. There are
    several forms of GARCH modeling. The GARCH process is often preferred
    by financial modeling professionals because it provides a more real-world
    context than other forms when trying to *predict* the prices and rates of
    financial instruments.
        The general process for a GARCH model involves three steps:

1) The first is to estimate a best-fitting autoregressive model.
2) The second is to compute autocorrelations of the error term.
3) The third is to test for significance.

    GARCH models are used by financial professionals in several areas,
including trading, investing, hedging, and dealing. Two other widely
used approaches to estimate and predict financial volatility are as follows:

- The classic *historical volatility* (VolSD) method
- The exponentially weighted *moving average volatility* (VolEWMA)
  method.

The GARCH Process   GARCH models help to describe financial markets in
which volatility can change, becoming more volatile during periods of financial
crises or world events and less volatile during periods of relatively calm and
steady economic growth. On a plot of returns, for example, stock returns may
look relatively uniform for the years leading up to a financial crisis such as the
one in 2007. In the time period following the onset of a crisis, however, returns
may swing wildly from negative to positive territory. Moreover, the increased
volatility may be predictive of volatility going forward. Volatility may then
return to levels resembling that of precrisis levels or be more uniform going
forward. A simple regression model does not account for this variation in
volatility exhibited in financial markets and is not representative of the "black
swan" events that occur more than one would predict.

GARCH Models for Asset Returns   GARCH processes *differ* from homoskedastic
models, which assume constant volatility and are used in basic ordinary least
squares (OLS) analysis. OLS aims to minimize the deviations between data
points and a regression line to fit those points. With asset returns, *volatility
seems to vary during certain periods of time and depends on past variance*,
making a homoskedastic model not optimal or suitable.

GARCH processes, being autoregressive, *depend on past squared observations and past variances to model for current variance.* They are widely used in finance owing to *their effectiveness in modeling asset returns and inflation.* A GARCH process aims to minimize errors in forecasting by accounting for errors in prior forecasting, enhancing the accuracy of ongoing predictions.

More on the GARCH Process: http://www.investopedia.com/terms/g/generalalizedautogregressiveconditionalheteroskedasticity.asp#ixzz4XwGOV3MI

And now consider a CRAN package based on GARCH:

Computational Finance

Description	Time series analysis and computational finance.
Depends	R (>= 2.10.0)
Imports	graphics, stats, utils, quadprog, zoo
License	GPL-2
NeedsCompilation	yes
Author	Adrian Trapletti [aut],
	Kurt Hornik [aut, cre],
	Blake LeBaron [ctb] (BDS test code)
Maintainer	Kurt Hornik <Kurt.Hornik@R-project.org>
Repository	CRAN
Date/Publication	2017-01-17 12:33:40
Description	Download historical financial data from a given data provider over the WWW.

Usage

```
get.hist.quote(instrument = "^gdax", start, end,
          quote = c("Open", "High", "Low", "Close"),
          provider = c("yahoo", "oanda"),
          method = NULL,
          origin = "1899-12-30",
          compression = "d",
          retclass = c("zoo", "ts"),
          quiet = FALSE,
          drop = FALSE)
```

Arguments

instrument	a character string giving the name of the quote symbol to download. See the webpage of the data provider for information about the available quote symbols.
start	an R object specifying the date of the start of the period to download. This must be in a form which is recognized by `as.POSIXct`, which includes R POSIX date/time objects, objects of class `"date"` (from package `date`) and `"chron"` and `"dates"` (from package `chron`), and character strings representing dates in ISO 8601 format. Defaults to 1992-01-02.
end	an R object specifying the end of the download period, see above. Defaults to yesterday.
quote	a character string or vector indicating whether to download opening, high, low, or closing quotes, or volume. For the default provider, this may be specified as `"Open"`, `"High"`, `"Low"`, `"Close"`, `"AdjClose"`, and `"Volume"`, respectively. For the provider `"oanda"`, this argument is ignored. Abbreviations are allowed.
provider	a character string with the name of the data provider. Currently, `"yahoo"` and `"oanda"` are implemented. See http://quote.yahoo.com/and https://www.oanda.com/for more information.
method	tool to be used for downloading the data. See `download.file` for the available download methods and the default settings.
origin	an R object specifying the origin of the Julian dates, see above. Defaults to 1899-12-30 (Popular spreadsheet programs internally also use Julian dates with this origin).
compression	Governs the granularity of the retrieved data; `"d"` for daily, `"w"` for weekly or `"m"` for monthly. Defaults to `"d"`. For the provider `"oanda"`, this argument is ignored.
retclass	character specifying which class the return value should have: can be either `"zoo"` (with `"Date"` index), or `"ts"` (with numeric index corresponding to days since `origin`).
quiet	logical. Should status messages (if any) be suppressed?
Drop	logical. If `TRUE` the result is coerced to the lowest possible dimension. Default is `FALSE`.

Value

A time series containing the data either as a "zoo" series (default) or as a "ts" series. The "zoo" series is created with zoo and has an index of class "Date". If a "ts" series is returned, the index is in physical time, that is, weekends, holidays, and missing days are filled with NAs if not available. The time scale is given in Julian dates (days since the origin).

Author(s)

A. Trapletti

See Also

zoo, ts, as.Date, as.POSIXct, download.file;
http://quote.yahoo.com/, https://www.oanda.com/

Examples

```
con <- url("http://quote.yahoo.com")
if(!inherits(try(open(con), silent = TRUE), "try-error")) {
  close(con)
  x <- get.hist.quote(instrument = "^gspc",
                start = "1998-01-01",
                quote = "Close")
  plot(x)
  x <- get.hist.quote(instrument = "ibm", quote = c("Cl", "Vol"))
  plot(x, main = "International Business Machines Corp")
  spc <- get.hist.quote(instrument = "^gspc", start = "1998-01-
                01", quote = "Close")
  ibm <- get.hist.quote(instrument = "ibm",
                start = "1998-01-01",
                quote = "AdjClose")
  require("zoo") # For merge() method.
  x <- merge(spc, ibm)
  plot(x, main = "IBM vs S&P 500")
}
con <- url("https://www.oanda.com")
if(!inherits(try(open(con), silent = TRUE), "try-error")) {
  close(con)
  x <- get.hist.quote(instrument = "EUR/USD",
                provider = "oanda",
                start = Sys.Date() - 500)
  plot(x, main = "EUR/USD")
}
```

5 Portfolio optimization using the CRAN packages: `PortfolioAnalytics` and `DEoptim`

Here, one may consider two CRAN packages for portfolio optimization:

1) `PortfolioAnalytics` enables one to obtain a numerical portfolio solution for a numerical solution for some given objectives and/or objective functions.

2) `DEoptim` is then evoked to provide the numerical optimization.

These two CRAN packages are separately described, each followed by an illustrative worked example.

(1) **Package "PortfolioAnalytics"** April 19, 2015

Type	Package
Title	Portfolio analysis, including numerical methods for optimization of portfolios
Version	1.0.3636
Date	2015-04-18
Maintainer	Brian G. Peterson <brian@braverock.com>
Description	Portfolio optimization and analysis routines and graphics.
Depends	R (>= 2.14.0), zoo, xts (>= 0.8), foreach, PerformanceAnalytics (>= 1.1.0)
Suggests	quantmod, DEoptim(>= 2.2.1), iterators, fGarch, Rglpk, quadprog, ROI (>= 0.1.0), ROI.plugin.glpk (>= 0.0.2), ROI.plugin.quadprog (>= 0.0.2), ROI.plugin.symphony (>= 0.0.2), pso, GenSA, corpcor, testthat, nloptr (>= 1.0.0), MASS, robustbase
License	GPL
Copyright (c)	2004-2015
NeedsCompilation	yes
Authors	Brian G. Peterson [cre, aut, cph], Peter Carl [aut, cph], Kris Boudt [ctb, cph], Ross Bennett [ctb, cph], Hezky Varon [ctb], Guy Yollin [ctb], R. Douglas Martin [ctb]
Repository	CRAN
Date/Publication	2015-04-19 07:38:57

6

Financial Risk Modeling and Portfolio Optimization Using R

6.1 Introduction to the Optimization Process

6.1.1 Classical Optimization Approach in Mathematics

Generally, an optimization process may be represented in the following way:

Given: a function $f: A \rightarrow R$ from some set A to the set of all real numbers R.

To find:

a) "Maximization": An element x_0 in A such that $f(x_0) \geq f(x)$, for all x in A, or
b) "Minimization": An element x_0 in A such that $f(x_0) \leq f(x)$, for all x in A.

Such a formulation is called an *optimization problem* or a *mathematical programming problem*. Many theoretical as well as real-world problems may be modeled in this general approach, for example:

- A is some subset of the Euclidean space R^n, often specified by a set of *constraints*, equalities, or inequalities that the members of A have to satisfy.
- A feasible solution that minimizes (or maximizes) the objective function is called an *optimal solution*.

6.1.1.1 Global and Local Optimal Values

A real-valued function f defined on a domain X has a *global* (or *absolute*) *maximum point* at x_1 if and only if

$$f(x_1) \geq f(x), \quad \text{for all } x \text{ in } X.$$

Similarly, the function has a *global* (or *absolute*) *minimum point* at x_2 if and only if

$$f(x_2) \leq f(x), \quad \text{for all } x \text{ in } X.$$

The value of the function at a maximum point is called the *maximum value* of the function, and the value of the function at a minimum point is called the *minimum value* of the function.

Applied Probabilistic Calculus for Financial Engineering: An Introduction Using R, First Edition. Bertram K. C. Chan.
© 2017 John Wiley & Sons, Inc. Published 2017 by John Wiley & Sons, Inc.
Companion website: www.wiley.com/go/chan/appliedprobabilisticcalculus

Remarks:

- If the domain X is a metric space, then f is said to have a *local* (or *relative*) *maximum point* at the point x^* if there exists some $\epsilon > 0$ such that

$$f(x^*) \geq f(x), \quad \text{for all } x \text{ in } X \text{ within a distance } \varepsilon \text{ of } x^*.$$

- Similarly, the function has a *local* (or *relative*) *minimum point* at x^* if

$$f(x^*) \leq f(x), \quad \text{for all } x \text{ in } X \text{ within a distance } \varepsilon \text{ of } x^*.$$

- A similar definition can be used when X is a topological space, since the definition just given can be re-expressed in terms of neighborhoods. Note that a global maximum point is always a local maximum point, and similarly for minimum points.
- In both the global and local cases, the concept of a *strict* extremum can be defined. For example, x^* is a *strict global maximum point* if, for all x in X with $x \neq x^*$, one has $f(x^*) > f(x)$, and x^* is a *strict local maximum point* if there exists some $\epsilon > 0$ such that, for all x in X within distance ϵ of x^* with $x \neq x^*$, one has $f(x^*) > f(x)$.
- A point is a strict global maximum point if and only if it is the unique global maximum point, and similarly for minimum points.
- Conventional optimization problems are usually stated in terms of minimization. Generally, unless both the objective function and the feasible region are convex in a minimization problem, there may be several local minima.
- If the domain X is a metric space, then f is said to have a *local* (or *relative*) *maximum point* at the point x^* if there exists some $\epsilon > 0$ such that

$$f(x^*) \geq f(x), \quad \text{for all } x \text{ in } X \text{ within distance } \varepsilon \text{ of } x^*.$$

- Similarly, the function has a *local minimum point* at x^* if

$$f(x^*) \leq f(x), \quad \text{for all } x \text{ in } X \text{ within distance } \varepsilon \text{ of } x^*.$$

6.1.1.2 Graphical Illustrations of Global and Local Optimal Value

The foregoing concepts of global and local maximum or minimum are illustrated in Figure 6.1.

6.1.2 Locating Functional Maxima and Minima

- Finding global maxima and minima is the aim of mathematical optimization. If a function is continuous on a closed interval, then by the extreme value theorem global maxima and minima exist.
- Also, a global maximum (or minimum) must either be a local maximum (or minimum) in the interior of the domain or must lie on the boundary of the domain. Hence, an approach for finding a global maximum (or minimum) is to examine all the local maxima (or minima) *in the interior*, and also look at

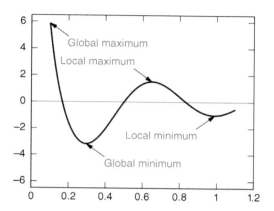

Figure 6.1 Local and global maxima and minima for $\cos(3\pi x)/x$, $0.1 \le x \le 1.1$.

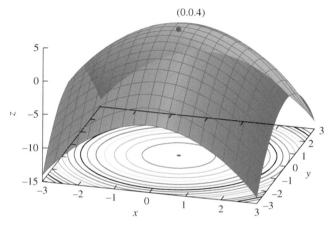

Figure 6.2 Graph of a paraboloid given by $z = f(x, y) = -(x^2 + y^2) + 4$. The global maximum at $(x, y, z) = (0, 0, 4)$ is indicated by a blue dot.

the maxima (or minima) of the points *on the boundary*, and select the largest (or smallest) one.

- For any function that is defined *piecewise*, one finds a maximum (or minimum) by finding the maximum (or minimum) of each piece separately, and then seeing which one is largest (or smallest).
- These concepts are illustrated in Figure 6.2.

6.2 Optimization Methodologies in Probabilistic Calculus for Financial Engineering

Preamble Since classical mathematical optimization techniques are generally built upon the approach of locating the minima at the turning points (at which the gradients are zeros), *it is implicitly assumed that these minima do exist!* However, in probabilistic calculus, such minima, or even such turning points, may *not* exist or may *not* be readily located. Heuristic approaches may have to be called upon to locate any existing and available minima.

Some useful heuristic methodology in locating minima, and the corresponding R programs for minimization are discussed in the following sections.

6.2.1 The Evolutionary Algorithms (EA)

Optimization algorithms inspired by the process of natural selection have been in use since the 1950s, and are often referred to as evolutionary algorithms. The genetic algorithm is one such method, and was invented by Holland in the 1960s. Genetic algorithms apply logical operations, usually on bit strings of fixed or variable lengths, in order to perform crossover, mutation, and selection on a population. Following successive generations, the members of the population are more likely to represent a minimum of an objective function. Genetic algorithms have shown to be useful heuristic methods for global optimization, in particular for combinatorial optimization problems. Evolution strategies are another variety of evolutionary algorithm. Genetic algorithms apply logical operations, usually on bit strings of fixed or variable length, in order to perform crossover, mutation, and selection on a population. Following successive generations, the members of the population are more likely to represent a minimum of an objective function.

Genetic algorithms are useful heuristic methods for global optimization, in particular for combinatorial optimization problems.

6.2.2 The Differential Evolution (DE) Algorithm

Another useful approach in portfolio optimization, particularly in assessing the equal risk contributions to the portfolio, is the differential evolution (DE) algorithm, introduced by Storn and Ulrich (1997), with its application demonstrated by Price, et al. (2006). It is considered to be a genetic algorithm, and is a derivative-free global optimizer.

The DEoptim implementation of DE was motivated by the objective of extending the set of algorithms available for global optimization in the R language and environment for statistical computing (R Development Core Team 2009). R enables rapid prototyping of objective functions, access to a wide array of tools for statistical modeling, and ability to generate customized plots of results with ease (which in many situations makes use of R preferable over the use of programs in languages like Java, MS Visual C++, Fortran 90, or Pascal). DEoptim is available at http://CRAN.R-project.org/package=DEoptim.

6.3   Financial Risk Modeling and Portfolio Optimization

In the real world of financial engineering, most private investors work through one or more "financial advisors', which, in turn, work with "wealth management corporations."

The following is an example of a typical financial advisor (Crystal Cove Advisors of Newport Beach, California), which invests via large professional organizations (such as the LPL Financial of San Diego, California).

6.3.1   An Example of a Typical Professional Organization in Wealth Management

6.3.1.1   LPL (Linsco Private Ledger) Financial
From Wikipedia (Figure 6.3):

LPL Financial is one of the largest organizations of independent financial advisors in the United States of America. Formed in 1989 through the

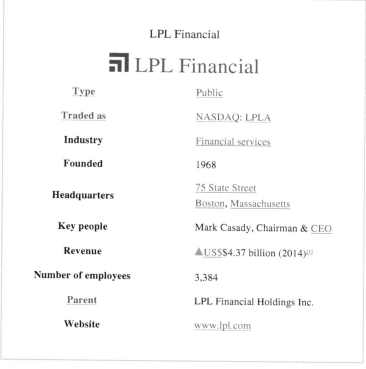

LPL Financial

🔝 LPL Financial

Type	Public
Traded as	NASDAQ: LPLA
Industry	Financial services
Founded	1968
Headquarters	75 State Street Boston, Massachusetts
Key people	Mark Casady, Chairman & CEO
Revenue	▲US$$4.37 billion (2014)[1]
Number of employees	3,384
Parent	LPL Financial Holdings Inc.
Website	www.lpl.com

Figure 6.3 LPL Financial.

merger of two brokerage companies – *Linsco* (established in 1968) and *Private Ledger* (established in 1973) – LPL has since expanded its number of independent financial advisors to more than 14,000. LPL Financial has main offices in

- Boston, MA,
- Charlotte, NC, and
- San Diego, CA

covering the eastern, southern, and western regions of the United States, respectively, as well as other associated territories.

Approximately 3500 employees support financial advisors, financial institutions, and technology, custody, and clearing service subscribers with enabling technology, comprehensive clearing, and compliance services, practice management programs and training, and independent research.

LPL Financial advisors support clients with a number of financial services, including equities, bonds, mutual funds, annuities, insurance, and fee-based programs. However, LPL Financial does *not* develop its own investment products, thus enabling the firm's investment professionals to offer financial ostensibly free from broker/dealer-inspired conflicts of interest.

Additional information on LPL

1) Recent Timeline
2) Some Important Statistics
3) External links

6.3.1.1.1 *Recent Timeline*

In 1989, LPL Financial was created through the merger of two small brokerage firms *Linsco* and *Private Ledger*.

In 2003, LPL Financial acquired Private Trust Company, which manages trusts and family assets for high-net-worth clients in all 50 states.

In 2004, LPL Financial acquired the broker/dealer operations of the Phoenix Companies, which offered Phoenix the chance to sell its products through the LPL Financial network.

In 2005, private equity firms Hellman & Friedman and Texas Pacific Group took a 60% stake in the firm.

In August 2006, LPL Financial expanded its client base following the purchase of UVEST Financial Services, which provided independent brokerage services to more than 300 regional and community banks and credit unions throughout the United States.

In June 2007, LPL Financial finalized its acquisition of several broker/dealers under the Pacific Select Group (aka Pacific Life) umbrella, which added 2000 financial advisors.

In September 2007, LPL Financial and Sun Life Financial announced a definitive agreement under which an affiliate of LPL Financial acquired Independent Financial Marketing Group, Inc. (IFMG) from Sun Life Financial. Sun Life Financial is a leading international financial services organization. The LPL Financial Institution Services business unit, which manages the IFMG business, is the nation's top provider of investment and insurance services to banks and credit unions. The deal closed in November 2007.

On January 1, 2008, Linsco/Private Ledger Corp. (LPL Financial Services) changed its brand name to "LPL Financial."

On November 18, 2010, LPL Investment Holdings Inc., the parent company of LPL Financial, become a publicly traded company on the NASDAQ Stock Market under ticker symbol LPLA.

In 2011, LPL Financial acquired New Jersey-based Concord Capital Partners, which helps trust companies automate the business practices of their internal investment management activities.

In January 2012, LPL Financial acquired Rockville-based Fortigent, which provides high-net-worth solutions and consulting services to RIAs, banks, and trust companies.

In November 2015, LPL Financial announced its first advisor online tool suite certification for Investment Support Services.

6.3.1.1.2 Some Important Statistics
Key facts regarding LPL Financial include the following:

- 4.5 million funded accounts
- 14,000 *financial advisors* supported
- Approximately 4500 technology, custody, and clearing service subscribers
- Approximately 700 financial institution partners
- $475 billion in advisory and brokerage assets

6.3.1.1.3 External links

- LPL Financial Web site
- LPL Financial recruiting Web site

As an example of a typical independent "Financial Advisor" serving the public (for a fee), through LPL, consider the following

Crystal Cove Advisors

- Crystal Cove Advisors is an investment management firm located in Newport Beach, California. Its team of professionals will provide clients with investment strategies that focus on *limiting risk while seeking profitable results*. It believes that avoiding serious loss is the best way to increase long-term returns.

- The financial markets are ever-changing with new challenges constantly presenting themselves. The reactions to these challenges are not always rational, but they are often volatile. For years, the investment community has championed a "buy and hold" strategy, which suggests that over the long term volatility is just noise and returns will be positive. At Crystal Cove Advisors one could not disagree more! The buy and hold philosophy has serious flaws in practice. It fails to account for the behavioral nature of the financial markets and the emotional reactions that follow. The majority of investment mistakes are made when an investor sees significant loss of principal, this stress turns "buy and hold" into "buy high and sell low – thus becoming victims of the BLASH (buy low and sell high) theory. " Crystal Cove Advisors recognizes the difficulties that negative market cycles present; thus one should place the focus on risk and employ tactical strategies that place *capital preservation* as the top priority.
- At Crystal Cove Advisors, a three-step process is used, specific to each client:
  1) Portfolio Analysis to determine risk tolerance and current exposure
  2) Portfolio Construction to establish allocation parameters and stress testing
  3) Portfolio Management to implement and adjust risk exposure as market cycles change.
- The goal of Crystal Cove Advisors is a service that provides the following:
  i) An asset allocation and portfolio optimization wealth-advising platform
  ii) Financial confidence to all clients so they can be totally free to focus on life and family.
- *Disclosures to All Clients*
  No strategy assures success, or can guarantee protection against loss. Tactical allocation may involve more frequent buying and selling of assets that can lead to higher transaction costs. Investors are advised to consider the tax consequences of moving positions more frequently.
- R version 3.2.2 (2015-08-14) -- "Fire Safety"
- Copyright (C) 2015 The R Foundation for Statistical ComputingPlatform: i386-w64-mingw32/i386 (32-bit)

R is free software and comes with ABSOLUTELY NO WARRANTYYou are welcome to redistribute it under certain conditions.Type 'license()' or 'licence()' for distribution details. Natural language support but running in an English locale. R is a collaborative project with many contributors. Type 'contributors()' for more information and 'citation()' on how to cite R or R packages in publications.Type 'demo()' for some demos, 'help()' for on-line help, or'help.start()' for an HTML browser interface to help.
Type 'q()' to quit R.

6.4 Portfolio Optimization Using R

6.4.1 Portfolio Optimization by Differential Evolution (DE) Using R

In 1997, Storn and Price developed an evolution strategy they termed differential evolution (DE). This approach searches the global optimum of a real-valued function of real-valued parameters, and does *not* require that the function be either continuous or differentiable. This technique has been successfully applied in a wide variety of fields, from computational physics to operations research.

The DEoptim implementation of DE extended the set of algorithms available for global optimization in the R environment for statistical computing.

In the DE algorithm, each generation transforms a set of parameter vectors, called the *population*, into another set of parameter vectors, the members of which are more likely to minimize the objective function, as follows:

1) To generate a new parameter vector NP, DE transforms an old parameter vector with the scaled difference of two randomly selected parameter vectors. The variable NP represents the number of parameter vectors in the population.

2) At generation 0, NP assumes that the optimal value of the parameter vector are made
   either using random values between upper and lower bounds for each parameter
   or using values given by the analyst.

   Each generation involves creating a new population from the current population members $x_{i,g}$, where
   i indicates the vectors that make up the population and
   g indicates the generation.

   This is achieved by the *differential mutation of the population members*:
   A trial mutant parameter vector $v_{i,g}$ is created by choosing *three* members of the population: $x_{r0,g}$, $x_{r1,g}$, and $x_{r2,g}$ at random. Then $v_{i,g}$ is generated as

   $$v_{i,g} = x_{r0,g} + K(x_{r1,g} - x_{r2,g})$$ (6.1)

   where K is a positive scaling factor. Effective values of K are typically less than 1.

3) This mutation is continued until either length(x) mutations have been made
   or
   $$r \text{ and} > CR$$ (6.2)
   where

   a) CR is a cross over probability
   $$CR \in [0, 1]$$ (6.3)
   and where rand denotes a random number from $U(0, 1)$.

4) The crossover probability CR controls the fraction of the parameter values that are copied from the mutant. CR approximates (but does not exactly represent) the probability that a parameter value will be inherited from the mutant, since at least one mutation always occurs. Mutation is applied in this way to each member of the population.

5) If an element v_j of the parameter vector is found to exceed the bounds after mutation and crossover, it is *reset*, where j is to index a parameter vector.

6) When using DEoptim

if $v_j >$ upper$_j$, it is reset as $v_j =$ upper$_j -$ rand $\cdot$ (upper$_j -$ lower$_j$) and
if $v_j <$ lower$_j$, it is reset as $v_j =$ lower$_j +$ rand $\cdot$ (upper$_j -$ lower$_j$).

This guarantees that candidate population members found to violate the bounds are set some random amount away from them, in such a way that the bounds are guaranteed to be satisfied. Then the objective function values associated with the children v are determined.

7) If a trial vector $v_{i,g}$ has equal or lower objective function value than the vector $x_{i,g}$, then $v_{i,g}$ replaces $x_{i,g}$ in the population; otherwise $x_{i,g}$ remains.

8) The algorithm automatically terminates after some set number of generations, or after the objective function value associated with the best member has been reduced below some preset threshold, or if it is unable to reduce the best member by a certain value over a specified number of iterations.

9) Applying the R code DEoptim in portfolio optimisation

Remarks:

a) At present, there are two versions of R codes for DE on the Comprehensive R Archive Network (CRAN):

1) DEoptim: Global Optimization by Differential Evolution – By implementing the differential evolution algorithm for global optimization of a real-valued function of a real-valued parameter vector: published on September 7, 2015, by D. Ardia, K. Mullen, B. Peterson, J. Ulrich, and K. Boudt.

2) DEoptimR: Differential Evolution Optimization in Pure R – An implementation n of a variant of the differential evolution stochastic algorithm for global optimization of nonlinear programming problems: published on July 1, 2016, by E. L. T. Conceicao and M. Maechler.

b) For the DEoptim code

Version:	2.2-3
:	foreach, iterators
Published:	2015-01-09

6.4.2 Portfolio Optimization by Special Numerical Methods

Numerical financial modeling is the building of an abstract representation or model of a real-world financial situation. This is a mathematical model for representing the performance of a financial asset or portfolio of an investment. Financial modeling is applicable to different areas of finance, including the following:

• Accounting and corporate finance applications
• Quantitative finance applications.

Financial modeling may mean an exercise in either asset pricing or corporate finance, of a quantitative nature. It is concerned with the modeling of a set of hypotheses about the behavior of markets or agents into numerical predictions. For example, a corporation's decisions about investments (the corporation will invest 25% of assets) or investment returns (returns on Stock X will be 20% higher than the market's returns).

In investment banking, corporate finance, and so on, *financial modeling* is largely synonymous with financial statement forecasting. This usually involves the preparation of detailed company-specific models used for decision-making purposes and financial analysis.

Applications of financial modeling include:

• Financial statement analysis
• Business valuation
• Management decision making
• Capital budgeting
• Cost of capital calculations
• Project finance
• Mergers and acquisitions
• Analyst buy/sell recommendations

These models are built around financial statements; calculations and outputs are monthly, quarterly, or annual. The inputs take the form of "assumptions," where the analyst *specifies* the values that will apply in each period for the following:

a) External/global variables (exchange rates, tax percentage, etc.; may be considered as the model parameters), and for internal *variables* (wages, costs, etc.) Correspondingly, both characteristics are reflected (at least implicitly) in the mathematical form of these models: first, the models are in discrete time; second, they are deterministic.
b) In many instances, such calculations are spreadsheet based, each institution having its own internal ground rules.

c) At a more sophisticated level is financial modeling, also known as quantitative finance, which is, in reality, mathematical modeling. A general distinctions are made among the following:

- *Quantitative Financial Management*: Models of the financial situation of a large, complex firm
- *Quantitative Asset Pricing*: Models of the returns of different stocks
- *Financial Engineering*: Models of the price or returns of derivative securities

These problems are generally probabilistic/stochastic and continuous in nature, and models here require complex algorithms, entailing finance. Modellers are generally referred to as "quants" (quantitative analysts), and typically have advanced academic degrees such as Master of Quantitative Finance, Master of Computational Finance, or Master of Financial Engineering. Not a few are at the Ph.D. level in quantitative disciplines such as mathematics, statistics, physics, engineering, computer science, operations research, and so on.

Alternatively, or in addition to their quantitative background, they complete a finance masters with a quantitative orientation.

6.4.3 Portfolio Optimization by the Black–Litterman Approach Using R

The Black–Litterman Model The goals of the Black–Litterrman model were

- to create a systematic method of specifying a portfolio, and then
- to incorporate the views of the analyst/portfolio manager into the estimation of market parameters.

Let,

$$A = \{a_1, a_2, a_3, \dots, a_n\}$$

be a set of random variables representing the returns of n assets. In the BL model, the joint distribution of A is taken to be multivariate normal, that is,

$$A \sim N(\mu, \Sigma)$$

The model then considers incorporating an analyst's views into the estimation of the market mean μ. If one considers

- μ to be a random variable which is itself normally distributed, and that
- its dispersion is proportional to that of the market,

then

$$\mu \sim N(\pi, \tau\Sigma)$$

where π is some parameter that may be determined by the analyst by some established procedure. On this point, Black and Litterman proposed (based on equilibrium considerations) that this should be obtainable from the intercepts of the capital asset pricing model.

Next, upon the consideration that the analyst has certain *subjective* views on the actual mean of the return for the holding period, this part of the BL model may allow the analyst to include *personal* views. BL suggested that such views should best be made as linear combinations, namely, *portfolios*, of the asset return variable mean μ:

- Each such personal view may be allocated a certain mean-and-error, (μ_i, ϵ_i), so that a typical view would take the form

$$p_{i1}\mu_1 + p_{i2}\mu_2 + p_{i3}\mu_3 + \cdots + p_{ij}\mu_j + \cdots + p_{in}\mu_n = q_i + \epsilon_i$$

where $\epsilon_i \sim N(0, \sigma_i^2)$.

The standard deviations σ_i^2 of each view may be assumed to control the confidence in each. Expressing these views in the form of a matrix, call the "pick" matrix, one obtains the "general" view specification

$$P_\mu \sim N(\mu, \Omega)$$

in which Ω is the diagonal matrix $\text{diag}(\sigma_2^1, \sigma_2^2, \sigma_3^2, \ldots, \sigma_n^2)$. It may be shown, using Bayes' law, that the posterior distribution of the market mean conditional on these views is

$$\mu|_{q,\Omega} \sim N(\mu_{\text{BL}}, \Sigma_{\text{BL}}^\mu)$$

where

$$\mu_{\text{BL}} = \{(\tau\Sigma)^{-1} + P^T\Omega^{-1}P\}^{-1}\{(\tau\Sigma)^{-1}\pi + P^T\Omega^{-1}q\}$$

$$\Sigma_{\text{BL}}^\mu = \{(\tau\Sigma)^{-1} + P^T\Omega^{-1}P\}^{-1}$$

One may obtain the posterior distribution of the market by taking

$$A|_{q,\Omega} = \mu|_{q,\Omega} + Z, \quad \text{and} \quad Z \sim N(0, \Sigma)$$

independent of μ.

One may then obtain

$$E[A] = \mu_{\text{BL}}$$

and

$$\Sigma_{\text{BL}} = \Sigma + \Sigma_{\text{BL}}^\mu$$

6.4.3.1 A Worked Example Portfolio Optimization by the Black–Litterman Approach Using R

Portfolio optimization using the R *code* BLCOP *that illustrates the implementation of the foregoing* BL *approach.*

6.4.3.1.1 *Introduction to the* R *Package* BLCOP

The R package BLCOP is an implementation of the Black–Litterman (BL) and copula opinion pooling frameworks. These two methods are suitably combined in this package.

6.4.3.1.2 *Overview of the Black–Litterman model*

The Black–Litterman model was devised in 1992 by Fisher Black and Robert Litterman with the objective of creating a systematic method of specifying and then incorporating analyst/portfolio manager views into the estimation of market parameters.

Let

$$A = \{a_1, a_2, a_3, \ldots, a_n\}$$

be a set of random variables representing the returns of n assets. In the BL model

a) the joint distribution of A is *assumed* to be multivariate normal, that is,

$$A \sim N(\mu, \Sigma)$$

b) to incorporate an analyst's views into the estimation of the market mean μ, it is further *assumed* that one may take itself to be a random variable which is itself normally distributed, and moreover that its dispersion is proportional to that of the market. Thus,

$$\mu \sim N(\pi, \tau\Sigma)$$

where π is some underlying parameter that may be determined by the analyst using some established procedure. Black and Litterman proposed, from equilibrium considerations, that this might be obtained from the intercepts of the capital asset pricing model.

c) Next, the analyst forms *subjective* views on the actual mean of the returns for the holding period. This is the assumption of the model that allows the analyst, or portfolio manager, to include his or her views. Moreover, BL proposed that views should be made on linear combinations (i.e., portfolios) of the asset return variable means μ.

Each view would take the form of a "mean plus error". For example, a typical view should be expressed as follows:

$$p_{i1}\mu_1 + p_{i2}\mu_2 + p_{i3}\mu_3 + \cdots + p_{in}\mu_n = q_i + \varepsilon_i$$

where $\varepsilon_i = N(0, \sigma_i)$.

The standard deviations σ_i^2 of each view could be taken as controlling the confidence in each view. Collecting these views into a matrix, call the "pick" matrix, one obtains the "general" view specification:

$$P_\mu \sim N(\mu, \Omega)$$

where Ω is the diagonal matrix diag $(\sigma_1^2, \sigma_2^2, \sigma_3^2, \ldots, \sigma_n^2)$.

Based on Bayes' law, it may be shown that the posterior distribution of the market mean conditional on this approach is

$$\mu|_{q,\Omega} \sim N(\mu_{BL}, \Sigma_{BL}^\mu)$$

where

$$\mu_{BL} = \{(\tau\Sigma)^{-1} + p^T\Omega^{-1}P\}^{-1}\{(\tau\Sigma)^{-1}\pi + p^T\Omega^{-1}q\}$$
$$\Sigma_{BL}^\mu = \{(\tau\Sigma)^{-1} + p^T\Omega^{-1}P\}^{-1}$$

These ideas are implemented in the BLCOP package, as shown in the following worked example.

Worked Example No. 1:

In the R domain:

```
> install.packages("BLCOP")
Installing package into 'C:/Users/Bert/Documents/R/win-
library/3.2'
(as 'lib' is unspecified)
--- Please select a CRAN mirror for use in this session ---
```

A CRAN mirror is selected.

```
trying URL 'https://stat.ethz.ch/CRAN/bin/windows/contrib/
3.2/BLCOP_0.3.1.zip'
Content type 'application/zip' length 646041 bytes (630 KB)
downloaded 630 KB
package 'BLCOP' successfully unpacked and MD5 sums checked
The downloaded binary packages are in
C:\Users\Bert\AppData\Local\Temp\RtmpmMX9H9\downloaded_
packages
> library(BLCOP)
Loading required package: MASS
Loading required package: quadprog
Warning message:
package 'BLCOP' was built under R version 3.2.4
> ls("package:BLCOP")
 [1] "addBLViews"          "addCOPViews"
 [3] "assetSet"            "BLCOPOptions"
 [5] "BLPosterior"         "BLViews"
 [7] "CAPMList"            "confidences"
 [9] "confidences<-"       "COPPosterior"
```

```
[11] "COPViews"                "createBLViews"
[13] "createCOPViews"          "deleteViews"
[15] "densityPlots"            "distribution"
[17] "getPosteriorEstim"       "getPriorEstim"
[19] "monthlyReturns"          "mvdistribution"
[21] "newPMatrix"              "numSimulations"
[23] "optimalPortfolios"       "optimalPortfolios.fPort"
[25] "PMatrix"                 "PMatrix<-"
[27] "posteriorEst"            "posteriorFeasibility"
[29] "posteriorMeanCov"        "posteriorSimulations"
[31] "priorViews"              "qv<-"
[33] "runBLCOPTests"           "sampleFrom"
[35] "show"                    "sp500Returns"
[37] "updateBLViews"           "US13wTB"
[39] "viewMatrix"
> pickMatrix <- matrix(c(1/2, -1, 1/2, rep(0, 3)), nrow = 1, ncol
+ = 6)
> views <- BLViews(P = pickMatrix, q = 0.06, confidences = 100,
+ assetNames = colnames(monthlyReturns))
> views # Outputting:
1 : 0.5*IBM+-1*MS+0.5*DELL=0.06  + eps. Confidence: 100
>
> priorMeans <- rep(0, 6)
> priorVarcov <- cov.mve(monthlyReturns)$cov
>
> marketPosterior <- posteriorEst(views = views, sigma =
+ priorVarcov, mu = priorMeans, tau = 1/2)
>
> marketPosterior # Outputting:
Prior means:
 IBM MS DELL C  JPM  BAC
  0   0   0    0    0    0
Posterior means:
     IBM       MS       DELL       C        JPM       BAC
0.0040334203 -0.0083696450 0.0114088881 -0.0008696187
-0.0031865767 0.0008424655
Posterior covariance:
        IBM       MS       DELL      C       JPM       BAC
IBM 0.014968077 0.010859214 0.011769151 0.010298180
0.007407937 0.003690800
MS  0.010859214 0.018563403 0.015853209 0.010405993 0.011377177
0.004699313
```

```
DELL 0.011769151 0.015853209 0.034133394 0.009431736
0.011381346 0.006756110
C    0.010298180 0.010405993 0.009431736 0.010811254 0.008246582
0.004926015
JPM  0.007407937 0.011377177 0.011381346 0.008246582
0.012875001 0.006741935
BAC  0.003690800 0.004699313 0.006756110 0.004926015
0.006741935 0.007630011
>
> finViews <- matrix(ncol = 4, nrow = 1, dimnames = list(NULL,
+ c("C","JPM","BAC","MS")))
> finViews[,1:4] <- rep(1/4,4)
> views <- addBLViews(finViews, 0.15, 90, views)
> views
1 : 0.5*IBM+-1*MS+0.5*DELL=0.06 + eps. Confidence: 100
2 : 0.25*MS+0.25*C+0.25*JPM+0.25*BAC=0.15
   + eps. Confidence: 90
>
> marketPosterior <- BLPosterior(as.matrix(monthlyReturns),
+ views, tau = 1/2,
+ marketIndex = as.matrix(sp500Returns),riskFree =
+ as.matrix(US13wTB))
>
>
> BLPosterior # Output:
function (returns, views, tau = 1, marketIndex, riskFree = NULL,
   kappa = 0, covEstimator = "cov")
{
   covEstimator <- match.fun(covEstimator)
   alphaInfo <- CAPMList(returns, marketIndex, riskFree =
riskFree)
   post <- posteriorEst(views, tau = tau, mu = alphaInfo
[["alphas"]],
     sigma = unclass(covEstimator(returns)), kappa = kappa)
   post
}
<environment: namespace:BLCOP>
>
> marketPosterior <- BLPosterior(as.matrix(monthlyReturns),
+ views, tau = 1/2,
+ marketIndex = as.matrix(sp500Returns),riskFree =
+ as.matrix(US13wTB))
> marketPosterior
```

```
Prior means :
    IBM         MS       DELL         C         JPM         BAC
0.020883598 0.059548398 0.017010062 0.014492325 0.027365230 0.002829908
Posterior means :
    IBM         MS       DELL         C         JPM         BAC
  0.06344562  0.07195806  0.07777653   0.04030821   0.06884519  0.02592776
Posterior covariance :
        IBM       MS      DELL       C       JPM       BAC
IBM 0.021334221 0.010575532 0.012465444 0.008518356
0.010605748 0.005281807
MS   0.010575532 0.031231768 0.017034827 0.012704758 0.014532900
0.008023646
DELL 0.012465444 0.017034827 0.047250599 0.007386821
0.009352949 0.005086150
C    0.008518356 0.012704758 0.007386821 0.016267422 0.010968240
0.006365457
JPM  0.010605748 0.014532900 0.009352949 0.010968240
0.028181136 0.011716834
BAC 0.005281807 0.008023646 0.005086150 0.006365457
0.011716834 0.011199343
>
#
# Optimization of Portfolios :
# The fPortfolio package, in CRAN, of the Rmetrics project
# RmCTWu09, has many functionalities available for portfolio
# optimization. The approach of this package will be used
# hereinafter.
>
> install.packages("fPortfolio")
Installing package into 'C:/Users/Bert/Documents/R/win-
library/3.2'
(as 'lib' is unspecified)
trying URL 'https://stat.ethz.ch/CRAN/bin/windows/contrib/
3.2/fPortfolio_3011.81.zip'
Content type 'application/zip' length 1066509 bytes (1.0 MB)
downloaded 1.0 MB
package 'fPortfolio' successfully unpacked and MD5 sums checked
The downloaded binary packages are in
C:\Users\Bert\AppData\Local\Temp\RtmpmMX9H9\downloaded_
packages
> library(fPortfolio)
Loading required package: timeDate
Loading required package: timeSeries
```

```
Loading required package: fBasics
Rmetrics Package fBasics
Analysing Markets and calculating Basic Statistics
Copyright (C) 2005-2014 Rmetrics Association Zurich
Educational Software for Financial Engineering and
Computational
Science
Rmetrics is free software and comes with ABSOLUTELY NO
WARRANTY.
https://www.rmetrics.org --- Mail to: info@rmetrics.org
Loading required package: fAssets
Rmetrics Package fAssets
Analysing and Modeling Financial Assets
Copyright (C) 2005-2014 Rmetrics Association Zurich
Educational Software for Financial Engineering and
Computational
Science
Rmetrics is free software and comes with ABSOLUTELY NO
WARRANTY.
https://www.rmetrics.org --- Mail to: info@rmetrics.org
Rmetrics Package fPortfolio
Portfolio Optimization
Copyright (C) 2005-2014 Rmetrics Association Zurich
Educational Software for Financial Engineering and
Computational Science
Rmetrics is free software and comes with ABSOLUTELY NO
WARRANTY.
https://www.rmetrics.org --- Mail to: info@rmetrics.org
Warning messages:
1: package 'fPortfolio' was built under R version 3.2.4
2: package 'timeDate' was built under R version 3.2.3
3: package 'timeSeries' was built under R version 3.2.3
4: package 'fBasics' was built under R version 3.2.3
5: package 'fAssets' was built under R version 3.2.3
> ls("package:fPortfolio") # Output:
  [1] "addRainbow"
  [2] "amplDataAdd"
  [3] "amplDataAddMatrix"
  [4] "amplDataAddValue"
  [5] "amplDataAddVector"
  [6] "amplDataOpen"
  [7] "amplDataSemicolon"
  [8] "amplDataShow"
```

```
 [9] "amplLP"
[10] "amplLPControl"
[11] "amplModelAdd"
[12] "amplModelOpen"
[13] "amplModelShow"
[14] "amplNLP"
[15] "amplNLPControl"
[16] "amplOutShow"
[17] "amplQP"
[18] "amplQPControl"
[19] "amplRunAdd"
[20] "amplRunOpen"
[21] "amplRunShow"
[22] "backtestAssetsPlot"
[23] "backtestDrawdownPlot"
[24] "backtestPlot"
[25] "backtestPortfolioPlot"
[26] "backtestRebalancePlot"
[27] "backtestReportPlot"
[28] "backtestStats"
[29] "backtestWeightsPlot"
[30] "bcpAnalytics"
[31] "bestDiversification"
[32] "budgetsModifiedES"
[33] "budgetsModifiedVAR"
[34] "budgetsNormalES"
[35] "budgetsNormalVAR"
[36] "budgetsSampleCOV"
[37] "cmlLines"
[38] "cmlPoints"
[39] "covEstimator"
[40] "covMcdEstimator"
[41] "covOGKEstimator"
[42] "covRisk"
[43] "covRiskBudgetsLinePlot"
[44] "covRiskBudgetsPie"
[45] "covRiskBudgetsPlot"
[46] "cvarRisk"
[47] "Data"
[48] "donlp2NLP"
[49] "donlp2NLPControl"
[50] "drawdownsAnalytics"
[51] "ECON85"
```

```
[52] "ECON85LONG"
[53] "efficientPortfolio"
[54] "emaSmoother"
[55] "eqsumWConstraints"
[56] "equalWeightsPoints"
[57] "equidistWindows"
[58] "feasibleGrid"
[59] "feasiblePortfolio"
[60] "frontierPlot"
[61] "frontierPlotControl"
[62] "frontierPoints"
[63] "garchAnalytics"
[64] "GCCINDEX"
[65] "GCCINDEX.RET"
[66] "getA"
[67] "getA.fPFOLIOSPEC"
[68] "getA.fPORTFOLIO"
[69] "getAlpha"
[70] "getAlpha.fPFOLIOSPEC"
[71] "getAlpha.fPFOLIOVAL"
[72] "getAlpha.fPORTFOLIO"
[73] "getConstraints"
[74] "getConstraints.fPORTFOLIO"
[75] "getConstraintsTypes"
[76] "getControl"
[77] "getControl.fPFOLIOSPEC"
[78] "getControl.fPORTFOLIO"
[79] "getCov"
[80] "getCov.fPFOLIODATA"
[81] "getCov.fPORTFOLIO"
[82] "getCovRiskBudgets"
[83] "getCovRiskBudgets.fPFOLIOVAL"
[84] "getCovRiskBudgets.fPORTFOLIO"
[85] "getData"
[86] "getData.fPFOLIODATA"
[87] "getData.fPORTFOLIO"
[88] "getEstimator"
[89] "getEstimator.fPFOLIODATA"
[90] "getEstimator.fPFOLIOSPEC"
[91] "getEstimator.fPORTFOLIO"
[92] "getMean"
[93] "getMean.fPFOLIODATA"
[94] "getMean.fPORTFOLIO"
```

```
[95]  "getMessages"
[96]  "getMessages.fPFOLIOBACKTEST"
[97]  "getMessages.fPFOLIOSPEC"
[98]  "getModel.fPFOLIOSPEC"
[99]  "getModel.fPORTFOLIO"
[100] "getMu"
[101] "getMu.fPFOLIODATA"
[102] "getMu.fPORTFOLIO"
[103] "getNAssets"
[104] "getNAssets.fPFOLIODATA"
[105] "getNAssets.fPORTFOLIO"
[106] "getNFrontierPoints"
[107] "getNFrontierPoints.fPFOLIOSPEC"
[108] "getNFrontierPoints.fPFOLIOVAL"
[109] "getNFrontierPoints.fPORTFOLIO"
[110] "getObjective"
[111] "getObjective.fPFOLIOSPEC"
[112] "getObjective.fPORTFOLIO"
[113] "getOptim"
[114] "getOptim.fPFOLIOSPEC"
[115] "getOptim.fPORTFOLIO"
[116] "getOptimize"
[117] "getOptimize.fPFOLIOSPEC"
[118] "getOptimize.fPORTFOLIO"
[119] "getOptions"
[120] "getOptions.fPFOLIOSPEC"
[121] "getOptions.fPORTFOLIO"
[122] "getParams"
[123] "getParams.fPFOLIOSPEC"
[124] "getParams.fPORTFOLIO"
[125] "getPortfolio"
[126] "getPortfolio.fPFOLIOSPEC"
[127] "getPortfolio.fPFOLIOVAL"
[128] "getPortfolio.fPORTFOLIO"
[129] "getRiskFreeRate"
[130] "getRiskFreeRate.fPFOLIOSPEC"
[131] "getRiskFreeRate.fPFOLIOVAL"
[132] "getRiskFreeRate.fPORTFOLIO"
[133] "getSeries"
[134] "getSeries.fPFOLIODATA"
[135] "getSeries.fPORTFOLIO"
[136] "getSigma"
[137] "getSigma.fPFOLIODATA"
```

```
[138] "getSigma.fPORTFOLIO"
[139] "getSmoother"
[140] "getSmoother.fPFOLIOBACKTEST"
[141] "getSmootherDoubleSmoothing"
[142] "getSmootherDoubleSmoothing.fPFOLIOBACKTEST"
[143] "getSmootherFun"
[144] "getSmootherFun.fPFOLIOBACKTEST"
[145] "getSmootherInitialWeights"
[146] "getSmootherInitialWeights.fPFOLIOBACKTEST"
[147] "getSmootherLambda"
[148] "getSmootherLambda.fPFOLIOBACKTEST"
[149] "getSmootherParams"
[150] "getSmootherParams.fPFOLIOBACKTEST"
[151] "getSmootherSkip"
[152] "getSmootherSkip.fPFOLIOBACKTEST"
[153] "getSolver"
[154] "getSolver.fPFOLIOSPEC"
[155] "getSolver.fPORTFOLIO"
[156] "getSpec"
[157] "getSpec.fPORTFOLIO"
[158] "getStatistics"
[159] "getStatistics.fPFOLIODATA"
[160] "getStatistics.fPORTFOLIO"
[161] "getStatus"
[162] "getStatus.fPFOLIOSPEC"
[163] "getStatus.fPFOLIOVAL"
[164] "getStatus.fPORTFOLIO"
[165] "getStrategy"
[166] "getStrategy.fPFOLIOBACKTEST"
[167] "getStrategyFun"
[168] "getStrategyFun.fPFOLIOBACKTEST"
[169] "getStrategyParams"
[170] "getStrategyParams.fPFOLIOBACKTEST"
[171] "getTailRisk"
[172] "getTailRisk.fPFOLIODATA"
[173] "getTailRisk.fPFOLIOSPEC"
[174] "getTailRisk.fPORTFOLIO"
[175] "getTailRiskBudgets"
[176] "getTailRiskBudgets.fPORTFOLIO"
[177] "getTargetReturn"
[178] "getTargetReturn.fPFOLIOSPEC"
[179] "getTargetReturn.fPFOLIOVAL"
[180] "getTargetReturn.fPORTFOLIO"
```

```
[181] "getTargetRisk"
[182] "getTargetRisk.fPFOLIOSPEC"
[183] "getTargetRisk.fPFOLIOVAL"
[184] "getTargetRisk.fPORTFOLIO"
[185] "getTrace"
[186] "getTrace.fPFOLIOSPEC"
[187] "getTrace.fPORTFOLIO"
[188] "getType"
[189] "getType.fPFOLIOSPEC"
[190] "getType.fPORTFOLIO"
[191] "getUnits.fPFOLIODATA"
[192] "getUnits.fPORTFOLIO"
[193] "getWeights"
[194] "getWeights.fPFOLIOSPEC"
[195] "getWeights.fPFOLIOVAL"
[196] "getWeights.fPORTFOLIO"
[197] "getWindows"
[198] "getWindows.fPFOLIOBACKTEST"
[199] "getWindowsFun"
[200] "getWindowsFun.fPFOLIOBACKTEST"
[201] "getWindowsHorizon"
[202] "getWindowsHorizon.fPFOLIOBACKTEST"
[203] "getWindowsParams"
[204] "getWindowsParams.fPFOLIOBACKTEST"
[205] "glpkLP"
[206] "glpkLPControl"
[207] "ipopQP"
[208] "ipopQPControl"
[209] "kendallEstimator"
[210] "kestrelQP"
[211] "kestrelQPControl"
[212] "lambdaCVaR"
[213] "listFConstraints"
[214] "lpmEstimator"
[215] "LPP2005"
[216] "LPP2005.RET"
[217] "markowitzHull"
[218] "maxBConstraints"
[219] "maxBuyinConstraints"
[220] "maxCardConstraints"
[221] "maxddMap"
[222] "maxFConstraints"
[223] "maxratioPortfolio"
[224] "maxreturnPortfolio"
```

```
[225] "maxsumWConstraints"
[226] "maxWConstraints"
[227] "mcdEstimator"
[228] "minBConstraints"
[229] "minBuyinConstraints"
[230] "minCardConstraints"
[231] "minFConstraints"
[232] "minriskPortfolio"
[233] "minsumWConstraints"
[234] "minvariancePoints"
[235] "minvariancePortfolio"
[236] "minWConstraints"
[237] "modifiedVaR"
[238] "monteCarloPoints"
[239] "mveEstimator"
[240] "nCardConstraints"
[241] "neosLP"
[242] "neosLPControl"
[243] "neosQP"
[244] "neosQPControl"
[245] "netPerformance"
[246] "nlminb2NLP"
[247] "nlminb2NLPControl"
[248] "nnveEstimator"
[249] "normalVaR"
[250] "parAnalytics"
[251] "pcoutAnalytics"
[252] "pfolioCVaR"
[253] "pfolioCVaRplus"
[254] "pfolioHist"
[255] "pfolioMaxLoss"
[256] "pfolioReturn"
[257] "pfolioTargetReturn"
[258] "pfolioTargetRisk"
[259] "pfolioVaR"
[260] "plot.fPORTFOLIO"
[261] "portfolioBacktest"
[262] "portfolioBacktesting"
[263] "portfolioConstraints"
[264] "portfolioData"
[265] "portfolioFrontier"
[266] "portfolioObjective"
[267] "portfolioReturn"
[268] "portfolioRisk"
```

```
[269] "portfolioSmoothing"
[270] "portfolioSpec"
[271] "print.solver"
[272] "quadprogQP"
[273] "quadprogQPControl"
[274] "ramplLP"
[275] "ramplNLP"
[276] "ramplQP"
[277] "rdonlp2"
[278] "rdonlp2NLP"
[279] "rglpkLP"
[280] "ripop"
[281] "ripopQP"
[282] "riskBudgetsPlot"
[283] "riskMap"
[284] "riskmetricsAnalytics"
[285] "riskSurface"
[286] "rkestrelQP"
[287] "rneosLP"
[288] "rneosQP"
[289] "rnlminb2"
[290] "rnlminb2NLP"
[291] "rollingCDaR"
[292] "rollingCmlPortfolio"
[293] "rollingCVaR"
[294] "rollingDaR"
[295] "rollingMinvariancePortfolio"
[296] "rollingPortfolioFrontier"
[297] "rollingSigma"
[298] "rollingTangencyPortfolio"
[299] "rollingVaR"
[300] "rollingWindows"
[301] "rquadprog"
[302] "rquadprogQP"
[303] "rsolnpNLP"
[304] "rsolveLP"
[305] "rsolveQP"
[306] "rsymphonyLP"
[307] "sampleCOV"
[308] "sampleVaR"
[309] "setAlpha<-"
[310] "setEstimator<-"
[311] "setNFrontierPoints<-"
[312] "setObjective<-"
```

```
[313] "setOptimize<-"
[314] "setParams<-"
[315] "setRiskFreeRate<-"
[316] "setSmootherDoubleSmoothing<-"
[317] "setSmootherFun<-"
[318] "setSmootherInitialWeights<-"
[319] "setSmootherLambda<-"
[320] "setSmootherParams<-"
[321] "setSmootherSkip<-"
[322] "setSolver<-"
[323] "setStatus<-"
[324] "setStrategyFun<-"
[325] "setStrategyParams<-"
[326] "setTailRisk<-"
[327] "setTargetReturn<-"
[328] "setTargetRisk<-"
[329] "setTrace<-"
[330] "setType<-"
[331] "setWeights<-"
[332] "setWindowsFun<-"
[333] "setWindowsHorizon<-"
[334] "setWindowsParams<-"
[335] "sharpeRatioLines"
[336] "shrinkEstimator"
[337] "singleAssetPoints"
[338] "slpmEstimator"
[339] "SMALLCAP"
[340] "SMALLCAP.RET"
[341] "solnpNLP"
[342] "solnpNLPControl"
[343] "solveRampl.CVAR"
[344] "solveRampl.MV"
[345] "solveRdonlp2"
[346] "solveRglpk.CVAR"
[347] "solveRglpk.MAD"
[348] "solveRipop"
[349] "solveRquadprog"
[350] "solveRquadprog.CLA"
[351] "solveRshortExact"
[352] "solveRsocp"
[353] "solveRsolnp"
[354] "spearmanEstimator"
[355] "SPISECTOR"
[356] "SPISECTOR.RET"
```

```
[357] "stabilityAnalytics"
[358] "summary.fPORTFOLIO"
[359] "surfacePlot"
[360] "SWX"
[361] "SWX.RET"
[362] "symphonyLP"
[363] "symphonyLPControl"
[364] "tailoredFrontierPlot"
[365] "tailRiskBudgetsPie"
[366] "tailRiskBudgetsPlot"
[367] "tangencyLines"
[368] "tangencyPoints"
[369] "tangencyPortfolio"
[370] "tangencyStrategy"
[371] "ternaryCoord"
[372] "ternaryFrontier"
[373] "ternaryMap"
[374] "ternaryPoints"
[375] "ternaryWeights"
[376] "turnsAnalytics"
[377] "twoAssetsLines"
[378] "varRisk"
[379] "waveletSpectrum"
[380] "weightedReturnsLinePlot"
[381] "weightedReturnsPie"
[382] "weightedReturnsPlot"
[383] "weightsLinePlot"
[384] "weightsPie"
[385] "weightsPlot"
[386] "weightsSlider"
>
> optPorts <- optimalPortfolios.fPort(marketPosterior, optimizer
+ = "tangencyPortfolio")
> optPorts # Output:
$priorOptimPortfolio
Title:
MV Tangency Portfolio
 Estimator:    getPriorEstim
 Solver:       solveRquadprog
 Optimize:     minRisk
 Constraints:  LongOnly
Portfolio Weights:
  IBM   MS DELL   C  JPM  BAC
```

IBM = International Business Machines Corporation
MS = Morgan Stanley
DELL = Dell Computers
C = Citigroup Incorporated
JPM = J P Morgan Chase and Company
BAC = Bank of America Corporation
0.0765 0.9235 0.0000 0.0000 0.0000 0.0000
Covariance Risk Budgets:
 IBM MS DELL   C JPM BAC
Target Returns and Risks:
 mean   mu Cov Sigma CVaR VaR
0.0000 0.0566   0.1460 0.0000 0.0000
Description:
 Wed Mar 23 07:38:59 2016 by user: Bert
$posteriorOptimPortfolio
Title:
 MV Tangency Portfolio
 Estimator: getPosteriorEstim
 Solver: solveRquadprog
 Optimize: minRisk
 Constraints: LongOnly
Portfolio Weights:
  IBM   MS   DELL   C   JPM   BAC
0.3633 0.1966 0.1622 0.0000 0.2779 0.0000

Covariance Risk Budgets:
 IBM MS DELL   C JPM BAC

Target Returns and Risks:
 mean   mu Cov Sigma CVaR VaR
0.0000 0.0689   0.1268 0.0000 0.0000

Description:
 Wed Mar 23 07:38:59 2016 by user: Bert

attr(,"class")
[1] "BLOptimPortfolios"
>
Portfolio Weights:
  IBM   MS   DELL   C   JPM   BAC
The fPortfolio package of the Rmetrics project ([RmCTWu09]), for
example, has a rich set of functionality available for portfolio
optimization. Their 0.0765 0.9235 0.0000 0.0000 0.0000 0.0000

```
#
# The selected portfolio consists of the following investments:
#
# IBM = International Business Machine
# MS = Morgan Stanley
# DELL = Dell Computer
# C = Citigroup Inc.
# JPM = JPMorgan Chase and Company
# BAC = Bank of America Corporation
#
Covariance Risk Budgets:
 IBM  MS  DELL  C  JPM  BAC

Target Returns and Risks:
 mean    mu  Cov Sigma  CVaR  VaR
0.0000 0.0566    0.1460 0.0000 0.0000

Description:
 Tue Mar 22 08:31:47 2016 by user: Bert

$posteriorOptimPortfolio
Title:
 MV Tangency Portfolio
 Estimator:     getPosteriorEstim
 Solver:        solveRquadprog
 Optimize:      minRisk
 Constraints:   LongOnly

Portfolio Weights:
    IBM      MS     DELL      C      JPM      BAC
 0.3633  0.1966  0.1622  0.0000  0.2779  0.0000

Covariance Risk Budgets:
 IBM  MS  DELL  C  JPM  BAC

Target Returns and Risks:
 mean    mu  Cov Sigma  CVaR  VaR
0.0000 0.0689    0.1268 0.0000 0.0000

Description:
 Tue Mar 22 08:31:47 2016 by user: Bert

attr(,"class")
```

Figure 6.4(a) BL-1: Preparing space for following plots.

Weights

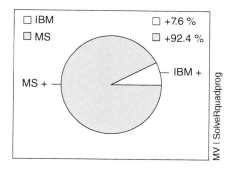

Figure 6.4(b) BL-2: Output of `weights pie (optPorts $priorOptimPortfolio)`.

```
[1] "BLOptimPortfolios"
> par(mfcol = c(2, 1))
> # Output: Figure 6.4a:  BL-1
>
(Providing space for following graphical outputs)
> weightsPie(optPorts$priorOptimPortfolio)
> # Output: Figure 6.4b:  BL-2
>
A piechart of the Relative Distributions of IBM and MS in the
Optimum Portfolio
>
> weightsPie(optPort8s$posteriorOptimPortfolio)
> # Output: Figure 6.5: BL-3
>
> optPorts2 <- optimalPortfolios.fPort(marketPosterior,
```

Weights   Figure 6.5 BL-3.

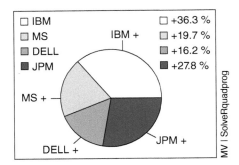

```
+ constraints = "minW[1:6]=0.1", optimizer =
"minriskPortfolio")
> optPorts2 # Output:
$priorOptimPortfolio
Title:
MV Minimum Risk Portfolio
 Estimator:      getPriorEstim
 Solver:         solveRquadprog
 Optimize:       minRisk
 Constraints:    minW

Portfolio Weights:
   IBM     MS     DELL    C     JPM     BAC
0.1137 0.1000 0.1000 0.1098 0.1000 0.4764

Covariance Risk Budgets:
 IBM  MS  DELL  C  JPM  BAC

Target Returns and Risks:
 mean      mu   Cov  Sigma   CVaR   VaR
0.0000 0.0157     0.0864 0.0000 0.0000

Description:
 Tue Mar 22 12:12:53 2016 by user: Bert

$posteriorOptimPortfolio
Title:
MV Minimum Risk Portfolio
 Estimator:      getPosteriorEstim
 Solver:         solveRquadprog
 Optimize:       minRisk
 Constraints:    minW
```

Portfolio Weights:
```
   IBM    MS    DELL    C     JPM     BAC
0.1000 0.1000 0.1000 0.1326 0.1000 0.4674
```

Covariance Risk Budgets:
```
 IBM  MS  DELL  C  JPM  BAC
```

Target Returns and Risks:
```
 mean    mu   Cov Sigma CVaR   VaR
0.0000 0.0457      0.1008 0.0000 0.0000
```

Description:
```
 Tue Mar 22 12:12:53 2016 by user: Bert
```

```
attr(,"class")
[1] "BLOptimPortfolios"
>
> densityPlots(marketPosterior, assetsSel = "JPM")
> # Output: Figure 6.6 BL-4
> dispersion c(.376,.253,.360,.333,.360,.600,.397,.396,.578,
+.775) / 1000
> sigma <- BLCOP:::.symmetricMatrix(dispersion, dim = 4)
> caps <- rep(1/4, 4)
> mu <- 2.5 * sigma %*% caps
> dim(mu) <- NULL
> marketDistribution <- mvdistribution("mt", mean = mu, S =
+ sigma, df = 5)
> class(marketDistribution)
> pick <- matrix(0, ncol = 4, nrow = 1,
> dimnames = list(NULL, c("SP", "FTSE", "CAC", "DAX")))
> pick[1,"DAX"] <- 1
> viewDist <- list(distribution("unif", min = -0.02, max = 0))
> views <- COPViews(pick, viewDist = viewDist, confidences =
+ 0.2, assetNames = c("SP", "FTSE", "CAC", "DAX"))
```

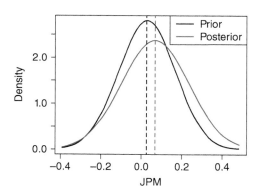

Figure 6.6 BL-4.

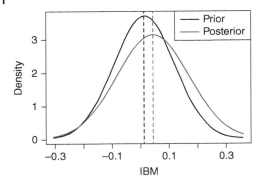

Figure 6.7 BL-5: Density Plot.

```
> newPick <- matrix(0, 1, 2)
> dimnames(newPick) <- list(NULL, c("SP", "FTSE"))
> newPick[1,] <- c(1, -1) # add a relative view
> views <- addCOPViews(newPick,
> list(distribution("norm", mean = 0.05, sd = 0.02)), 0.5, views)
> marketPosterior <- COPPosterior(marketDistribution, views,
+ numSimulations = 50000)
> densityPlots(marketPosterior, assetsSel = 4)
> # Output: Figure 6.7 BL-5
```

6.4.4 More Worked Examples of Portfolio Optimization Using R

6.4.4.1 Worked Examples of Portfolio Optimization – No. 1 Portfolio Optimization by `PerformanceAnalytics` in CRAN

The R package `PerformanceAnalytics`, is a collection of econometric functions for performance and risk analysis, designed to analyze nonnormal return streams in asset allocation and portfolio optimization. It is mostly tested on return data, rather than price, but most functions will work with irregular return data as well, and increasing numbers of functions will work with price data where possible.

`PortfolioAnalytics`-**package**

Numeric Methods for Optimization of Portfolios

Description `PortfolioAnalytics` is an R package providing numerical solutions for portfolio problems with complex constraints and objective sets. The goal of the package is to aid practitioners and researchers in solving portfolio optimization problems with complex constraints and objectives that mirror real-world applications.

One of the goals of the packages is to provide a common interface to specify constraints and objectives that can be solved by any supported solver (i.e., optimization method). Currently, supported optimization methods include the following:

- Random portfolios
- Differential evolution
- Particle swarm optimization
- Generalized simulated annealing
- Linear and quadratic programming routines

The solver can be specified with the optimize_method argument in optimize. portfolio and optimize.portfolio.rebalancing. The optimize_method argument must be one of random, DEoptim, pso, GenSA, ROI, quadprog, glpk, or symphony.

Additional information on random portfolios is provided further. The differential evolution algorithm is implemented via the DEoptim package, the particle swarm optimization algorithm via the pso package, the generalized simulated annealing via the GenSA package, and linear and quadratic programming are implemented via the ROI package that acts as an interface to the Rglpk, Rsymphony.

Worked Example No. 2

PortfolioAnalytics-package
Numeric methods for optimization of portfolios
indexes   **Six Major Economic Indexes**

Description
Monthly data of five indexes beginning on 2000 January, 31 and ending 2009 December, 31.

The following are the five financial indexes, together with inflation factors:

1) U.S. bonds
2) U.S. equities
3) International equities
4) Commodities
5) U.S. T-Bills
6) Inflation factors

Usage

```
data(indexes)
```

Format
CSV converted into xts object with monthly observations.

Examples

```
data(indexes)
#preview the data
```

```
head(indexes)
#summary period statistics
summary(indexes)
```

In the R domain

```
>
> data(indexes)
>
> #preview the data
> head(indexes) # Output:
US Bonds  US Equities   Int'l Equities Commodities  US T-bill
Inflation
1980-01-31 -0.0272    0.0610     0.0462    0.0568  0.0104    0.0149
1980-02-29 -0.0669    0.0031    -0.0040   -0.0093  0.0106    0.0146
1980-03-31  0.0053   -0.0987    -0.1188   -0.1625  0.0121    0.0120
1980-04-30  0.0992    0.0429     0.0864    0.0357  0.0137    0.0095
1980-05-31  0.0000    0.0562     0.0446    0.0573  0.0106    0.0095
1980-06-30  0.0605    0.0296     0.0600    0.0533  0.0066    0.0000
>
> #summary period statistics
> summary(indexes) # Output:
>

          Index          US Bonds          US Equities
Min.    :1980-01-31   Min.    :-0.066900   Min.    :-0.215200
1st Qu.:1987-07-23   1st Qu.:-0.000550   1st Qu.:-0.016000
Median :1995-01-15   Median: 0.005850   Median : 0.012950
Mean    :1995-01-14   Mean    : 0.006825   Mean    : 0.008546
3rd Qu.:2002-07-07   3rd Qu.: 0.014025   3rd Qu.: 0.037625
Max.    :2009-12-31   Max.    : 0.128800   Max.    : 0.134300

  Int'l Equities      Commodities        US Tbill           Inflation
Min.    :-0.25270  Min.    :-0.392700  Min. N
```

6.4.4.2 Worked Example for Portfolio Optimization – No. 2 Portfolio Optimization using the R code DEoptim

Worked Example No. 3

DEoptim was published on the CRAN website (The Comprehensive R Archive Network) as the package DEoptim:Global Optimization by Differential Evolution:

Version: 2.2-3

Suggests: foreach, iterators, parallel

Published:	2015-01-09
Author:	David Ardia [aut], Katharine Mullen [aut, cre], Brian Peterson [aut], Joshua Ulrich [aut], Kris Boudt [ctb]
Maintainer:	Katharine Mullen <mullenkate@gmail.com>
BugReports:	NA
License:	GPL-2 | GPL-3 [expanded from: GPL (≥ 2)]
URL:	NA
NeedsCompilation:	Yes
Citation:	DEoptim citation info
Materials:	README NEWS
In views:	Optimization
CRAN checks:	DEoptim results

It implements the differential evolution algorithm for global optimization of a real-valued function of a real-valued parameter vector.

The following is a detailed description of this R code:

DEoptim **Differential Evolution Optimization**

Description

Performs evolutionary global optimization via the *differential evolution* algorithm.

Usage

```
DEoptim(fn, lower, upper, control = DEoptim.control(), ...,
    fnMap=NULL)
```

Arguments

fn   the function to be optimized (minimized). The function should have as its first argument the vector of real-valued parameters to optimize, and return a scalar real result. NA and NaN values are not allowed.

lower, upper   two vectors specifying scalar real lower and upper bounds on each parameter to be optimized, so that the i-th element of lower and upper applies to the i-th

> parameter. The implementation searches between lower and upper for the global optimum/minimum of fn.

control   a list of control parameters; see DEoptim.control.

fnMap   an optional function that will be run after each population is created, but before the population is passed to the objective function. This allows the user to impose integer/cardinality constriants.

...   further arguments to be passed to fn.

Details

DEoptim performs optimization (minimization) of fn.

The control argument is a list; see the help file for DEoptim.control for details.

The R implementation of differential evolution (DE):

DEoptim, was first published on the Comprehensive R Archive Network (CRAN) in 2005 by David Ardia. Early versions were written in pure R. Since version 2.0-0 (published in CRAN in 2009), the package has relied on an interface to a C implementation of DE, which is significantly faster for most problems as compared to the implementation in pure R. The C interface is in many respects similar to the MS Visual C++ v5.0 implementation of the differential evolution algorithm distributed with the book Differential Evolution: "A Practical Approach to Global Optimization" by Price et al. and may be available at http://www.icsi.berkeley.edu/~storn/. Since version 2.0-3, the C implementation dynamically allocates the memory required to store the population, removing limitations on the number of members in the population and length of the parameter vectors that may be optimized. Since version 2.2-0, the package allows for parallel operation, so that the evaluations of the objective function may be performed using all available cores. This is accomplished using either the built-in parallel package or the foreach package. If parallel operation is desired, the user should set parallelType and make sure that the arguments and packages needed by the objective function are available; see DEoptim.control, the example below and examples in the sandbox directory for details.

Since becoming publicly available, the package DEoptim has been used to solve optimization problems arising in diverse domains.

- To perform a maximization (instead of minimization) of a given function, simply define a new function that is the opposite of the function to maximize and apply DEoptim to it.
- To integrate additional constraints (other than box constraints) on the parameter x of fn(x), for instance $x[1] + x[2]^2 < 2$, integrate the constraint within the function to optimize, for instance

```
fn <- function(x) {
if (x[1] + x[2]^2 >= 2) {
r <- Inf
```

```
else{ ... }
return(r)
}
```

This simplistic strategy usually does not work all that well for gradient-based or Newton-type methods. It is likely to be acceptable when the solution is in the interior of the feasible region, but when the solution is on the boundary, optimization algorithm would have a difficult time converging.

Furthermore, when the solution is on the boundary, this strategy would make the algorithm converge to an inferior solution in the interior. However, for methods such as DE, which are not gradient based, this strategy might not be that bad.

Note that DEoptim stops if any NA or NaN value is obtained. Under these conditions: redefine the function to handle these values (for instance, set NA to Inf in ones objective function).

It is important to emphasize that the result of DEoptim is a random variable, that is, different results may be obtained when the algorithm is run repeatedly with the same settings. Hence, the user should set the random seed if they want to reproduce the results, for example, by setting set.seed(1234) before the call of DEoptim.

DEoptim *depends on repeated evaluation of the objective function in order to move the population toward a global minimum.* Users interested in making DEoptim run as fast as possible should consider using the package in parallel mode (so that all CPUs available are used), and also ensure that evaluation of the objective function is as efficient as possible (e.g., by using vectorization in pure R code, or writing parts of the objective function in a lower level language like C or Fortran).

Further details and examples of the R package DEoptim can be found in Mullen et al. (2011) and Ardia et al. (2011a, 2011b) or look at the package's vignette by typing vignette("DEoptim").

Also, an illustration of the package usage for a high-dimensional nonlinear portfolio optimization problem is available by typing vignette("DEoptim PortfolioOptimization").

Remarks:

A) Differential evolution is a search heuristic code introduced by Storn and Price (1997). Its remarkable performance as a global optimization algorithm on continuous numerical minimization problems has been extensively explored. DE belongs to the class of genetic algorithms that use biology-inspired operations of crossover, mutation, and selection on a population in order to minimize an objective function over the course of successive generations.

B) As with other evolutionary algorithms, DE solves optimization problems by evolving a population of candidate solutions using alteration and selection operators. DE uses floating-point instead of bit string encoding of

population members, and arithmetic operations instead of logical operations in mutation. DE is particularly well suited to find the global optimum
of a real-valued function of real-valued parameters, and does not require
that the function be either continuous or differentiable.

Examples

```
## Rosenbrock Banana function
## The function has a global minimum f(x) = 0 at the point
## (1,1).
## Note that the vector of parameters to be optimized must be ##
the first
## argument of the objective function passed to DEoptim.
Rosenbrock <- function(x){
x1 <- x[1]
x2 <- x[2]
100 * (x2 - x1 * x1)^2 + (1 - x1)^2
}
## DEoptim searches for minima of the objective function
## between lower and upper bounds on each parameter to be ##
optimized. Therefore in the call to DEoptim we specify vectors ##
that comprise the lower and upper bounds; these vectors are ## the
same length as the parameter vector.
lower <- c(-10,-10)
upper <- -lower
## run DEoptim and set a seed first for replicability
set.seed(1234)
DEoptim(Rosenbrock, lower, upper)
## increase the population size
DEoptim(Rosenbrock, lower, upper, DEoptim.control(NP = 100))
## change other settings and store the output
outDEoptim <- DEoptim(Rosenbrock, lower, upper, DEoptim.control
(NP = 80,
itermax = 400, F = 1.2, CR = 0.7))
## plot the output
plot(outDEoptim)
## 'Wild' function, global minimum at about -15.81515
Wild <- function(x)
DEoptim-methods 7
10 * sin(0.3 * x) * sin(1.3 * x^2) +
0.00001 * x^4 + 0.2 * x + 80
plot(Wild, -50, 50, n = 1000, main = "'Wild function'")
outDEoptim <- DEoptim(Wild, lower = -50, upper = 50,
```

```
control = DEoptim.control(trace = FALSE))
plot(outDEoptim)
DEoptim(Wild, lower = -50, upper = 50,
control = DEoptim.control(NP = 50))
## The below examples shows how the call to DEoptim can be
## parallelized.
## Note that if your objective function requires packages to be
## loaded or has arguments supplied via \code{...}, these
## should be specified using the \code{packages} and
## \code{parVar} arguments in control.
## Not run: Genrose <- function(x) {
## One generalization of the Rosenbrock banana valley function ##
(n parameters)
## n <- length(x)
## make it take some time ...
Sys.sleep(.001)
1.0 + sum (100 * (x[-n]^2 - x[-1])^2 + (x[-1] - 1)^2)
}
# get some run-time on simple problems
maxIt <- 250
n <- 5
oneCore <- system.time( DEoptim(fn=Genrose, lower=rep(-25, n),
upper=rep(25, n),
control=list(NP=10*n, itermax=maxIt)))
withParallel <- system.time( DEoptim(fn=Genrose, lower=rep
(-25, n), upper=rep(25, n),
control=list(NP=10*n, itermax=maxIt, parallelType=1)))
## Compare timings
(oneCore)
(withParallel)
## End(Not run)
```

In the R domain:

```
>
> install.packages("DEoptim")
Installing package into 'C:/Users/Bert/Documents/R/win-
library/3.2'
(as 'lib' is unspecified)
--- Please select a CRAN mirror for use in this session ---
```

```
# A CRAN mirror is selected.

trying URL
'https://cran.cnr.berkeley.edu/bin/windows/contrib/3.2/
DEoptim_2.2-3.zip'
Content type 'application/zip' length 633185 bytes (618 KB)
downloaded 618 KB
package 'DEoptim' successfully unpacked and MD5 sums checked
The downloaded binary packages are in
C:\Users\Bert\AppData\Local\Temp\RtmpIj26Xj\downloaded_pac
kages
> library(DEoptim)
DEoptim package
Differential Evolution algorithm in R
Authors: D. Ardia, K. Mullen, B. Peterson and J. Ulrich
Warning message:
package 'DEoptim' was built under R version 3.2.5
> ls("package:DEoptim")
[1] "DEoptim"     "DEoptim.control"
> ## Rosenbrock Banana function> ## The function has a global
minimum f(x) = 0 at the point
> ## (1,1).
> ## Note that the vector of parameters to be optimized must be
> ## the first
> ## argument of the objective function passed to DEoptim.
> Rosenbrock <- function(x){
+ x1 <- x[1]
+ x2 <- x[2]
+ 100 * (x2 - x1 * x1)^2 + (1 - x1)^2
+ }
> ## DEoptim searches for minima of the objective function
> ## between lower and upper bounds on each parameter to be
> ## optimized.
> ## Therefore, in the call to DEoptim we specify vectors that
> ## comprise the lower and upper bounds; these vectors are
> ## the same length as the parameter vector.
> lower <- c(-10,-10)
> upper <- -lower
> ## run DEoptim and set a seed first for replicability
> set.seed(1234)
> DEoptim(Rosenbrock, lower, upper) # Outputting:
Iteration: 1 bestvalit: 231.182756 bestmemit:   2.298807   6.799427
Iteration: 2 bestvalit: 147.937835 bestmemit:  -2.864559   7.052430
Iteration: 3 bestvalit: 147.937835 bestmemit:  -2.864559   7.052430
Iteration: 4 bestvalit: 12.976620 bestmemit:  -2.155809   4.821222
Iteration: 5 bestvalit: 12.976620 bestmemit:  -2.155809   4.821222
```

```
------------------------------------------------------------------
Iteration: 198 bestvalit: 0.000000 bestmemit:   1.000000   1.000000
Iteration: 199 bestvalit: 0.000000 bestmemit:   1.000000   1.000000
Iteration: 200 bestvalit: 0.000000 bestmemit:   1.000000   1.000000
$optim
$optim$bestmem
par1 par2
  1   1

$optim$bestval
[1] 3.010254e-16

$optim$nfeval
[1] 402

$optim$iter
[1] 200

$member
$member$lower
par1 par2
 -10 -10

$member$upper
par1 par2
  10  10

$member$bestmemit
         par1        par2
1   2.8062121   6.21197105
2   2.2988072   6.79942674
3  -2.8645593   7.05243041
4  -2.8645593   7.05243041
5  -2.1558090   4.82122176
------------------------------------
198 1.0000000 0.99999997
199 1.0000000 0.99999997
200 1.0000000 0.99999997

$member$bestvalit
 [1] 2.797712e+02 2.311828e+02 1.479378e+02 1.479378e+02 1.297662e+01
 ----------------------------------------------------------
 [196] 3.010254e-16 3.010254e-16 3.010254e-16 3.010254e-16
3.010254e-16
$member$pop
           [,1]       [,2]
 [1,]  1.0000000 1.0000001
 [2,]  1.0000000 0.9999999
 [3,]  1.0000000 0.9999999
```

```
 [4,]  1.0000000 1.0000000
 [5,] -1.6667868 2.8245124
 [6,]  1.0000001 1.0000001
 [7,]  1.0000001 1.0000002
 [8,]  1.0000000 1.0000001
 [9,]  1.0000000 1.0000000
[10,]  1.0000000 1.0000000
[11,]  1.0000000 1.0000000
[12,]  1.0000000 1.0000000
[13,]  1.0000000 1.0000000
[14,]  1.0000000 1.0000001
[15,]  1.0000000 1.0000000
[16,]  1.0000000 1.0000001
[17,]  1.0000000 1.0000001
[18,]  0.9999999 0.9999999
[19,]  1.0000000 1.0000000
[20,] -1.6708498 2.9124751

$member$storepop
list()
attr(,"class")
[1] "DEoptim"
> ## increase the population size
> DEoptim(Rosenbrock, lower, upper, DEoptim.control(NP =
> 100))
Iteration: 1 bestvalit: 0.831267 bestmemit:   0.993395   1.078004
Iteration: 2 bestvalit: 0.831267 bestmemit:   0.993395   1.078004
Iteration: 3 bestvalit: 0.136693 bestmemit:   1.238842   1.506507
Iteration: 4 bestvalit: 0.136693 bestmemit:   1.238842   1.506507
Iteration: 5 bestvalit: 0.136693 bestmemit:   1.238842   1.506507
--------------------------------------------------------------
--------------------------------------------------------------
Iteration: 198 bestvalit: 0.000000 bestmemit:   1.000000   1.000000
Iteration: 199 bestvalit: 0.000000 bestmemit:   1.000000   1.000000
Iteration: 200 bestvalit: 0.000000 bestmemit:   1.000000   1.000000
$optim
$optim$bestmem
par1 par2
  1   1

$optim$bestval
[1] 6.981784e-23

$optim$nfeval
[1] 402
```

```
$optim$iter
[1] 200

$member
$member$lower
par1 par2
 -10 -10

$member$upper
par1 par2
  10  10

$member$bestmemit
          par1          par2
1  -0.05055801   -0.3587592
2   0.99339454    1.0780042
3   0.99339454    1.0780042
4   1.23884195    1.5065075
5   1.23884195    1.5065075
----------------------------------
198 1.00000000 1.0000000
199 1.00000000 1.0000000
200 1.00000000 1.0000000

$member$bestvalit
[1] 1.415855e+01 8.312672e-01 8.312672e-01 1.366929e-01
1.366929e-01
  -------------------------------------------------------
  -------------------------------------------------------
[196] 1.241771e-21 6.981784e-23 6.981784e-23 6.981784e-23
6.981784e-23

$member$pop
           [,1]       [,2]
 [1,] 1.000000  1.000000
 [2,] 1.000000  1.000000
 [3,] 1.000000  1.000000
 [4,] 1.000000  1.000000
 [5,] 1.000000  1.000000
 -----------------------------
[98,] 1.000000   1.000000
[99,] 1.000000   1.000000
[100,] 1.000000  1.000000

$member$storepop
list()
```

```
attr(,"class")
[1] "DEoptim"
> ## change other settings and store the output
> outDEoptim <- DEoptim(Rosenbrock, lower, upper,
> DEoptim.control(NP = 80, itermax = 400, F = 1.2, CR = 0.7))
Iteration:  1 bestvalit: 3.635506 bestmemit: 1.317411 1.547563
Iteration:  2 bestvalit: 3.635506 bestmemit: 1.317411 1.547563
Iteration:  3 bestvalit: 3.635506 bestmemit: 1.317411 1.547563
Iteration:  4 bestvalit: 3.635506 bestmemit: 1.317411 1.547563
Iteration:  5 bestvalit: 3.635506 bestmemit: 1.317411 1.547563
------------------------------------------------------------
Iteration: 398 bestvalit: 0.000000 bestmemit:  1.000000
1.000000
Iteration: 399 bestvalit: 0.000000 bestmemit:  1.000000
1.000000
Iteration: 400 bestvalit: 0.000000 bestmemit:  1.000000
1.000000
> ## plot the output
> plot(outDEoptim)
> # Outputting: Figure 6.3 DEoptim-1
>
> ## 'Wild' function, global minimum at about -15.81515
> Wild <- function(x)
+ 10 * sin(0.3 * x) * sin(1.3 * x^2) +
```

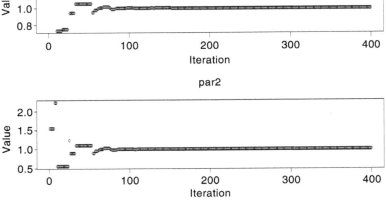

Figure 6.8 DEoptim-1.

'Wild function'

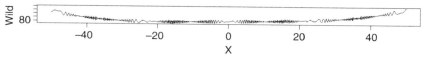

Figure 6.9 DEoptim-2.

```
+ 0.00001 * x^4 + 0.2 * x + 80
> plot(Wild, -50, 50, n = 1000, main = "'Wild function'")
> # Outputting: Figure 6.4 DEoptim-2
>
> outDEoptim <- DEoptim(Wild, lower = -50, upper = 50,
> # Outputting: Figure 6.5 DEoptim-3
>
> DEoptim(Wild, lower = -50, upper = 50,
+ control = DEoptim.control(NP = 50))
Iteration:   1 bestvalit: 68.834199 bestmemit: -15.802280
Iteration:   2 bestvalit: 68.834199 bestmemit: -15.802280
Iteration:   3 bestvalit: 67.562828 bestmemit: -15.509441
Iteration:   4 bestvalit: 67.562828 bestmemit: -15.509441
Iteration:   5 bestvalit: 67.562828 bestmemit: -15.509441
-----------------------------------------------------------
Iteration: 198 bestvalit: 67.467735 bestmemit: -15.815151
Iteration: 199 bestvalit: 67.467735 bestmemit: -15.815151
Iteration: 200 bestvalit: 67.467735 bestmemit: -15.815151
$optim
$optim$bestmem
   par1
-15.81515
```

par1

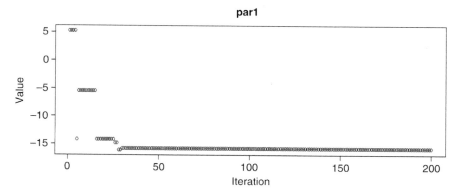

Figure 6.10 DEoptim-3.

```
$optim$bestval
[1] 67.46773

$optim$nfeval
[1] 402

$optim$iter
[1] 200

$member
$member$lower
par1
 -50

$member$upper
par1
  50

$member$bestmemit
        par1
1    -25.66072
2    -15.80228
3    -15.80228
4    -15.50944
5    -15.50944
--------------------------------
198 -15.81515
199 -15.81515
200 -15.81515$member$bestvalit

[1] 69.34871 68.83420 68.83420 67.56283 67.56283 67.56283
-------------------------------------------------------
[196] 67.46773 67.46773 67.46773 67.46773 67.46773
$member$pop
       [,1]
 [1,] -15.66161
 [2,] -15.66161
 [3,] -15.81515
 [4,] -15.66161
 [5,] -15.81515
 ----------------------
[48,] -15.66162
[49,] -15.66161
[50,] -15.66161
```

```
$member$storepop
list()

attr(,"class")
[1] "DEoptim"
>
>
> Genrose <- function(x) {
+ ## One generalization of the Rosenbrock banana valley function
(n parameters)
+ n <- length(x)
+ ## make it take some time ...
+ Sys.sleep(.001)
+ 1.0 + sum (100 * (x[-n]^2 - x[-1])^2 + (x[-1] - 1)^2)
+ }
>
> # get some run-time on simple problems
> maxIt <- 250
> n <- 5
> oneCore <- system.time( DEoptim(fn=Genrose, lower=rep(-25, n),
upper=rep(25, n),
+ control=list(NP=10*n, itermax=maxIt)))
Iteration: 1 bestvalit: 67180.992768 bestmemit:    0.787009
-4.017866   4.091383  -2.065631  -7.987533
Iteration: 2 bestvalit: 67180.992768 bestmemit:    0.787009
-4.017866   4.091383  -2.065631  -7.987533
Iteration: 3 bestvalit: 67180.992768 bestmemit:    0.787009
-4.017866   4.091383  -2.065631  -7.987533
Iteration: 4 bestvalit: 67180.992768 bestmemit:    0.787009
-4.017866   4.091383  -2.065631  -7.987533
Iteration: 5 bestvalit: 42308.544031 bestmemit:    0.787009
-4.017866   0.493543  -2.065631  -7.987533
-------------------------------------------------------------
Iteration: 248 bestvalit: 1.079131 bestmemit:   -1.005728
0.991545   0.993995   1.004214   1.004883
Iteration: 249 bestvalit: 1.079131 bestmemit:   -1.005728
0.991545   0.993995   1.004214   1.004883
Iteration: 250 bestvalit: 1.079131 bestmemit:   -1.005728
0.991545   0.993995   1.004214   1.004883
> withParallel <- system.time( DEoptim(fn=Genrose, lower=rep
(-25, n), upper=rep(25, n),
+ control=list(NP=10*n, itermax=maxIt, parallelType=1)))
Iteration: 1 bestvalit: 1036700.619060 bestmemit:    8.705266
3.108165  -7.886303  -3.630207  -7.869868
```

```
Iteration: 2 bestvalit: 1036700.619060 bestmemit:   8.705266
3.108165  -7.886303  -3.630207  -7.869868
Iteration: 3 bestvalit: 1021360.089135 bestmemit:  -7.209533
-0.069989  -4.584420  -9.804176  15.312698
Iteration: 4 bestvalit: 312699.104689 bestmemit:  -3.456957
5.695299  -6.193019  -1.552886   0.895982
Iteration: 5 bestvalit: 157145.882276 bestmemit:   1.055869
-5.262289   5.023291  -1.470140  -15.176823
-------------------------------------------------------------
Iteration: 248 bestvalit: 1.664090 bestmemit:  -0.933494
0.874286   0.780743   0.630968   0.421854
Iteration: 249 bestvalit: 1.664090 bestmemit:  -0.933494
0.874286   0.780743   0.630968   0.421854
Iteration: 250 bestvalit: 1.664090 bestmemit:  -0.933494
0.874286   0.780743   0.630968   0.421854
> ## Compare timings
> (oneCore)
  user system elapsed
  0.00  0.00  12.58
> (withParallel)
  user system elapsed
  0.59  0.11   7.07
> ## End
```

6.4.4.3   Worked Example for Portfolio Optimization – No. 3 Portfolio Optimization Using the R Code PortfolioAnalytics in CRAN

A numerical portfolio solution for rather complex constraints or objective functions may be obtained by using a function that returns the specific risk measure as the objective function.

This numerical optimization is performed either with the differential evolution optimizer contained in the package DEoptim (see Worked Example No. 3) or by means of randomly generated portfolios satisfying the given constraints.

Worked Example No. 4: Portfolio Optimization using the R code PortfolioAnalytics

The PortfolioAnalytics-package

Numeric methods for optimization of portfolios

Description

PortfolioAnalytics is an R package to compute numerical solutions for portfolio problems with complex constraints and objective sets. The goal of the package is to aid practitioners and researchers in solving portfolio optimization problems with complex constraints and objectives that reflect real-world applications.

One of the goals of this package is to provide a common interface to specify constraints and objectives that can be solved by any supported optimization method. The supported optimization methods include the following:

- Random portfolios
- Differential evolution
- Particle swarm optimization
- Generalized simulated annealing
- Linear and quadratic programming routines

The solver can be specified with the optimize-method argument in `optimize.portfolio` and `optimize.portfolio.rebalancing`. The optimize-method argument must be one of `random`, `DEoptim`, `pso`, `GenSA`, `ROI`, `quadprog`, `glpk`, or performed with the differential evolution optimizer.

This package `PortfolioAnalytics` allows one to obtain a numerical portfolio solution for rather complex constraints or objective functions. For risk measure, one may use a function that returns the specific risk measure as the objective function.

The numerical optimization is performed either with the differential evolution optimizer contained in the package `DEoptim` (see Worked Example No. 2) or by means of randomly generated portfolios satisfying the given constraints.

Selected Example:
From the `PortfolioAnalytics`-package

Numeric methods for optimization of portfolios

`ac.ranking`	**Asset Ranking**
Description	Compute the first moment from a single complete sort
Usage	`ac.ranking(R, order, ...)`

Arguments

`R`	xts object of asset returns
`order`	a vector of indexes of the relative ranking of expected asset returns in ascending order. For example, order = c(2, 3, 1, 4) means that the expected returns of R[,2] < R[,3], < R[,1] < R[,4].
`...`	any other passthrough parameters

Details
This function computes the estimated centroid vector from a single complete sort using the analytical approximation as described in Almgren and Chriss, Portfolios

from Sorts. The centroid is estimated and then scaled such that it is on a scale similar to the asset returns. By default, the centroid vector is scaled according to the median of the asset mean returns.

Value

The estimated first moments based on ranking views.

See Also

```
centroid.complete.mc
centroid.sectors
centroid.sign
centroid.buckets
```

Examples

```
data(edhec)
R <- edhec[,1:4]
ac.ranking(R, c(2, 3, 1, 4))
```

In the R **domain:**

```
> install.packages("PortfolioAnalytics")
Installing package into 'C:/Users/Bert/Documents/R/win-
library/3.2'
(as 'lib' is unspecified)
trying URL 'https://cran.cnr.berkeley.edu/bin/windows/
contrib/3.2/PortfolioAnalytics_1.0.3636.zip'
Content type 'application/zip' length 1552727 bytes (1.5 MB)
downloaded 1.5 MB
package 'PortfolioAnalytics' successfully unpacked and MD5 sums
checked
The downloaded binary packages are in
C:\Users\Bert\AppData\Local\Temp\Rtmpk9Dosw
\downloaded_packages
> library(PortfolioAnalytics)
Loading required package: zoo
Attaching package: 'zoo'
The following objects are masked from 'package:base':
  as.Date, as.Date.numeric
Loading required package: xts
Loading required package: foreach
foreach: simple, scalable parallel programming from Revolution
Analytics
```

```
Use Revolution R for scalability, fault tolerance and more.
http://www.revolutionanalytics.com
Loading required package: PerformanceAnalytics
Package PerformanceAnalytics (1.4.3541) loaded.
Copyright (c) 2004-2014 Peter Carl and Brian G. Peterson, GPL-2 |
GPL-3
http://r-forge.r-project.org/projects/returnanalytics/
Attaching package: 'PerformanceAnalytics'
The following object is masked from 'package:graphics':
  legend
Warning messages:
1: package 'PortfolioAnalytics' was built under R version 3.2.4
2: package 'xts' was built under R version 3.2.3
3: package 'PerformanceAnalytics' was built under R version 3.2.3
> ls("package:PortfolioAnalytics")
 [1] "ac.ranking"                "add.constraint"
 [3] "add.objective"             "add.objective_v1"
 [5] "add.objective_v2"          "add.sub.portfolio"
 [7] "applyFUN"                  "black.litterman"
 [9] "box_constraint"            "CCCgarch.MM"
[11] "center"                    "centroid.buckets"
[13] "centroid.complete.mc"      "centroid.sectors"
[15] "centroid.sign"             "chart.Concentration"
[17] "chart.EF.Weights"          "chart.EfficientFrontier"
[19] "chart.EfficientFrontierOverlay" "chart.GroupWeights"
[21] "chart.RiskBudget"          "chart.RiskReward"
[23] "chart.Weights"             "combine.optimizations"
[25] "combine.portfolios"        "constrained_objective"
[27] "constrained_objective_v1"  "constrained_objective_v2"
[29] "constraint"                "constraint_ROI"
[31] "create.EfficientFrontier"  "diversification"
[33] "diversification_constraint" "EntropyProg"
[35] "equal.weight"              "extractCokurtosis"
[37] "extractCoskewness"         "extractCovariance"
[39] "extractEfficientFrontier"  "extractGroups"
[41] "extractObjectiveMeasures"  "extractStats"
[43] "extractWeights"            "factor_exposure_constraint"
[45] "fn_map"                    "generatesequence"
[47] "group_constraint"          "HHI"
[49] "insert_objectives"         "inverse.volatility.weight"
[51] "is.constraint"             "is.objective"
[53] "is.portfolio"              "leverage_exposure_constraint"
[55] "meanetl.efficient.frontier" "meanvar.efficient.frontier"
```

```
[57] "meucci.moments"              "meucci.ranking"
[59] "minmax_objective"           "mult.portfolio.spec"
[61] "objective"                   "optimize.portfolio"
[63] "optimize.portfolio.parallel" "optimize.portfolio.rebalancing"
[65] "optimize.portfolio.rebalancing_v1" "optimize.portfolio_v1"
[67] "optimize.portfolio_v2"       "portfolio.spec"
[69] "portfolio_risk_objective"    "pos_limit_fail"
[71] "position_limit_constraint"   "quadratic_utility_objective"
[73] "random_portfolios"          "random_portfolios_v1"
[75] "random_portfolios_v2"        "random_walk_portfolios"
[77] "randomize_portfolio"         "randomize_portfolio_v1"
[79] "randomize_portfolio_v2"      "regime.portfolios"
[81] "return_constraint"           "return_objective"
[83] "risk_budget_objective"       "rp_grid"
[85] "rp_sample"                   "rp_simplex"
[87] "rp_transform"                "scatterFUN"
[89] "set.portfolio.moments"       "statistical.factor.model"
[91] "trailingFUN"                 "transaction_cost_constraint"
[93] "turnover"                    "turnover_constraint"
[95] "turnover_objective"          "update_constraint_v1tov2"
[97] "var.portfolio"               "weight_concentration_objective"
[99] "weight_sum_constraint"
> ac.ranking
function (R, order, ...)
{
  if (length(order) != ncol(R))
    stop("The length of the order vector must equal the number of
assets")
  nassets <- ncol(R)
  if (hasArg(max.value)) {
    max.value <- match.call(expand.dots = TRUE)$max.value
  }
  else {
    max.value <- median(colMeans(R))
  }
  c_hat <- scale.range(centroid(nassets), max.value)
  out <- vector("numeric", nassets)
  out[rev(order)] <- c_hat
  return(out)
}
<environment: namespace:PortfolioAnalytics>
> data(edhec)
> edhec
```

	Convertible Arbitrage	CTA Global	Distressed Securities
1997-01-31	0.0119	0.0393	0.0178
1997-02-28	0.0123	0.0298	0.0122
1997-03-31	0.0078	-0.0021	-0.0012
1997-04-30	0.0086	-0.0170	0.0030
1997-05-31	0.0156	-0.0015	0.0233
----------	----------	----------	----------
2009-06-30	0.0241	-0.0147	0.0198
2009-07-31	0.0611	-0.0012	0.0311
2009-08-31	0.0315	0.0054	0.0244

	Emerging Markets	Equity Market	Neutral Event Driven
1997-01-31	0.0791	0.0189	0.0213
1997-02-28	0.0525	0.0101	0.0084
1997-03-31	-0.0120	0.0016	-0.0023
1997-04-30	0.0119	0.0119	-0.0005
1997-05-31	0.0315	0.0189	0.0346
----------	----------	----------	----------
2009-06-30	0.0013	0.0036	0.0123
2009-07-31	0.0451	0.0042	0.0291
2009-08-31	0.0166	0.0070	0.0207

	Fixed Income	Arbitrage Global Macro	Long/Short Equity
1997-01-31	0.0191	0.0573	0.0281
1997-02-28	0.0122	0.0175	-0.0006
1997-03-31	0.0109	-0.0119	-0.0084
1997-04-30	0.0130	0.0172	0.0084
1997-05-31	0.0118	0.0108	0.0394
----------	----------	----------	----------
2009-06-30	0.0126	-0.0076	0.0009
2009-07-31	0.0322	0.0166	0.0277
2009-08-31	0.0202	0.0050	0.0157

	Merger Arbitrage	Relative Value	Short Selling	Funds of Funds
1997-01-31	0.0150	0.0180	-0.0166	0.0317
1997-02-28	0.0034	0.0118	0.0426	0.0106
1997-03-31	0.0060	0.0010	0.0778	-0.0077
1997-04-30	-0.0001	0.0122	-0.0129	0.0009
1997-05-31	0.0197	0.0173	-0.0737	0.0275

```
- - - - - - - - - - - - - - - - - - - - - - - - - - - - - - - - - - - - - - - - - - - - -
2009-06-30      0.0104      0.0101     -0.0094      0.0024
2009-07-31      0.0068      0.0260     -0.0596      0.0153
2009-08-31      0.0102      0.0162     -0.0165      0.0113
> R <- edhec[,1:4]
> ac.ranking(R, c(2, 3, 1, 4))
[1]  0.01432457 -0.05000000 -0.01432457 0.05000000
>
> R # R = edhec [, 1:4], viz. Column Vectors 1 thru 4 of edhec:
#
#      1. Convertible Arbitrage
#      2. CTA Global
#      3. Distressed Securities
#      4. Emerging Markets
#
# but NOT Columns 5 thru 6:
#      5. Equity Market
#      6. Neutral Event Driven
# and, those 4 columns are:
#
                Convertible  CTA          Distressed
                Arbitrage    Global       Securities
1997-01-31        0.0119     0.0393         0.0178
1997-02-28        0.0123     0.0298         0.0122
1997-03-31        0.0078    -0.0021        -0.0012
1997-04-30        0.0086    -0.0170         0.0030
1997-05-31        0.0156    -0.0015         0.0233
- - - - - - - - - - - - - - - - - - - - - - - - - - - - - - - - - - - - - - - - -
2009-06-30        0.0241    -0.0147         0.0198
2009-07-31        0.0611    -0.0012         0.0311
2009-08-31        0.0315     0.0054         0.0244
                Emerging
                Markets
1997-01-31        0.0791
1997-02-28        0.0525
1997-03-31       -0.0120
1997-04-30        0.0119
1997-05-31        0.0315
- - - - - - - - - - - - - - - - - - - - - -
2009-06-30        0.0013
2009-07-31        0.0451
2009-08-31        0.0166
>
```

```
> ac.ranking(R, c(2, 3, 1, 4))
> # Outputting:
[1]  0.01432457 -0.05000000 -0.01432457  0.05000000
```

Remark:

order a vector of indexes of the relative ranking of expected asset returns in ***ascending*** order. Thus, the order = c(2, 3, 1, 4) means that the **expected**

returns of R[,2] < R[,3], < R[,1] < R[,4].

6.4.4.4   Worked Example for Portfolio Optimization – Portfolio Optimization by AssetsM in CRAN

Worked Example No. 8

In the R **domain:**

```
>
> library(urca)
Warning message:
package 'urca' was built under R version 3.2.3
> library(vars)
Loading required package: MASS
Loading required package: strucchange
Loading required package: zoo
Attaching package: 'zoo'
The following objects are masked from 'package:base':
  as.Date, as.Date.numeric
Loading required package: sandwich
Loading required package: lmtest
Warning messages:
1: package 'vars' was built under R version 3.2.3
2: package 'MASS' was built under R version 3.2.4
3: package 'strucchange' was built under R version 3.2.3
4: package 'lmtest' was built under R version 3.2.3
> ## Loading data set and converting to zoo
> data(EuStockMarkets)
> Assets <- as.zoo(EuStockMarkets)
> ## Aggregating as month's-end series
> AssetsM <- aggregate(Assets, as.yearmon,tail, 1)
> head(AssetsM)
> # Outputting:
            DAX   SMI    CAC    FTSE
```

```
Jun 1991 1628.75 1678.1 1772.8 2443.6
Jul 1991 1619.29 1727.2 1754.7 2588.8
Aug 1991 1649.88 1736.6 1864.3 2645.7
Sep 1991 1605.47 1664.7 1880.6 2621.7
Nov 1991 1571.06 1622.6 1739.7 2420.2
> ## Applying unit root tests for subsample
> AssetsMsub <- window(AssetsM, start = start(AssetsM), end =
+                "Jun 1996")
> ## Levels
> ADF <- lapply(AssetsMsub, ur.df, type = "drift", selectlags =
+            "AIC")
> ERS <- lapply(AssetsMsub, ur.ers)
> ## Differences
> DADF <- lapply(diff(AssetsMsub), ur.df, selectlags = "AIC")
> DERS <- lapply(diff(AssetsMsub), ur.ers)
> ## VECM
> VEC <- ca.jo(AssetsMsub, ecdet = "none", spec = "transitory")
> summary(VEC)
> # Outputting:
######################
#     Johansen-Procedure    #
######################
Test type: maximal eigenvalue statistic (lambda max), with
linear trend
Eigenvalues (lambda):
[1] 0.431634123 0.245497723 0.091715832 0.001757365
Values of teststatistic and critical values of test:
        test   10pct   5pct   1pct
r <= 3 |  0.10   6.50   8.18   11.65
r <= 2 |  5.68  12.91  14.90   19.19
r <= 1 | 16.62  18.90  21.07   25.75
 r = 0 | 33.33  24.78  27.14   32.14
Eigenvectors, normalised to first column:
   (These are the co-integration relations)
        DAX.11     SMI.11      CAC.11     FTSE.11
DAX.11  1.000000  1.00000000  1.000000  1.0000000
SMI.11 -6.031316 -0.35450941 -1.797697  0.4074565
CAC.11  4.545907  0.04603755 -2.607493 -0.7843438
FTSE.11 7.547658 -0.29885304  2.049960 -0.4496086
Weights W:
(This is the loading matrix)
```

```
            DAX.11       SMI.11       CAC.11         FTSE.11
DAX.d -0.01371706 -0.3917460 -0.0004852363  0.001711701
SMI.d  0.01145194 -0.2074083  0.0591189734  0.005281964
CAC.d -0.04923235 -0.2558996  0.0307056371 -0.002976012
FTSE.d -0.07672776 -0.1903755  0.0233982140  0.006038698
>
```

6.4.4.5 Worked Examples from *Pfaff*

Forecast

```
> install.packages("forecast")
Installing package into 'C:/Users/Bert/Documents/R/win-
library/3.2'
(as 'lib' is unspecified)
--- Please select a CRAN mirror for use in this session ---
also installing the dependency 'ggplot2'
> A CRAN mirror is selected
trying URL 'https://cran.cnr.berkeley.edu/bin/windows/
contrib/3.2/ggplot2_2.1.0.zip'
Content type 'application/zip' length 2002561 bytes (1.9 MB)
downloaded 1.9 MB
trying URL 'https://cran.cnr.berkeley.edu/bin/windows/
contrib/3.2/forecast_7.1.zip'
Content type 'application/zip' length 1356674 bytes (1.3 MB)
downloaded 1.3 MB
package 'ggplot2' successfully unpacked and MD5 sums checked
package 'forecast' successfully unpacked and MD5 sums checked
The downloaded binary packages are in
    C:\Users\Bert\AppData\Local\Temp\Rtmp6X3C4j
\downloaded_packages
> library(forecast)
Loading required package: zoo
Attaching package: 'zoo'
The following objects are masked from 'package:base':
  as.Date, as.Date.numeric
Loading required package: timeDate
This is forecast 7.1
> ls("package:forecast")
 [1] "accuracy"          "Acf"             "arfima"
 [4] "Arima"             "arima.errors"    "arimaorder"
 [7] "auto.arima"        "bats"            "bizdays"
[10] "BoxCox"            "BoxCox.lambda"   "Ccf"
[13] "croston"           "CV"              "dm.test"
```

```
[16] "dshw"                "easter"             "ets"
[19] "findfrequency"       "fitted.Arima"       "forecast"
[22] "forecast.ar"          "forecast.Arima"   "forecast.bats"
[25] "forecast.ets"        "forecast.fracdiff" "forecast.HoltWinters"
[28] "forecast.lm"          "forecast.nnetar"   "forecast.stl"
[31] "forecast.stlm"        "forecast.StructTS" "forecast.tbats"
[34] "fourier"              "fourierf"           "gas"
[37] "geom_forecast"        "GeomForecast"       "getResponse"
[40] "ggAcf"               "ggCcf"             "ggmonthplot"
[43] "ggPacf"              "ggseasonplot"       "ggtaperedacf"
[46] "ggtaperedpacf"        "ggtsdisplay"        "gold"
[49] "holt"                "hw"               "InvBoxCox"
[52] "is.acf"              "is.Arima"          "is.bats"
[55] "is.constant"         "is.ets"            "is.forecast"
[58] "is.mforecast"        "is.nnetar"          "is.nnetarmodels"
[61] "is.splineforecast"   "is.stlm"            "logLik.ets"
[64] "ma"                  "meanf"             "monthdays"
[67] "msts"                "na.interp"          "naive"
[70] "ndiffs"              "nnetar"             "nsdiffs"  .
[73] "Pacf"                "plot.ar"            "plot.Arima"
[76] "plot.bats"            "plot.ets"           "plot.forecast"
[79] "plot.splineforecast" "plot.tbats"         "rwf"
[82] "seasadj"             "seasonaldummy"    "seasonaldummyf"
[85] "seasonplot"          "ses"               "simulate.ar"
[88] "simulate.Arima"      "simulate.ets"       "simulate.fracdiff"
[91] "sindexf"             "snaive"            "splinef"
[94] "StatForecast"        "stlf"              "stlm"
[97] "subset.ts"            "taperedacf"         "taperedpacf"
[100] "taylor"             "tbats"             "tbats.components"
[103] "thetaf"             "tsclean"            "tsdisplay"
[106] "tslm"               "tsoutliers"         "wineind"
[109] "woolyrnq"
```

Example 1: forecast-accuracy

```
> fit1 <- rwf(EuStockMarkets[1:200,1],h=100)
> fit2 <- meanf(EuStockMarkets[1:200,1],h=100)
> accuracy(fit1)
            ME   RMSE   MAE    MPE    MAPE  MASE    ACF1
Training set 0.4393467 15.8461 8.959648 0.02128535 0.5533582
1 -0.007760885
```

```
> accuracy(fit2)
                ME      RMSE      MAE        MPE    MAPE MASE
Training set 6.24991e-14 57.83723 46.98108 -0.1240662 2.865357    1
> accuracy(fit1,EuStockMarkets[201:300,1])
            ME     RMSE     MAE       MPE       MAPE    MASE
Training set 0.4393467 15.8461  8.959648  0.02128535 0.5533582 1.00000
Test set    0.8900000 78.1811 63.311200 -0.16763307 3.7897316 7.06626
            ACF1
Training set  -0.007760885
Test set         NA
> accuracy(fit2,EuStockMarkets[201:300,1])
                ME      RMSE     MAE       MPE     MAPE     MASE
Training set 6.249910e-14   57.83723   46.98108 -0.1240662
2.865357 1.000000
Test set     8.429485e+01 114.96571 105.65361  4.7004265  6.075968
2.248855
> plot(fit1)
> # Outputting: Figure 6.10  Forecast-Accuracy-1
>
> lines(EuStockMarkets[1:300,1])
>
> # Outputting: Figure 6.11  Forecast-Accuracy-2, and
                Figure 6.12  Forecast-Accuracy-2
```

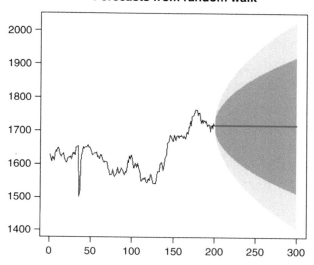

Forecasts from random walk

Figure 6.11 Forecast-Accuracy-1.

Forecasts from random walk

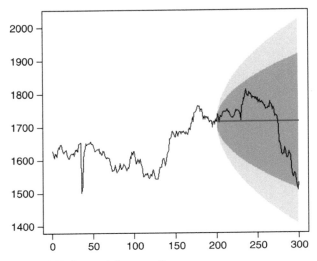

Figure 6.12 Forecast-Accuracy-2.

Example 2: forecast-arfima

```
>
> library(fracdiff)
> x <- fracdiff.sim( 100, ma=-.4, d=.3)$series
> fit <- arfima(x)
> tsdisplay(residuals(fit))
> # Outputting: Figure 6.13 Forecast-arfirma
```

Example 3: forecast-Arima

To fit the ARIMA model to univariate time series data.

Description

Largely a wrapper for the Arima function in the stats package. The main difference is that this function permits a drift term. It is also possible to take an arima model from a previous call to Arima and re-apply it to the data x.

Usage

```
Arima(x, order=c(0,0,0), seasonal=c(0,0,0),
      xreg=NULL, include.mean=TRUE, include.drift=FALSE,
      include.constant, lambda=model$lambda,
      transform.pars=TRUE, fixed=NULL, init=NULL,
```

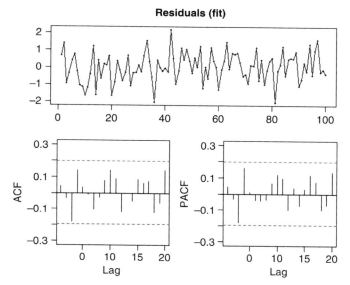

Figure 6.13 Forecast-arfirma.

```
        method=c("CSS-ML","ML","CSS"), n.cond,
        optim.control=list(), kappa=1e6, model=NULL)
>
> fit <- Arima(WWWusage,order=c(3,1,0))
> plot(forecast(fit,h=20))
> # Outputting: Figure 6.14 ARIMA-1
```

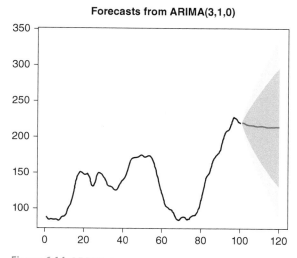

Figure 6.14 ARIMA-1.

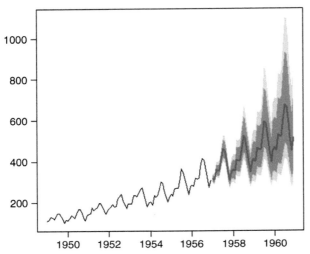

Figure 6.15 ARIMA-2.

```
>
> # Fit model to first few years of AirPassengers data
> air.model <-
+Arima(window(AirPassengers,end=1956+11/12),order=c(0,1,1)
+, seasonal=list(order=c(0,1,1),period=12),lambda=0)
> plot(forecast(air.model,h=48))
> # Outputting: Figure 6.15 ARIMA-2
>
> lines(AirPassengers)
> # Figure 6.16 ARIMA-3
>
> # Apply fitted model to later data
> air.model2 <-
+ Arima(window(AirPassengers,start=1957),model=air.model)
>
> # Forecast accuracy measures on the log scale.
> # in-sample one-step forecasts.
> accuracy(air.model)
> Outputting:
          ME    RMSE    MAE   MPE   MAPE   MASE    ACF1
Training set 0.5159268 12.13132 8.14054 0.07949083 1.900931
0.2266508 -0.2166661
```

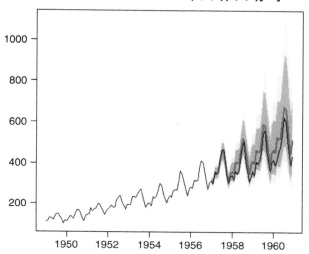

Forecasts from ARIMA(0,1,1)(0,1,1)[12]

Figure 6.16 ARIMA-3.

```
>
> # out-of-sample multi-step forecasts
> accuracy(forecast(air.model,h=48,lambda=NULL),
+        log(window(AirPassengers,start=1957)))

                      ME       RMSE      MAE      MPE      MAPE      MASE
Training set 0.35762533 7.8973404 5.78834425 0.1458472 2.670181
0.1982148

Test set    -0.08403416 0.1031891 0.08801596 -1.3982000 1.463555
0.0030140

Theil's U
Training set 0.05807465        NA
Test set     0.75730561    0.9290965
>
```

Review Questions and Exercises

1 Maxima and minima for a function of several variables
   For a function of several variables, such as the probable value, at any given
   time, of a financial investment portfolio of many stocks, bonds, and so on,
   the value of the portfolio at any given time is, therefore, a function of many

variables. Mathematically, the value of this portfolio may be expressed as a function of future time. With respect to this portfolio value,

a Describe mathematically, what is meant by its
  i) local maximum and local minimum
  ii) global maximum and global minimum of the portfolio function.
b How may each of these values be determined?

2 Optimization methodologies in probabilistic calculus for financial engineering
  This subject encompasses a very wide field of applied mathematics – as different approaches may measure risks that may not be robust, other measures include the CVAR (conditional value at risk). Other approaches often consider portfolio optimization in two stages: optimizing weights of asset classes to hold and optimizing weights of assets within the same asset class. An example of the former would be selecting the proportions in bonds versus stocks, while an example of the latter would be choosing the proportions of the stock subportfolio placed in stocks A, B, C, . . . , Z. Bonds and shares have different financial characteristics. These important differences call for diversification of optimization approaches for different classes of investments. With this introduction in mind, write an outline to form a practical optimization methodology for a portfolio which is invested in the following portfolio that consists of each of the following assets:

  1 Cash (including cash loans from financial institutions)
  2 Termed fixed interest deposits
  3 Portfolio of mixed equities of stocks and high-yield corporate bonds in the United States
  4 Same as (3), but with extensions to overseas financial products
  5 Private companies in the construction of retirement homes for special markets, such as the retiring "Baby Boomers"
  6 Cash income producing real estates: private and commercial
  7 Noncash income producing real estates: private and commercial
  8 Commodities and precious metals
  9 Collectibles
  10 American, European, and Asian options

3 The following paper, available on the Internet, provides a number of models for global optimization:

  1 The standard Markowitz model
  2 A model with risk-free asset (Tobin model)
  3 A multiobjective model for portfolio optimization
  4 A model based on Minkowski absolute metric of risk estimation
  5 A model based on Minkowski semiabsolute metric of risk estimation
  6 A model based on Chebyshev metric of risk estimation (maxmin and minimax models)
  7 The Sharpe model with fractional criteria

8 Linear models of returns

9 A model with limited number assets (cardinality constrained)

10 A model with buy-in thresholds

11 Models with transaction costs

12 Models with integral (lot) assets

13 Models with submodular constraints of diversification of risks

14 Models using fuzzy expected return

Study the following paper on this survey – available at http://www.math.uni-magdeburg.de/~girlich/preprints/preprint0906.pdf Consider other available survey papers on this subject, and comment on your findings.

Some further remarks on optimization methodologies in probabilistic calculus for financial engineering are as follows:

A The package `PortfolioAnalytics` provides a numerical solution of the portfolio with complex constraints and objective functions.

B Following that, the numerical optimization may then be performed either with the differential evolution optimizer in the package `DEoptim` or randomly.

C Carefully study the next two examples selected from the CRAN package: Portfolio Optimization Using the CRAN Package and Portfolio Optimization Using the CRAN Packages

Portfolio Optimization Using the CRAN **Package** `PortfolioAnalytics: ac. ranking` **for Portfolio Optimization with Asset Ranking** The following is a description of this CRAN package: `PortfolioAnalytics:` **Portfolio Analysis, Including Numerical Methods for Optimization of Portfolios**

Portfolio optimization and analysis routines and graphics.

Version:	1.0.3636
Depends:	R ($\geq$ 2.14.0), zoo, xts ($\geq$ 0.8), foreach, PerformanceAnalytics ($\geq$ 1.1.0)
Suggests:	quantmod, DEoptim ($\geq$ 2.2.1), iterators, fGarch, Rglpk, quadprog, ROI ($\geq$ 0.1.0), ROI.plugin.glpk ($\geq$ 0.0.2), ROI. plugin.quadprog ($\geq$ 0.0.2), ROI.plugin.symphony ($\geq$ 0.0.2), pso, GenSA, corpcor, testthat, nloptr ($\geq$ 1.0.0), MASS, robustbase
Published:	2015-04-19
Author:	Brian G. Peterson [cre, aut, cph], Peter Carl [aut, cph], Kris Boudt [ctb, cph], Ross Bennett [ctb, cph], Hezky Varon [ctb], Guy Yollin [ctb], R. Douglas Martin [ctb]
Maintainer:	Brian G. Peterson <brian@braverock.com>

License: GPL-2 | GPL-3 [expanded from: GPL]

Copyright: 2004–2015

NeedsCompilation: yes

Materials: README

CRAN checks: PortfolioAnalytics results

Downloads:

Reference manual: PortfolioAnalytics.pdf

Vignettes: Design Thoughts of the PortfolioAnalytics Package

                     Portfolio Optimization with ROI in PortfolioAnalytics

                     Custom Moment and Objective Functions

                     An Introduction to Portfolio Optimization with PortfolioAnalytics

                     Portfolio Optimization with CVaR budgets in PortfolioAnalytics

Package source: PortfolioAnalytics_1.0.3636.tar.gz

Windows binaries: r-devel: PortfolioAnalytics_1.0.3636.zip, r-release: PortfolioAnalytics_1.0.3636.zip, r-oldrel: PortfolioAnalytics_1.0.3636.zip

OS X Mavericks binaries: r-release: PortfolioAnalytics_1.0.3636.tgz, r-oldrel: PortfolioAnalytics_1.0.3636.tgz

Within this `CRAN` package, consider the following function program: `ac.ranking`. This function computes the estimated centroid vector from a single complete sort using the analytical approximation as described in Almgren and Chriss, "Portfolios from Sorts". The centroid is estimated and then scaled such that it is on a scale similar to the asset returns. By default, the centroid vector is scaled according to the median of the asset mean returns.

This program is described further:

`ac.ranking`   **Asset Ranking**

Description

Compute the first moment from a single complete sort.

Usage

`ac.ranking(R, order, ...)`

Arguments

`R`   xts object of asset returns

`order`   a vector of indexes of the relative ranking of expected asset returns in ascending order. For example, order = c

```
(2, 3, 1, 4) means that the expected returns of R[,2]
< R[,3], < R[,1] < R[,4].
```
. . .   any other passthrough parameters

Details

This function computes the estimated centroid vector from a single complete sort using the analytical approximation as described in Almgren and Chriss, Portfolios from Sorts. The centroid is estimated and then scaled such that it is on a scale similar to the asset returns. By default, the centroid vector is scaled according to the median of the asset mean returns.

Value

The estimated first moments based on ranking views

See Also

centroid.complete.mc centroid.sectors centroid.sign centroid.
buckets

Examples

```
**
data(edhec)
R <- edhec[,1:4]
ac.ranking(R, c(2, 3, 1, 4))
#
data(edhec)
R <- edhec[,1:10]
ac.ranking(R, c(1, 2, 3, 4, 5, 6, 7, 8; 9, 10))
```

Portfolio Optimization Using the CRAN **Packages** DEoptim:DEoptim-methods **for Portfolio Optimization Using the** DEoptim **Approach**

A) DEoptim: **Global optimization by differential evolution**
B) DEoptim-methods : DEoptim-**methods**

A) DEoptim: **Global optimization by differential evolution**

This introduction to the R package DEoptim is an abbreviated version of the manuscript published in the *Journal of Statistical Software*. DEoptim implements the Differential Evolution algorithm for global optimization of a real-valued function of a real-valued parameter vector. The implementation of differential evolution in DEoptim interfaces with C code for efficiency. Moreover, the package is self-contained and does not depend on any other packages.

Implements the differential evolution algorithm for global optimization of a real-valued function of a real-valued parameter vector.

Version:	2.2-4
Depends:	parallel
Suggests:	foreach, iterators, colorspace, lattice
Published:	2016-12-19
Author:	David Ardia [aut], Katharine Mullen [aut, cre], Brian Peterson [aut], Joshua Ulrich [aut], Kris Boudt [ctb]
Maintainer:	Katharine Mullen <mullenkate@gmail.com>
License:	GPL-2 | GPL-3 [expanded from: GPL (≥ 2)]
NeedsCompilation:	yes
Citation:	DEoptim citation info
Materials:	README NEWS
In views:	Optimization
CRAN checks:	DEoptim results

Downloads:

Reference manual:	DEoptim.pdf
Vignettes:	DEoptim: An R Package for Differential Evolution Large scale portfolio optimization with DEoptim
Package source:	DEoptim_2.2-4.tar.gz
Windows binaries:	r-devel: DEoptim_2.2-4.zip, r-release: DEoptim_2.2-4.zip, r-oldrel: DEoptim_2.2-4.zip
OS X Mavericks binaries:	r-release: DEoptim_2.2-4.tgz, r-oldrel: DEoptim_2.2-4.tgz
Old sources:	DEoptim archive

Reverse Dependencies:

Reverse depends:	EcoHydRology, galts, IBHM, likeLTD, micEconCES, quickpsy, selectMeta
Reverse imports:	BBEST, CEGO, covmat, DstarM, FuzzyStatProb, MSGARCH, SpaDES
Reverse suggests:	BayesianTools, MSCMT, nanop, npsp, PortfolioAnalytics, RcppDE, SACOBRA, SPOT

Examples from CRAN:
Use the following two examples, in R, to illustrate the use of the aforementioned approach in portfolio optimization:

DEoptim **Differential Evolution Optimization**

Description
Performs evolutionary global optimization via the differential evolution algorithm.

Usage
```
DEoptim(fn, lower, upper, control = DEoptim.control(), ...,
                        fnMap=NULL)
```

Arguments

fn	the function to be optimized (minimized). The function should have as its first argument the vector of real-valued parameters to optimize, and return a scalar real result. NA and NaN values are not allowed.
lower, upper	two vectors specifying scalar real lower and upper bounds on each parameter to be optimized, so that the i-th element of lower and upper applies to the ith parameter. The implementation searches between lower and upper for the global optimum (minimum) of fn.
control	a list of control parameters; see DEoptim.control.
fnMap	an optional function that will be run after each population is created, but before the population is passed to the objective function. This allows the user to impose integer/cardinality constriants.
...	further arguments to be passed to fn.

Details
DEoptim performs optimization (minimization) of fn.

The control argument is a list; see the help file for DEoptim.control for details.

The R implementation of DE, DEoptim, was first published on the Comprehensive R Archive Network (CRAN) in 2005 by David Ardia. Early versions were written in pure R. Since version 2.0-0 (published on CRAN in 2009), the package has relied on an interface to a C implementation of DE, which is significantly faster on most problems as compared to the implementation in pure R. The C interface is in many respects similar to the MS Visual C++ v5.0 implementation of the differential evolution algorithm distributed with the book *Differential Evolution – A Practical Approach to Global Optimization* by Price et al. and found online at http://www1.icsi.berkeley.edu/~storn/code. html. Since version 2.0-3, the C implementation dynamically allocates the memory required to store the population, removing limitations on the number of members in the population and length of the parameter vectors that may be optimized. Since version 2.2-0, the package allows for parallel operation, so that the evaluations of the objective function may be performed using all available cores. This is accomplished using either the built-in parallel package or the foreach package. If parallel operation is desired, the user should set paral- lelType and make sure that the arguments and packages needed by the objective function are available; see DEoptim.control, the example below and examples in the sandbox directory for details.

Since becoming publicly available, the package DEoptim has been used by several authors to solve optimization problems arising in diverse domains.

To perform a maximization (instead of minimization) of a given function, simply define a new function which is the opposite of the function to maximize and apply DEoptim to it.

To integrate additional constraints (other than box constraints) on the parameters x of fn(x), for instance x[1] + x[2]^2 < 2, integrate the constraint within the function to optimize, for instance:

```
fn <- function(x){
if (x[1] + x[2]^2 >= 2){
r <- Inf
else{
...
}
return(r)
}
```

This simplistic strategy usually does not work all that well for gradient-based or Newton-type methods. It is likely to be alright when the solution is in the interior of the feasible region, but when the solution is on the boundary, optimization algorithm would have a difficult time converging. Furthermore, when the solution is on the boundary, this strategy would make the algorithm converge to an inferior solution in the interior. However, for methods such as DE that are not gradient based, this strategy might not be that bad.

Note that DEoptim stops if any NA or NaN value is obtained. You have to redefine your function to handle these values (for instance, set NA to Inf in your objective function).

It is important to emphasize that the result of DEoptim is a random variable, that is, different results may be obtained when the algorithm is run repeatedly with the same settings. Hence, the user should set the random seed if they want to reproduce the results, for example, by setting set.seed(1234) before the call of DEoptim.

DEoptim relies on repeated evaluation of the objective function in order to move the population toward a global minimum. Users interested in making DEoptim run as fast as possible should consider using the package in parallel mode (so that all CPUs available are used), and also ensure that evaluation of the objective function is as efficient as possible (e.g., by using vectorization in pure R code, or writing parts of the objective function in a lower level language like C or Fortran).

Further details and examples of the R package DEoptim can be found in Mullen et al. (2011) and Ardia et al. (2011a, 2011b) or look at the package's vignette by typing vignette("DEoptim"). Also, an illustration of the package usage for a high-dimensional nonlinear portfolio optimization problem is available by typing vignette("DEoptimPortfolioOptimization").

Please cite the package in publications. Use citation("DEoptim").

Value

The output of the function DEoptim is a member of the S3 class DEoptim. More precisely, this is a list (of length 2) containing the following elements: optim, a list containing the following elements:

- bestmem: the best set of parameters found.
- bestval: the value of fn corresponding to bestmem.
- nfeval: number of function evaluations.
- iter: number of procedure iterations.

member, a list containing the following elements:

- lower: the lower boundary.
- upper: the upper boundary.
- bestvalit: the best value of fn at each iteration.
- bestmemit: the best member at each iteration.
- pop: the population generated at the last iteration.
- storepop: a list containing the intermediate populations.

Members of the class DEoptim have a plot method that accepts the argument plot.type.

plot.type = "bestmemit" results in a plot of the parameter values that represent the lowest value of the objective function in each generation.

plot.type = "bestvalit" plots the best value of the objective function in each generation. Finally, `plot.type = "storepop"` results in a plot of stored populations (which are only available if these have been saved by setting the control argument of `DEoptim` appropriately). Storing intermediate populations allows us to examine the progress of the optimization in detail. A summary method also exists and returns the best parameter vector, the best value of the objective function, the number of generations optimization ran, and the number of times the objective function was evaluated.

Note

DE is a search heuristic introduced by Storn and Price (1997). Its remarkable performance as a global optimization algorithm on continuous numerical minimization problems has been extensively explored. DE belongs to the class of genetic algorithms that use biology-inspired operations of crossover, mutation, and selection on a population in order to minimize an objective function over the course of successive generations. As with other evolutionary algorithms, DE solves optimization problems by evolving a population of candidate solutions using alteration and selection operators. DE uses floating-point instead of bit string encoding of population members, and arithmetic operations instead of logical operations in mutation. DE is particularly well suited to find the global optimum of a real-valued function of real-valued parameters, and does not require that the function be either continuous or differentiable.

Let NP denote the number of parameter vectors (members) $x \in R^d$ in the population. In order to create the initial generation, NP guesses for the optimal value of the parameter vector are made, either using random values between lower and upper bounds (defined by the user) or using values given by the user. Each generation involves creation of a new population from the current population members $\{x_i \mid i = 1, \ldots, \text{NP}\}$, where i indexes the vectors that make up the population. This is accomplished using differential mutation of the population members. An initial mutant parameter vector v_i is created by choosing three members of the population, x_{r0}, x_{r1}, and x_{r2}, at random. Then v_i is generated as

$$v_i = x_{r0} + F(x_{r1} - x_{r2})$$

where F is the differential weighting factor, effective values for which are typically between 0 and 1. After the first mutation operation, mutation is continued until d mutations have been made, with a crossover probability $CR \in [0, 1]$. The crossover probability CR controls the fraction of the parameter values that are copied from the mutant. If an element of the trial parameter vector is found to violate the bounds after mutation and crossover, it is reset in such a way that the bounds are respected (with the specific protocol depending on the implementation). Then, the objective function values associated with the children are determined. If a trial vector has equal or lower objective function value than the previous vector, it replaces the previous vector in the population;

otherwise the previous vector remains. Variations of this scheme have also been proposed.

Intuitively, the effect of the scheme is that the shape of the distribution of the population in the search space is converging with respect to size and direction toward areas with high fitness. The closer the population gets to the global optimum, the more the distribution will shrink and, therefore, reinforce the generation of smaller difference vectors.

As a general advice regarding the choice of *NP*, *F*, and CR, Storn et al. (2006) state the following:

Set the number of parents NP to 10 times the number of parameters, select differential weighting factor $F = 0.8$ and crossover constant $CR = 0.9$. Make sure that you initialize your parameter vectors by exploiting their full numerical range, that is, if a parameter is allowed to exhibit values in the range $[-100, 100]$, it is a good idea to pick the initial values from this range instead of unnecessarily restricting diversity. If you experience misconvergence in the optimization process you usually have to increase the value for NP, but often you only have to adjust *F* to be a little lower or higher than 0.8. If you increase NP and simultaneously lower *F* a little, convergence is more likely to occur but generally takes longer, that is, DE is getting more robust (there is always a convergence speed/robustness trade-off).

DE is much more sensitive to the choice of *F* than it is to the choice of CR. CR is more like a fine tuning element. High values of CR, like $CR = 1$, give faster convergence if convergence occurs. Sometimes, however, you have to go down as much as $CR = 0$ to make DE robust enough for a particular problem. For more details on the DE strategy, we refer the reader to Storn and Price (1997) and Price et al. (2006).

Author(s)

David Ardia, Katharine Mullen <mullenkate@gmail.com>, Brian Peterson, and Joshua Ulrich.

See Also

`DEoptim.control` for control arguments, `DEoptim-methods` for methods on DEoptim objects, including some examples in plotting the results; `optim` or `constrOptim` for alternative optimization algorithms.

Examples

```
## Rosenbrock Banana function
## The function has a global minimum f (x)=0 at the point (1,1) .
## Note that the vector of parameters to be optimized must be ##
the first argument of the objective function passed to
## DEoptim.
Rosenbrock <- function(x) {
```

```
x1 <- x[1]
x2 <- x[2]
100 * (x2 - x1 * x1)^2 + (1 - x1)^2
}
## DEoptim searches for minima of the objective function
## between lower and upper bounds on each parameter to be ##
optimized. Therefore in the call to DEoptim we specify vectors ##
that comprise the lower and upper bounds; these vectors are ## the
same length as the parameter vector.
lower <- c(-10,-10)
upper <- -lower
## run DEoptim and set a seed first for replicability
set.seed(1234)
DEoptim(Rosenbrock, lower, upper)
## increase the population size
DEoptim(Rosenbrock, lower, upper, DEoptim.control(NP = 100))
## change other settings and store the output
outDEoptim <- DEoptim(Rosenbrock, lower, upper,
                 DEoptim.control(NP = 80,
                    itermax = 400, F = 1.2, CR = 0.7))
## plot the output
plot(outDEoptim)
## 'Wild' function, global minimum at about -15.81515
Wild <- function(x)
10 * sin(0.3 * x) * sin(1.3 * x^2) +
   0.00001 * x^4 + 0.2 * x + 80
plot(Wild, -50, 50, n = 1000, main = "'Wild function'")
outDEoptim <- DEoptim(Wild, lower = -50, upper = 50,
                 control = DEoptim.control(trace = FALSE))
plot(outDEoptim)
DEoptim(Wild, lower = -50, upper = 50,
      control = DEoptim.control(NP = 50))
## The below examples shows how the call to DEoptim can be
## parallelized.
## Note that if your objective function requires packages to be
## loaded or has arguments supplied via \code{...}, these
## should be specified using the \code{packages} and
## \code{parVar} arguments
## in control.
## Not run:
Genrose <- function(x) {
## One generalization of the Rosenbrock banana valley function ##
(n parameters)
```

```
n <- length(x)
## make it take some time ...
Sys.sleep(.001)
1.0 + sum (100 * (x[-n]^2 - x[-1])^2 + (x[-1] - 1)^2)
}
# get some run-time on simple problems
maxIt <- 250
n <- 5
oneCore <- system.time( DEoptim(fn=Genrose, lower=rep(-25, n),
upper=rep(25, n),
control=list(NP=10*n, itermax=maxIt)))
withParallel <- system.time( DEoptim(fn=Genrose, lower=rep
(-25, n), upper=rep(25, n),
control=list(NP=10*n, itermax=maxIt, parallelType=1)))
## Compare timings
(oneCore)
(withParallel)
## End(Not run)
```

B) DEoptim-methods **DEoptim-methods**

Description
Methods for DEoptim objects.

Usage

```
## S3 method for class 'DEoptim'
summary(object, ...)
## S3 method for class 'DEoptim'
plot(x, plot.type = c("bestmemit", "bestvalit", "storepop"),
...)
```

Arguments

object   an object of class DEoptim; usually, a result of a call
            to DEoptim.
x   an object of class DEoptim; usually, a result of a call
            to DEoptim.
plot.type   should we plot the best member at each iteration, the
            best value at each iteration or the intermediate
            populations?
...   further arguments passed to or from other methods.

Details

Members of the class DEoptim have a plot method that accepts the argument plot.type.
plot.type = "bestmemit" results in a plot of the parameter values that represent the lowest value of the objective function each generation.
plot.type = "bestvalit" plots the best value of the objective function each generation. Finally,
plot.type = "storepop" results in a plot of stored populations (which are only available if these have been saved by setting the control argument of DEoptim appropriately). Storing intermediate populations allows us to examine the progress of the optimization in detail. A summary method also exists and returns the best parameter vector, the best value of the objective function, the number of generations optimization ran, and the number of times the objective function was evaluated.

Note

Further details and examples of the R package DEoptim can be found in Mullen et al. (2011) and Ardia et al. (2011a, 2011b) or look at the package's vignette by typing vignette("DEoptim").

Please cite the package in publications. Use citation("DEoptim").

Author(s)

David Ardia, Katharine Mullen <mullenkate@gmail.com>, Brian Peterson, and Joshua Ulrich.

See Also

DEoptim and DEoptim.control.

Examples

```
## Rosenbrock Banana function
## The function has a global minimum f(x)=0 at the point (1,1).
## Note that the vector of parameters to be optimized must be ##
the first argument of the objective function passed to
## DEoptim.
Rosenbrock <- function(x){
x1 <- x[1]
x2 <- x[2]
100 * (x2 - x1 * x1)^2 + (1 - x1)^2
}
```

```
lower <- c(-10, -10)
upper <- -lower
set.seed(1234)
outDEoptim <- DEoptim(Rosenbrock, lower, upper)
## print output information
summary(outDEoptim)
## plot the best members
plot(outDEoptim, type = 'b')
## plot the best values
dev.new()
plot(outDEoptim, plot.type = "bestvalit", type = 'b', col =
'blue')
## rerun the optimization, and store intermediate populations
outDEoptim <- DEoptim(Rosenbrock, lower, upper,
DEoptim.control(itermax = 500,
storepopfrom = 1, storepopfreq = 2))
summary(outDEoptim)
## plot intermediate populations
dev.new()
plot(outDEoptim, plot.type = "storepop")
## Wild function
Wild <- function(x)
10 * sin(0.3 * x) * sin(1.3 * x^2) +
0.00001 * x^4 + 0.2 * x + 80
outDEoptim = DEoptim(Wild, lower = -50, upper = 50,
DEoptim.control(trace = FALSE, storepopfrom = 50,
storepopfreq = 1))
plot(outDEoptim, type = 'b')
dev.new()
plot(outDEoptim, plot.type = "bestvalit", type = 'b')
## Not run:
## an example with a normal mixture model: requires package ##
mvtnorm
library(mvtnorm)
## neg value of the density function
negPdfMix <- function(x) {
tmp <- 0.5 * dmvnorm(x, c(-3, -3)) + 0.5 * dmvnorm(x, c(3, 3))
-tmp
}
## wrapper plotting function
plotNegPdfMix <- function(x1, x2)
negPdfMix(cbind(x1, x2))
## contour plot of the mixture
```

```
x1 <- x2 <- seq(from = -10.0, to = 10.0, by = 0.1)
thexlim <- theylim <- range(x1)
z <- outer(x1, x2, FUN = plotNegPdfMix)
contour(x1, x2, z, nlevel = 20, las = 1, col = rainbow(20),
xlim = thexlim, ylim = theylim)
set.seed(1234)
outDEoptim <- DEoptim(negPdfMix, c(-10, -10), c(10, 10),
  DEoptim.control(NP = 100, itermax = 100, storepopfrom = 1,
  storepopfreq = 5))
## convergence plot
dev.new()
plot(outDEoptim)
## the intermediate populations indicate the bi-modality of the
## function
dev.new()
plot(outDEoptim, plot.type = "storepop")
## End(Not run)
```

The CRAN Package `PortfolioAnalytics`

In the R domain:

```
>
> install.packages("PortfolioAnalytics")
Installing package into 'C:/Users/Bert/Documents/R/win-
library/3.3'
(as 'lib' is unspecified)
--- Please select a CRAN mirror for use in this session ---
```

A CRAN mirror is selected.

```
also installing the dependencies 'iterators', 'foreach'
> library(PortfolioAnalytics)
Attaching package: 'PerformanceAnalytics'
>
> ls("package:PortfolioAnalytics")
 [1] "ac.ranking"                "add.constraint"
 [3] "add.objective"             "add.objective_v1"
 [5] "add.objective_v2"          "add.sub.portfolio"
 [7] "applyFUN"                  "black.litterman"
 [9] "box_constraint"            "CCCgarch.MM"
[11] "center"                    "centroid.buckets"
[13] "centroid.complete.mc"      "centroid.sectors"
[15] "centroid.sign"             "chart.Concentration"
```

```
[17] "chart.EF.Weights"                  "chart.EfficientFrontier"
[19] "chart.EfficientFrontierOverlay"    "chart.GroupWeights"
[21] "chart.RiskBudget"                  "chart.RiskReward"
[23] "chart.Weights"                     "combine.optimizations"
[25] "combine.portfolios"               "constrained_objective"
[27] "constrained_objective_v1"         "constrained_objective_v2"
[29] "constraint"                        "constraint_ROI"
[31] "create.EfficientFrontier"         "diversification"
[33] "diversification_constraint"       "EntropyProg"
[35] "equal.weight"                      "extractCokurtosis"
[37] "extractCoskewness"                 "extractCovariance"
[39] "extractEfficientFrontier"          "extractGroups"
[41] "extractObjectiveMeasures"          "extractStats"
[43] "extractWeights"                    "factor_exposure_constraint"
[45] "fn_map"                            "generatesequence"
[47] "group_constraint"                  "HHI"
[49] "insert_objectives"                 "inverse.volatility.weight"
[51] "is.constraint"                     "is.objective"
[53] "is.portfolio"                      "leverage_exposure_constraint"
[55] "meanetl.efficient.frontier"        "meanvar.efficient.frontier"
[57] "meucci.moments"                    "meucci.ranking"
[59] "minmax_objective"                  "mult.portfolio.spec"
[61] "objective"                         "optimize.portfolio"
[63] "optimize.portfolio.parallel"      "optimize.portfolio.rebalancing"
[65] "optimize.portfolio.rebalancing_v1" "optimize.portfolio_v1"
[67] "optimize.portfolio_v2"            "portfolio.spec"
[69] "portfolio_risk_objective"         "pos_limit_fail"
[71] "position_limit_constraint"        "quadratic_utility_objective"
[73] "random_portfolios"                "random_portfolios_v1"
[75] "random_portfolios_v2"             "random_walk_portfolios"
[77] "randomize_portfolio"              "randomize_portfolio_v1"
[79] "randomize_portfolio_v2"           "regime.portfolios"
[81] "return_constraint"                "return_objective"
[83] "risk_budget_objective"            "rp_grid"
[85] "rp_sample"                         "rp_simplex"
[87] "rp_transform"                      "scatterFUN"
[89] "set.portfolio.moments"            "statistical.factor.model"
[91] "trailingFUN"                       "transaction_cost_constraint"
[93] "turnover"                          "turnover_constraint"
[95] "turnover_objective"               "update_constraint_v1tov2"
[97] "var.portfolio"                     "weight_concentration_objective"
[99] "weight_sum_constraint"
```

```
>
> ac.ranking
```

```
function (R, order, ...)
{
  if (length(order) != ncol(R))
    stop("The length of the order vector must equal the number of
assets")
  nassets <- ncol(R)
  if (hasArg(max.value)) {
    max.value <- match.call(expand.dots = TRUE)$max.value
  }
  else {
    max.value <- median(colMeans(R))
  }
  c_hat <- scale.range(centroid(nassets), max.value)
  out <- vector("numeric", nassets)
  out[rev(order)] <- c_hat
  return(out)
}
<environment: namespace:PortfolioAnalytics>
>
> # Run 1
> data(edhec)
> R <- edhec[,1:4]
> ac.ranking(R, c(2, 3, 1, 4)) # Outputting:
[1]  0.01432457 -0.05000000 -0.01432457  0.05000000
>
> # Run 2
> data(edhec)
> R <- edhec[,1:10]
> ac.ranking(R, c(1, 2, 3, 4, 5, 6, 7, 8, 9, 10)) # Outputting:
 [1] -0.050000000 -0.032377611 -0.021216605 -0.012155458
-0.003968706
 [6]  0.003968706  0.012155458  0.021216605  0.032377611
 0.050000000
>
```

The CRAN **Package** DEoptim **Portfolio Optimization Using the** CRAN
Packages: DEoptim:DEoptim-methods **for Portfolio Optimization using
the** DEoptim **approach.** DEoptim: Global Optimization by Differential Evolution

In the R **domain:**

```
>
> install.packages("DEoptim")
```

```
Installing package into 'C:/Users/Bert/Documents/R/win-
library/3.3'
(as 'lib' is unspecified)
--- Please select a CRAN mirror for use in this session ---
```

A CRAN mirror is selected.

```
trying URL 'https://cran.ism.ac.jp/bin/windows/contrib/3.3/
DEoptim_2.2-4.zip'
Content type 'application/zip' length 680535 bytes (664 KB)
downloaded 664 KB
package 'DEoptim' successfully unpacked and MD5 sums checked
The downloaded binary packages are in
C:\Users\Bert\AppData\Local\Temp\RtmpkPO86K
\downloaded_packages
> library(DEoptim)
Loading required package: parallel
DEoptim package
Differential Evolution algorithm in R
Authors: D. Ardia, K. Mullen, B. Peterson and J. Ulrich
> ls("package:DEoptim")
[1] "DEoptim"      "DEoptim.control"
> DEoptim
function (fn, lower, upper, control = DEoptim.control(), ...,
  fnMap = NULL)
{
  if (length(lower) != length(upper))
    stop("'lower' and 'upper' are not of same length")
  if (!is.vector(lower))
    lower <- as.vector(lower)
  if (!is.vector(upper))
    upper <- as.vector(upper)
  if (any(lower > upper))
    stop("'lower' > 'upper'")
  if (any(lower == "Inf"))
    warning("you set a component of 'lower' to 'Inf'. May imply
'NaN' results",
      immediate. = TRUE)
  if (any(lower == "-Inf"))
    warning("you set a component of 'lower' to '-Inf'. May imply
'NaN' results",
      immediate. = TRUE)
  if (any(upper == "Inf"))
```

```
    warning("you set a component of 'upper' to 'Inf'. May imply
'NaN' results",
        immediate. = TRUE)
  if (any(upper == "-Inf"))
    warning("you set a component of 'upper' to '-Inf'. May imply
'NaN' results",
        immediate. = TRUE)
  if (!is.null(names(lower)))
    nam <- names(lower)
  else if (!is.null(names(upper)) & is.null(names(lower)))
    nam <- names(upper)
  else nam <- paste("par", 1:length(lower), sep = "")
  ctrl <- do.call(DEoptim.control, as.list(control))
  ctrl$npar <- length(lower)
  if (is.na(ctrl$NP))
    ctrl$NP <- 10 * length(lower)
  if (ctrl$NP < 4) {
    warning("'NP' < 4; set to default value 10*length(lower)\n",
        immediate. = TRUE)
    ctrl$NP <- 10 * length(lower)
  }
  if (ctrl$NP < 10 * length(lower))
    warning("For many problems it is best to set 'NP' (in
'control') to be at least ten times the length of the parameter
vector. \n",
        immediate. = TRUE)
  if (!is.null(ctrl$initialpop)) {
    ctrl$specinitialpop <- TRUE
    if (!identical(as.numeric(dim(ctrl$initialpop)), as.
numeric(c(ctrl$NP,
        ctrl$npar))))
        stop("Initial population is not a matrix with dim. NP x
length(upper).")
  }
  else {
    ctrl$specinitialpop <- FALSE
    ctrl$initialpop <- 0
  }
  ctrl$trace <- as.numeric(ctrl$trace)
  ctrl$specinitialpop <- as.numeric(ctrl$specinitialpop)
  ctrl$initialpop <- as.numeric(ctrl$initialpop)
  if (!is.null(ctrl$cluster)) {
    if (!inherits(ctrl$cluster, "cluster"))
```

```
      stop("cluster is not a 'cluster' class object")
    parallel::clusterExport(cl, ctrl$parVar)
    fnPop <- function(params, ...) {
      parallel::parApply(cl = ctrl$cluster, params, 1,
        fn, ...)
    }
  }
  else if (ctrl$parallelType == 2) {
    if (!foreach::getDoParRegistered()) {
      foreach::registerDoSEQ()
    }
    args <- ctrl$foreachArgs
    fnPop <- function(params, ...) {
      my_chunksize <- ceiling(NROW(params)/foreach::
getDoParWorkers())
      my_iter <- iterators::iter(params, by = "row", chunksize =
my_chunksize)
      args$i <- my_iter
      args$.combine <- c
      if (!is.null(args$.export))
        args$.export = c(args$.export, "fn")
      else args$.export = "fn"
      if (is.null(args$.errorhandling))
        args$.errorhandling = c("stop", "remove", "pass")
      if (is.null(args$.verbose))
        args$.verbose = FALSE
      if (is.null(args$.inorder))
        args$.inorder = TRUE
      if (is.null(args$.multicombine))
        args$.multicombine = FALSE
      foreach::"%dopar%"(do.call(foreach::foreach, args),
        apply(i, 1, fn, ...))
    }
  }
  else if (ctrl$parallelType == 1) {
    cl <- parallel::makeCluster(parallel::detectCores())
    packFn <- function(packages) {
      for (i in packages) library(i, character.only = TRUE)
    }
    parallel::clusterCall(cl, packFn, ctrl$packages)
    parallel::clusterExport(cl, ctrl$parVar)
    fnPop <- function(params, ...) {
      parallel::parApply(cl = cl, params, 1, fn, ...)
```

```
      }
    }
    else {
      fnPop <- function(params, ...) {
        apply(params, 1, fn, ...)
      }
    }
    if (is.null(fnMap)) {
      fnMapC <- function(params, ...) params
    }
    else {
      fnMapC <- function(params, ...) {
        mappedPop <- t(apply(params, 1, fnMap))
        if (all(dim(mappedPop) != dim(params)))
          stop("mapping function did not return an object with ",
            "dim NP x length(upper).")
        dups <- duplicated(mappedPop)
        np <- NCOL(mappedPop)
        tries <- 0
        while (tries < 5 && any(dups)) {
          nd <- sum(dups)
          newPop <- matrix(runif(nd * np), ncol = np)
          newPop <- rep(lower, each = nd) + newPop * rep(upper -
            lower, each = nd)
          mappedPop[dups,] <- t(apply(newPop, 1, fnMap))
          dups <- duplicated(mappedPop)
          tries <- tries + 1
        }
        if (tries == 5)
          warning("Could not remove ", sum(dups), " duplicates from
the mapped ",
            "population in 5 tries. Evaluating population with
duplicates.",
            call. = FALSE, immediate. = TRUE)
        storage.mode(mappedPop) <- "double"
        mappedPop
      }
    }
    outC <- .Call("DEoptimC", lower, upper, fnPop, ctrl, new.env
(),
      fnMapC, PACKAGE = "DEoptim")
    if (ctrl$parallelType == 1)
      parallel::stopCluster(cl)
```

```
    if (length(outC$storepop) > 0) {
      nstorepop <- floor((outC$iter - ctrl$storepopfrom)/ctrl
$storepopfreq)
      storepop <- list()
      cnt <- 1
      for (i in 1:nstorepop) {
        idx <- cnt:((cnt - 1) + (ctrl$NP * ctrl$npar))
        storepop[[i]] <- matrix(outC$storepop[idx], nrow = ctrl
$NP,
          ncol = ctrl$npar, byrow = TRUE)
        cnt <- cnt + (ctrl$NP * ctrl$npar)
        dimnames(storepop[[i]]) <- list(1:ctrl$NP, nam)
      }
    }
    else {
      storepop = NULL
    }
    names(outC$bestmem) <- nam
    iter <- max(1, as.numeric(outC$iter))
    names(lower) <- names(upper) <- nam
    bestmemit <- matrix(outC$bestmemit[1:(iter * ctrl$npar)],
      nrow = iter, ncol = ctrl$npar, byrow = TRUE)
    dimnames(bestmemit) <- list(1:iter, nam)
    storepop <- as.list(storepop)
    outR <- list(optim = list(bestmem = outC$bestmem, bestval =
outC$bestval,
      nfeval = outC$nfeval, iter = outC$iter), member = list
(lower = lower,
      upper = upper, bestmemit = bestmemit, bestvalit = outC
$bestvalit,
      pop = t(outC$pop), storepop = storepop))
    attr(outR, "class") <- "DEoptim"
    return(outR)
}
<environment: namespace:DEoptim>
> ## Rosenbrock Banana function
> ## The function has a global minimum f(x) = 0 at the point
+ ## (1,1).
> ## Note that the vector of parameters to be optimized must be
> ## the first
> ## argument of the objective function passed to DEoptim.
> Rosenbrock <- function(x) {
+ x1 <- x[1]
```

```
+ x2 <- x[2]
+ 100 * (x2 - x1 * x1)^2 + (1 - x1)^2
+ }
>
> ## DEoptim searches for minima of the objective function
> ## between
> ## lower and upper bounds on each parameter to be
> ## optimized. Therefore
> ## in the call to DEoptim we specify vectors that comprise the
> ## lower and upper bounds; these vectors are the same length
> ## as the
> ## parameter vector.
> lower <- c(-10,-10)
> upper <- -lower
>
> ## run DEoptim and set a seed first for replicability
> set.seed(1234)
> DEoptim(Rosenbrock, lower, upper)
Iteration: 1 bestvalit: 231.182756 bestmemit:   2.298807  6.799427
Iteration: 2 bestvalit: 147.937835 bestmemit:  -2.864559  7.052430
Iteration: 3 bestvalit: 147.937835 bestmemit:  -2.864559  7.052430
 . . . . . . . . . . . . . . . . . . . . . . . . . . . . . . . .
Iteration: 199 bestvalit: 0.000000 bestmemit:  1.000000  1.000000
Iteration: 200 bestvalit: 0.000000 bestmemit:  1.000000  1.000000
$optim
$optim$bestmem
par1 par2
  1   1
$optim$bestval
[1] 3.010666e-16
$optim$nfeval
[1] 402
$optim$iter
[1] 200
$member
$member$lower
par1 par2
 -10 -10
$member$upper
par1 par2
  10  10
```

```
$member$bestmemit
     par1      par2
1   2.8062121 6.21197105
2   2.2988072 6.79942674
3  -2.8645593 7.05243041
 . . . . . . . . . . . . ...
199 1.0000000 0.99999997
200 1.0000000 0.99999997
$member$bestvalit
  [1] 2.797712e+02 2.311828e+02 1.479378e+02 1.479378e+02 1.297662e+01
  [6] 1.297662e+01 7.484302e+00 4.177053e+00 1.522458e+00 1.522458e+00
 [11] 1.522458e+00 1.522458e+00 1.505673e+00 1.505673e+00 1.505673e+00
 . . . . . . . . . . . . . . . . . . . . . . . . . . . . . . ..
[191] 7.122327e-16 7.122327e-16 7.122327e-16 7.122327e-16
7.122327e-16
[196] 3.010666e-16 3.010666e-16 3.010666e-16 3.010666e-16
3.010666e-16
$member$pop
        [,1]         [,2]
 [1,]  1.0000000 1.0000001
 [2,]  1.0000000 0.9999999
 [3,]  1.0000000 0.9999999
 [4,]  1.0000000 1.0000000
 [5,] -1.6667868 2.8245124
 [6,]  1.0000001 1.0000001
 [7,]  1.0000001 1.0000002
 [8,]  1.0000000 1.0000001
 [9,]  1.0000000 1.0000000
[10,]  1.0000000 1.0000000
[11,]  1.0000000 1.0000000
[12,]  1.0000000 1.0000000
[13,]  1.0000000 1.0000000
[14,]  1.0000000 1.0000001
[15,]  1.0000000 1.0000000
[16,]  1.0000000 1.0000001
[17,]  1.0000000 1.0000001
[18,]  0.9999999 0.9999999
[19,]  1.0000000 1.0000000
[20,] -1.6708498 2.9124751
$member$storepop
list()
attr(,"class")
[1] "DEoptim"
```

```
>
>
> ## increase the population size
> DEoptim(Rosenbrock, lower, upper, DEoptim.control(NP = 100))
Iteration: 1 bestvalit: 0.831267 bestmemit:   0.993395   1.078004
Iteration: 2 bestvalit: 0.831267 bestmemit:   0.993395   1.078004
Iteration: 3 bestvalit: 0.136693 bestmemit:   1.238842   1.506507
. . . . .  . . .  . .  . . . . .  . . .  . . .  . . . . . . .
Iteration: 199 bestvalit: 0.000000 bestmemit:   1.000000   1.000000
Iteration: 200 bestvalit: 0.000000 bestmemit:   1.000000   1.000000
$optim
$optim$bestmem
par1 par2
  1  1
$optim$bestval
[1] 1.561559e-22
$optim$nfeval
[1] 402
$optim$iter
[1] 200
$member
$member$lower
par1 par2
 -10 -10
$member$upper
par1 par2
 10  10
$member$bestmemit
    par1          par2
1  -0.05055801 -0.3587592
2   0.99339454  1.0780042
3   0.99339454  1.0780042
. . . . .  . . .  . . . .  . . .  . . .  . . . . . .  . . .
199 1.00000000 1.0000000
200 1.00000000 1.0000000
$member$bestvalit
  [1] 1.415855e+01 8.312672e-01 8.312672e-01 1.366929e-01
1.366929e-01
  [6] 1.366929e-01 1.366929e-01 1.366929e-01 1.366929e-01
5.009681e-02
. . . . .  . . .  . . .  . . .  . . .  . . .  . . .  . . .
[196] 1.273210e-21 1.273210e-21 3.241491e-22 3.241491e-22
1.561559e-22
```

```
$member$pop
       [,1]    [,2]
 [1,]  1.000000 1.000000
 [2,]  1.000000 1.000000
 [3,]  1.000000 1.000000
 [4,]  1.000000 1.000000
 [5,]  1.000000 1.000000
 . . .  . . . . . . . . . .  . . . . . .  . . . . .  . .
[100,]  1.000000 1.000000
$member$storepop
list()
attr(,"class")
[1] "DEoptim"
>
>
> ## change other settings and store the output
> outDEoptim <- DEoptim(Rosenbrock, lower, upper,
+ DEoptim.control(NP = 80,
+ itermax = 400, F = 1.2, CR = 0.7))
Iteration: 1 bestvalit: 3.635506 bestmemit:   1.317411   1.547563
Iteration: 2 bestvalit: 3.635506 bestmemit:   1.317411   1.547563
Iteration: 3 bestvalit: 3.635506 bestmemit:   1.317411   1.547563
Iteration: 4 bestvalit: 3.635506 bestmemit:   1.317411   1.547563
Iteration: 5 bestvalit: 3.635506 bestmemit:   1.317411   1.547563
 . . . . . .  . . . . . . . . . . .  . . . . . . .  . . . . . . . .
Iteration: 399 bestvalit: 0.000000 bestmemit:   1.000000
1.000000
Iteration: 400 bestvalit: 0.000000 bestmemit:   1.000000
1.000000
>
> ## plot the output
> plot(outDEoptim)
> # Outputting: Figure 6.17
```

Figure 6.17 > plot(outDEoptim).

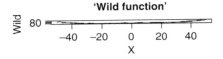

'Wild function'

Figure **6.18** > plot (Wild, -50, 50, n = 1000, main = "'Wild function'").

```
>
> ## 'Wild' function, global minimum at about -15.81515
> Wild <- function(x)
+ 10 * sin(0.3 * x) * sin(1.3 * x^2) +
+ 0.00001 * x^4 + 0.2 * x + 80
>
> plot(Wild, -50, 50, n = 1000, main = "'Wild function'")
> # Outputting: Figure 6.18
>
> outDEoptim <- DEoptim(Wild, lower = -50, upper = 50,
+ control = DEoptim.control(trace = FALSE))
> plot(outDEoptim)
> # Outputting: Figure 6.19
>
> DEoptim(Wild, lower = -50, upper = 50,
+ control = DEoptim.control(NP = 50))
Iteration: 1 bestvalit: 68.834199 bestmemit: -15.802280
Iteration: 2 bestvalit: 68.834199 bestmemit: -15.802280
Iteration: 3 bestvalit: 67.562828 bestmemit: -15.509441
Iteration: 4 bestvalit: 67.562828 bestmemit: -15.509441
Iteration: 5 bestvalit: 67.562828 bestmemit: -15.509441

. . . . . . . . . . . . . . . . . . . . . . . . . . . . . .
Iteration: 199 bestvalit: 67.467735 bestmemit: -15.815151
Iteration: 200 bestvalit: 67.467735 bestmemit: -15.815151
```

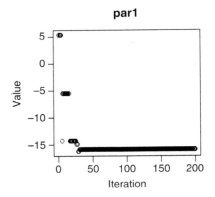

par1

Figure **6.19** > plot (outDEoptim).

```
$optim
$optim$bestmem
    par1
-15.81515
$optim$bestval
[1] 67.46773
$optim$nfeval
[1] 402
$optim$iter
[1] 200
$member
$member$lower
par1
 -50
$member$upper
par1
 50
$member$bestmemit
      par1
1  -25.66072
2  -15.80228
3  -15.80228
4  -15.50944
5  -15.50944
. . . . . . . .  . . . . . . . . . . . . . . . . . .
199 -15.81515
200 -15.81515
$member$bestvalit
 [1] 69.34871 68.83420 68.83420 67.56283 67.56283 67.56283
67.48859 67.48859
. . . . . . . .  . . . . . . . . . . . . . . . . . . .
[193] 67.46773 67.46773 67.46773 67.46773 67.46773 67.46773
67.46773 67.46773
$member$pop
      [,1]
[1,] -15.66161
[2,] -15.66161
[3,] -15.81515
. . . . . . . .  . . . . . . . . . . . . . . . .
[49,] -15.66161
[50,] -15.66161
$member$storepop
list()
```

```
attr(,"class")
[1] "DEoptim"
> ## The below examples shows how the call to DEoptim can be
> ## parallelized.
> ## Note that if your objective function requires packages to be
> ## loaded or has arguments supplied via \code{...}, these should be
> ## specified using the \code{packages} and \code{parVar}
arguments
> ## in control.
> ## Not run:
> Genrose <- function(x) {
+ ## One generalization of the Rosenbrock banana valley function
(n parameters)
+ n <- length(x)
+ ## make it take some time ...
+ Sys.sleep(.001)
+ 1.0 + sum (100 * (x[-n]^2 - x[-1])^2 + (x[-1] - 1)^2)
+ }
>
> # get some run-time on simple problems
> maxIt <- 250
> n <- 5
> oneCore <- system.time( DEoptim(fn=Genrose, lower=rep(-
+ 25, n), upper=rep(25, n),
+ control=list(NP=10*n, itermax=maxIt)))
Iteration: 1 bestvalit: 67180.992768 bestmemit:   0.787009
-4.017866   4.091383  -2.065631  -7.987533
Iteration: 2 bestvalit: 67180.992768 bestmemit:   0.787009
-4.017866   4.091383  -2.065631  -7.987533
Iteration: 3 bestvalit: 67180.992768 bestmemit:   0.787009
-4.017866   4.091383  -2.065631  -7.987533
Iteration: 4 bestvalit: 67180.992768 bestmemit:   0.787009
-4.017866   4.091383  -2.065631  -7.987533
Iteration: 5 bestvalit: 42308.544031 bestmemit:   0.787009
-4.017866   0.493543  -2.065631  -7.987533
  . . . . . . . . . . . . . . . . . . . . . . . . . . . . . .
Iteration: 249 bestvalit: 1.079131 bestmemit:  -1.005728
0.991545   0.993995   1.004214   1.004883
Iteration: 250 bestvalit: 1.079131 bestmemit:  -1.005728
0.991545   0.993995   1.004214   1.004883
> withParallel <- system.time( DEoptim(fn=Genrose, lower=rep(-
+ 25, n), upper=rep(25, n),
+ control=list(NP=10*n, itermax=maxIt, parallelType=1)))
```

```
Iteration: 1 bestvalit: 1036700.619060 bestmemit:   8.705266
3.108165  -7.886303  -3.630207  -7.869868
Iteration: 2 bestvalit: 540605.725273 bestmemit:   6.552843
-2.202845   6.752675   6.739165   2.361229
Iteration: 3 bestvalit: 134358.516530 bestmemit:  -0.173169
2.578942  -5.957804   1.230374   3.368074
Iteration: 4 bestvalit: 134358.516530 bestmemit:  -0.173169
2.578942  -5.957804   1.230374   3.368074
Iteration: 5 bestvalit: 40126.427537 bestmemit:   0.397502
-3.122870   1.249653   4.073320  -1.059661
Iteration: 6 bestvalit: 40126.427537 bestmemit:   0.397502
-3.122870   1.249653   4.073320  -1.059661
. . . . . . . . . . . . . . . . . . . . . . . . . . . . . . . .
Iteration: 249 bestvalit: 2.071510 bestmemit:   0.904367
0.836285   0.691696   0.466066   0.219700
Iteration: 250 bestvalit: 2.071510 bestmemit:   0.904367
0.836285   0.691696   0.466066   0.219700
> ## Compare timings
> (oneCore)
  user  system elapsed
  0.00   0.00  12.58
> (withParallel)
  user  system elapsed
  0.42   0.21   5.86
>
```

References

Adler, J. (2010) *R in a Nutshell: A Desktop Quick Reference*, O'Reilly Media, Inc., Sebastopol, CA.

Allen, E. (2007) *Modeling with Ito Stochastic Differential Equations*, Springer, New York.

ANOVA http://en.wikipedia.org/wiki/Analysis_of_variance

Aragon, T.J. (2011) *Applied Epidemiology Using R (epir)*, UC Berkeley School of Public Health, and San Francisco Department of Public Health, Berkeley, California.

Ardia, D., Mullen, K., Peterson, B.G., and Ulrich, J. (2011) DEoptim: differential evolution optimization in R. Available at http://CRAN.R-project.org/ package=DEoptim. C); http://www.icsi.berkeley.edu/~storn/code.html.

Ardia, D., Boudt, K., Carl, P., Mullen, K.M., Peterson, B.G. (2011) Differential evolution with EoptimD. an application to non-convex portfolio optimization. *The R Journal*, **3** (1), 27–34.

Ardia, D., Ospina, A.J.D., Giraldo G.N.D. (2011) Jump-diffusion calibration using differential evolution. *Wilmott Magazine*, **55** (September), 76–79.

Assets Allocation for Efficient Portfolios [http://www.kellogg.northwestern.edu/ faculty/papanikolaou/htm/finc460/ln/lecture1.pdf]

Baxter, M. and Rennie, A. (1996) *Financial Calculus, An Introduction to Derivative Pricing*, Cambridge University Press, Cambridge, UK.

Beach, S. and Orlov, A. (2006) An Application of the Black–Litterman Model with EGARCH-M-Derived views for international portfolio management. Working paper.

Bevan and Winkelmann (1998), Using the Black–Litterman Global Asset Allocation Model: Three Years of Practical Experience. *Goldman Sachs Fixed Income Research paper*.

Biostatistician job search http://jobs.amstat.org/jobs/4627784/biostatistician-1

Applied Probabilistic Calculus for Financial Engineering: An Introduction Using R, First Edition. Bertram K. C. Chan.
© 2017 John Wiley & Sons, Inc. Published 2017 by John Wiley & Sons, Inc.
Companion website: www.wiley.com/go/chan/appliedprobabilisticcalculus

Black, F. and Litterman, R. (1990) Asset Allocation: Combining Investor Views with Market Equilibrium. *Goldman Sachs Fixed Income Research Note*, September, 1990.

Black, F. and Litterman, R. (1991a) Global portfolio optimization. *Journal of Fixed Income*, **1** (2), 7–18.

Black, F. and Litterman, R. (1991b) Global Asset Allocation with Equities, Bonds and Currencies. *Goldman Sachs Fixed Income Research Note*, October, 1991.

Black, F. and Litterman, R. (1992) Global portfolio optimization. *Financial Analysts Journal*, **48** (5), 28–43.

Blanchett, D., and Kaplan, P. (2013) Alpha-Beta- and-Now . . . Gamma. *Institutional Investor Journals*, **1** (2), 29–45.

Blanchett, D. and Ratner, H. (2015) http://corporate.morningstar.com/us/documents/MethodologyDocuments/Morningstar_Asset_Allocation_Optimization_Methodology.pdf

Blanchett, D., Finke, M., and Guillemette, M. (2013) Variable-Risk-Preference-Bias, published by Morningstar, Inc. https://corporate.morningstar.com/.../Variable-Risk- Preference-Bias . . .

Blanchett, D., Finke, M., and Guillemette, M. (2014) Morningstar Working Draft, Washington, Chicago, IL, July 7.

BLCOP Package, (2015) Geometric brownian motion: Portfolio optimization using the R code BLCOP which illustrates the implementation of the foregoing BL approach. Figure 2.4 (from Wikipedia), Gochez, February 4, Available at https://cran.r-project.org/web/packages/BLCOP/vignettes/BLCOP.pdf

BMI Notes (2012) Body Mass Index. http://en.wikipedia.org/wiki/Body_mass_index

Braga, M. D. and Natale, F. P. (2007) TEV Sensitivity to Views in Black–Litterman Model.

Centers for Disease Control and Prevention. (2005) Antiretroviral postexposure, prophylaxis after sexual, injection-drug use, or other non-occupational exposure to HIV in the United States: recommendations from the U.S. Department of Health and Human Services. MMWR Recom. Rep. 2005 Jan;54 (RR-2):1-20. Available at http://www.cdc.gov/mmwr/preview/mmwrhtml/rr5402a1.htm.

Chambers, J. M., Cleveland, W. S., Kleiner, B. and Tukey, P. A. (1983) *Graphical Methods for Data Analysis*, Wadsworth, Belmont, CA.

Chan, Bertram (1978) A New School Mathematics for Hong Kong, vols1A, 1B, 2A, 2B, 3A, 3B, 4A, 4B, 5A, and 5B (also exercise volumes), Ling Kee Publishing Company, Hong Kong.

Chan, Bertram K. C. (2015) *Biostatistics for Epidemiology and Public Health Using R*, Springer Publishing, New York.

Chan, K. C., Karolyi, G. A., Longstaff, F. A., and Sanders, A. B. (1992) An empirical comparison of alternative models of the short-term interest rate. *Journal of Finance*, **47** (3), 1209–1227.

Chandra, S. (2003) Regional economy size and the growth-instability frontier: evidence from Europe. *Journal of Regional Science* **43** (1), 95–122. doi: 10.1111/1467-9787.00291

Chandra, S. and Shadel, W.G. (2007) Crossing disciplinary boundaries: applying financial portfolio theory to model the organization of the self-concept. *Journal of Research in Personality*, **41** (2), 346–373. doi: 10.1016/j.jrp.2006.04.007

Chicago Board of Options Exchange (CBOE) index: https://www.janus.com/advisor/funds/janus-balanced-fund?gclid=CKvJx6yOsdACFQIdaQodzxkNVA 2017.

Christadoulakis (2002) Bayesian optimal portfolio selection: the Black–Litterman approach. Class notes.

Coelen, N. (2000) Black–Scholes Option Pricing Model, Trinity University. Available at http://ramanujan.math.trinity.edu/tumath/research/studpapers/s11.pdfBlack-

Coelen, N. (2002) Introduction: finance is one of the most rapidly changing and fastest growing areas.

CRAN, The Comprehensive R Archive Network. http://cran.r-project.org/ 2017.

Da, Z. and Jagnannathan, R. (2005) Teaching note on Black–Litterman model.

Dalgaard, P. (2002) *Introductory Statistics with R*. Springer Statistics and Computing Series, Springer Science+Business Media, New York, NY.

Damghani B. M. (2013) The non-misleading value of inferred correlation: an introduction to the cointelation model. *Wilmott Magazine*. doi: 10.1002/wilm.102

Daniel, W.W. (2005) *Biostatistics: A Foundation for Analysis in the Health Sciences*, John Wiley & Sons, Inc., New York, NY.

Daroczi, G. et al. (2013) *"Introduction to R for Quantitative Finance"*, PACKT Publishing, Birmingham, U.K. www.packtpub.com/introduction-to-r-for quantitative finance/book

DeGroot, M. H. (1970) *Optimal Statistical Decisions*. Wiley Interscience.

Doganoglu, T., Hartz, C., and Mittnik, S. (2007) Portfolio optimization when risk factors are conditionally varying and heavy tailed. *Computational Economics* **29**, 333–354. doi: 10.1007/s10614-006-9071-1

Dow Jones Industrial Average (DJIA) and Inflation http://thesovereigninvestor.com/exclusives/wall-street-legend-predicts-dow-50000/?z=590252 2017.

Dowd, K. (2007) *Measuring Market Risk*, 2nd Edition, John Wiley& Sons, Ltd, Hoboken, NJ.

Everitt, B.S. and Hothorn, T. (2006) *A Handbook of Statistical Analysis Using R*, Chapman & Hall/CRC, Boca Raton, FL.

Fantazzinni, D. (2009) The effects of misspecified marginals and copulas on computing the value at risk: a Monte Carlo study. *Computational Statistics Data Analysis*, **53** (6), 2168–2188.

Financial Engineering 101 https://en.wikipedia.org/wiki/LPL_Financial; 1.4: Modern Portfolio Theory: https://en.m.wikipedia.org/wiki/Modern_portfolio_theory 2017.

Financial News Media (2017) 5 reasons why stocks may keep going higher, CNN report available at http://money.cnn.com

Firoozye N. and Blamont, D. (2003) Asset Allocation Model. Global Markets Research, Deutsche Bank, July.

Frost, P. and Savarino, J. (1986) An empirical bayes approach to efficient portfolio selection. *Journal of Financial and Quantitative Analysis*, **21** (3), 293–305.

Fusai, G. and Meucci, A. (2003) Assessing Views. *Risk Magazine*, **16** (3), S18–S21.

Georgakopoulos, H. (2015) *Quantitative Trading with R: Understanding Mathematical and Computational Tools from a Quant's Perspective*. Palgrave Macmillan.

Giacometti, R., Bertocchi, M., Rachev, S. T., and Fabozzi, F. J. (2007) Stable distributions in the Black–Litterman approach to asset allocation. *Quantitative Finance*, **7** (4), 423–433.

Hardy, G. H. (1908, 1967) *A Course of Pure Mathematics*, 10th Edition, Cambrideg University Press, Cambridge, U.K.

Hardy, G.H. (2009) *A Course of Pure Mathematics*, 3rd Edition Cambridge University Press, Cambridge, UK (First edition 1908).

He, G. and Litterman, R. (1999) The Intuition Behind Black-Litterman Model Portfolios. Goldman Sachs Asset Management Working paper. (Copyright 2007–2014, Jay Walters.)

Henderson, D. and Plaschko, P. (2006) *Stochastic Differential Equations In Science And Engineering*, World Scientific.

Herold, U. (2003) Portfolio construction with qualitative forecasts. *Journal of Portfolio Management*, **30** (Fall 2003), 161–172.

Herold, U. (2005) Computing implied returns in a meaningful way. *Journal of Asset Management*, **6** (1), 53–64.

Holland, J. H. (1975) *Adaptation in Natural Artificial Systems*, University of Michigan Press, Ann Arbor, MI.

Hubbard, D. (2007) *How to Measure Anything: Finding the Value of Intangibles in Business*, John Wiley & Sons Ltd, Hoboken, NJ.

Hull, J. C. (2000) *Options, Futures, and Other Derivatives*, Prentice Hall, Upper Saddle River, NJ.

Idzorek, T. (2005) A Step-By-Step Guide to the Black–Litterman Model, Incorporating User-Specified Confidence Levels. Working Paper.

Idzorek, T. (2006) Strategic Asset Allocation and Commodities. Ibbotson White Paper.

Ito Lemma Reference: http://www.sjsu.edu/faculty/watkins/ito.htm

Jedrzejewski, F. (2009) *Modeles aleatoires et physique probabiliste*. Springer, New York.

Klebaner, F. C. (1998) *Introduction to Stochastic Calculus with Applications*, Imperial College Press, London.

Koch, W. (2004) Consistent Asset Return Estimates. The Black–Litterman Approach. Cominvest presentation, October 2004.

Krishnan, H. and Mains, N. (2005) The two-factor Black–Litterman model. Risk Magazine, July.

Lamberton, D. and Lapeyre, B. (1991, 1996) *Introduction to Stochastic Calculus Applied to Finance* (Translated from French by Rabeau, N. and Mantion, F.), Chapman & Hall, London, U.K.

Lintner, J. (1965) The valuation of risk assets and the selection of risky investments in stock portfolios and capital budgets. *The Review of Economics and Statistics (The MIT Press)*, **47** (1), 13–39. doi: 10.2307/1924119.

Litterman, B. and Quantitative Research Group (2003) *Modern Investment Management: An Equilibrium Approach*, John Wiley & Sons, Inc., New Jersey.

Lo, A. W. (1999) *A Non-Random Walk Down Wall Street*, Princeton University Press.

Lo, A. W. (2004) The adaptive markets hypothesis: market efficiency from an evolutionary perspective. Journal of Portfolio Management.

Lo, A. W. and Mackinlay, A. C. (2002) *A Non-Random Walk Down Wall Street*, 5th Edition, Princeton University Press. pp. 4–47.

Low, R.K.Y., Faff, R., Aas, K. (2016) Enhancing mean–variance portfolio selection by modeling distributional asymmetries. *Journal of Economics and Business*, doi: 10.1016/j.jeconbus.2016.01.003

Lyuu, Y.-D. (2002) *Financial Engineering & Computation: Principles, Mathematics, Algorithms*, Cambridge University Press, UK.

Mankert, C. (2006) The Black–Litterman model: mathematical and behavioral finance approaches towards its use in practice. Licentiate thesis.

Markowitz, H.M. (1952) Portfolio Selection. *The Journal of Finance*, **7** (1), 77–91. doi: 10.2307/2975974. JSTOR 2975974.

Markowitz, H.M. (1952) Portfolio selection. *The Journal of Finance*, **7** (1), 77–91. doi: 10.2307/2975974. JSTOR 2975974. (The MPT was proposed by Economist Harry Markowitz in 1952, for which he was later awarded a Nobel Memorial Prize in Economic Sciences.)

Martellini, L. and Ziemann, V. (2007) Extending Black–Litterman analysis beyond the mean-variance framework. *Journal of Portfolio Management*, **33** (4), 33–44.

Merton, R. (1972) An analytic derivation of the efficient portfolio frontier. *Journal of Financial and Quantitative Analysis*, 7, 1851–1872.

Meucci, A. (2005) *Risk and Asset Allocation*, Springer Finance.

Meucci, A. (2006) Beyond Black–Litterman in Practice: A Five-Step Recipe to Input Views on Non-Normal Markets. Working paper available on SSRN.

Meucci, A. (2008) The Black–Litterman approach: original model and extensions. April. Available at SSRN: http://ssrn.com/abstract=1117574

Meucci, A. (2008) Fully Flexible Views: Theory and Practice. Working paper available on SSRN.

Meucci, A. (2008) The Black–Litterman approach: original model and extensions. Available at SSRN: http://ssrn.com/abstract=1117574

Meucci, A. (2010) The Black–Litterman Approach: Original Model and Extensions. Working paper available on SSRN.

Michaud R. (1989) The Markowitz optimization enigma: is optimized optimal? *Financial Analysts Journal*, **45** (1), 31–42.

Michaud R. (1998) *Efficient Asset Management*, Harvard Business School Press, Boston, MA.

Midas Group (2005) Black–Litterman Asset Allocation Model Portfolio Theory of Information Retrieval, July 11, 2009

Mitchell, M. (1998) *An Introduction to Genetic Algorithms*, The MIT Press, Cambridge, MA.

Mittal, H.V. (2011) *R Graphics Cookbook*, PACKT Publiching Ltd., Birmingham, UK.

Mullen, K.M., Ardia, D., Gil, D., Windover, D., Cline, J. (2011) DEoptim: an R package for global optimization by differential evolution. *Journal of Statistical Software*, **40** (6), 1–26.

Murrell, P. (2006) *R Graphics*, Computer Science and Data Analysis Series, Chapman & Hall/CRC, Boca Raton, FL.

Peleg, D. (2014) *Fundamental Models in Financial Theory*, The MIT Press, Cambridge, MA.

Peng, R.D. and Dominici, F. (2008) *Statistical Methods for Environmental Epidemiology with R – A Case Study in Air Pollution and Health*, Springer, NY.

Pfaff, B. (2013) Financial risk modeling and portfolio optimization using R. (Suitable distributions for returns: p. 53–83; Modeling dependence: p. 136–138; are illustrations of the implementation of the foregoing BL approach using the BLCOP package in R; Portfolio optimization: 160, 161, 189–215, 217–254, 160–165; R Packages: 197–201, etc.; Robust portfolio optimization: 155–188; R packages: 166–171, 317–319).

Price, K.V., Storn, R.M., and Lampinen J.A. (2006) *Differential Evolution: A Practical Approach to Global Optimization*. Springer, Berlin Heidelberg.

Qian, E. and Gorman, S. (2001) Conditional distribution in portfolio theory. *Financial Analysts Journal*, **57** (2), 44–51.

Rachev, S.T. and Mittnik, S. (2000) *Stable Paretian Models in Finance*. John Wiley & Sons New Approaches for Portfolio Optimization: Parting with the Bell Curve: Interview with Prof. Rachev and Prof. Mittnik. Available at https://statistik.econ.kit.edu/download/doc_secure1/RM-Interview-Rachev-Mittnik-EnglishTranslation.pdf (Risk Manager Journal (2006) New Approaches for Portfolio Optimization: Parting with the Bell Curve: Interview with Prof. Svetlozar Rachev and Prof. Mittnik.)

Risk Manager Journal (2006) New Approaches for Portfolio Optimization: Parting with the Bell Curve (Interview with Professor Svetlozar Rachev and Prof.Stefan Mittnik;. A Simple Example of a Portfolio Consisting of Two Risky Assets.

Available at http://faculty.washington.edu/ezivot/econ424/introductionPortfolioTheory.pdf).

Ross, S. (1999) *An Introduction to Mathematical Finance*, Cambridge University Press, UK.

Rupert, D. and Matteson, D.S. (2016) *Statistics and Data Analysis for Financial Engineering*, 2nd Edition, Springer, NY, p. 491. Available at http://people.orie.cornell.edu/davidr/SDAFE2/index.html (Solutions to Selected Computer Lab Problems and Exercises in Chapter 16 of Statistics and Data Analysis for Financial Engineering, 2nd ed., by D. Ruppert and D.S. Matteson © 2016 David Ruppert and David S. Matteson).

Sabbadini, T. (2010) *Manufacturing Portfolio Theory (PDF)*, International Institute for Advanced Studies in Systems Research and Cybernetics.

Salomons, A. (2007) The Black–Litterman model, hype or improvement? Thesis.

Satchell, S. and Scowcroft, A. (2000) A demystification of the Black–Litterman model: managing quantitative and traditional portfolio construction. *Journal of Asset Management*, **1** (2), 138–150.

Schottenfeld, D. (2002) Epidemiology: an introduction. *American Journal of Epidemiology*, **156** (2), 188–190.

Seefeld, K. and Linder, L. (2005) *Statistics Using* R *with Biological Examples*, Department of Mathematics & Statistics, University of New Hampshire, Durham, NH http://cran.r-project.org/doc/contrib/Seefeld_StatsRBio.pdf

Sharpe, W. F. (1964) Capital asset prices: a theory of market equilibrium under conditions of risk. *Journal of Finance*, **19** (3), 425–442. doi: 10.2307/2977928. JSTOR 2977928

Sharpe, W. F. Macro-Investment Analysis, Stanford University.

Sokolnikoff, I.S. and Redheffer, R.M. (1966) *Mathematics of Physics and Modern Engineering*, 2nd Edition, McGraw-Hill Book Company, N.Y (First Edition 1958).

Statistics Canada: http://www.statcan.gc.ca/edu/power-pouvoir/ch3/5214785-eng.htm#a1

Storn R., and Price K. (1997) Differential evolution: a simple and efficient heuristic for global optimization over continuous spaces. *Journal of Global Optimization*, **11** (4), 341–359.

Storn, R., and Ulrich, I. (1997) Differential evolution: simple and efficient heuristic for global optimization over continuous spaces. *Journal of Global Optimization*, **11** (4), 341–359.

Syvertsen, C.O. (2013) The Black–Litterman model. 24. 1. BLCOP – portfolio of 5 assets 2. Gochez, F.

Taleb, N. N. (2007) *The Black Swan: The Impact of the Highly Improbable*. Random House.

Teetor, P. (2011) R *Cookbook: Proven Recipes for Data Analysis, Statistics, and Graphics*, O'Reilly Media, Sebastopol, CA.

The canonical Black–Litterman reference model references (Many of these references are available at www.blacklitterman.org) with links to many of these papers. 2017.

Theil, H. (1971) *Principles of Econometrics*, John Wiley & Sons, Inc.

Tobin, J. (1958) Liquidity preference as behavior towards risk. *The Review of Economic Studies*, **25** (2), 65–86. doi: 10.2307/2296205. JSTOR 2296205

Venables, W.N., and Smith, D.M., and the R Development Core Team (2004) *An Introduction to R*, Network Theory, Ltd., Bristol, UK.

Verzani, J. (2005) *Using R for Introductory Statistics*, Chapman & Hall/CRC, Boca Raton, FL.

Virasakdi, C., (2007) *Analysis of Epidemiological Data Using R and Epicalc*, Epidemiology Unit, Prince of Songkla University, Thailand.

Walters, J. (2007) Black–Litterman Model Available at http://www.blacklitterman. org/Black-Litterman.pdf; http://papers.ssrn.com/sol3/papers.cfm? abstract_id=1314585 Also see The Black–Litterman Model in Detail, June 20, 2014; Copyright 2007-14: Jay Walters, CFA, jwalters@blacklitterman.org Me08, p. 5).

Wang, J. (2009) Portfolio Theory of Information Retrieval. Proceedings of the 32nd international ACM SIGIR conference on research and development in information retrieval, Boston, MA.

Wikipedia (2017) LPL (Linsco Private Ledger) Financial.

Wilmott, P., Howison, S., and Dewynne, J. (1995) *The Mathematics of Financial Derivatives: A Student Introduction*, Cambridge University Press, UK.

Xiong, J.X. and Idzorek, T.M. (2011) The impact of skewness and fat tails on the asset allocation decision. *Financial Analysts Journal*, **67**, 23–35.

http://corporate.morningstar.com/us/documents/MethodologyDocuments/ Morning Star_Asset_Allocation_Optimization_Methodology.pdf

https://en.wikipedia.org/wiki/Modern_portfolio_theory

https://investor.vanguard.com/mutual-funds/managed- payout/#/

Index

Applied Probabilistic Calculus for Financial Engineering: An Introduction Using R, First Edition.
Bertram K. C. Chan.
© 2017 John Wiley & Sons, Inc. Published 2017 by John Wiley & Sons, Inc.
Companion website: www.wiley.com/go/chan/appliedprobabilisticcalculus